Clothesline Math

The Master Number Sense Maker

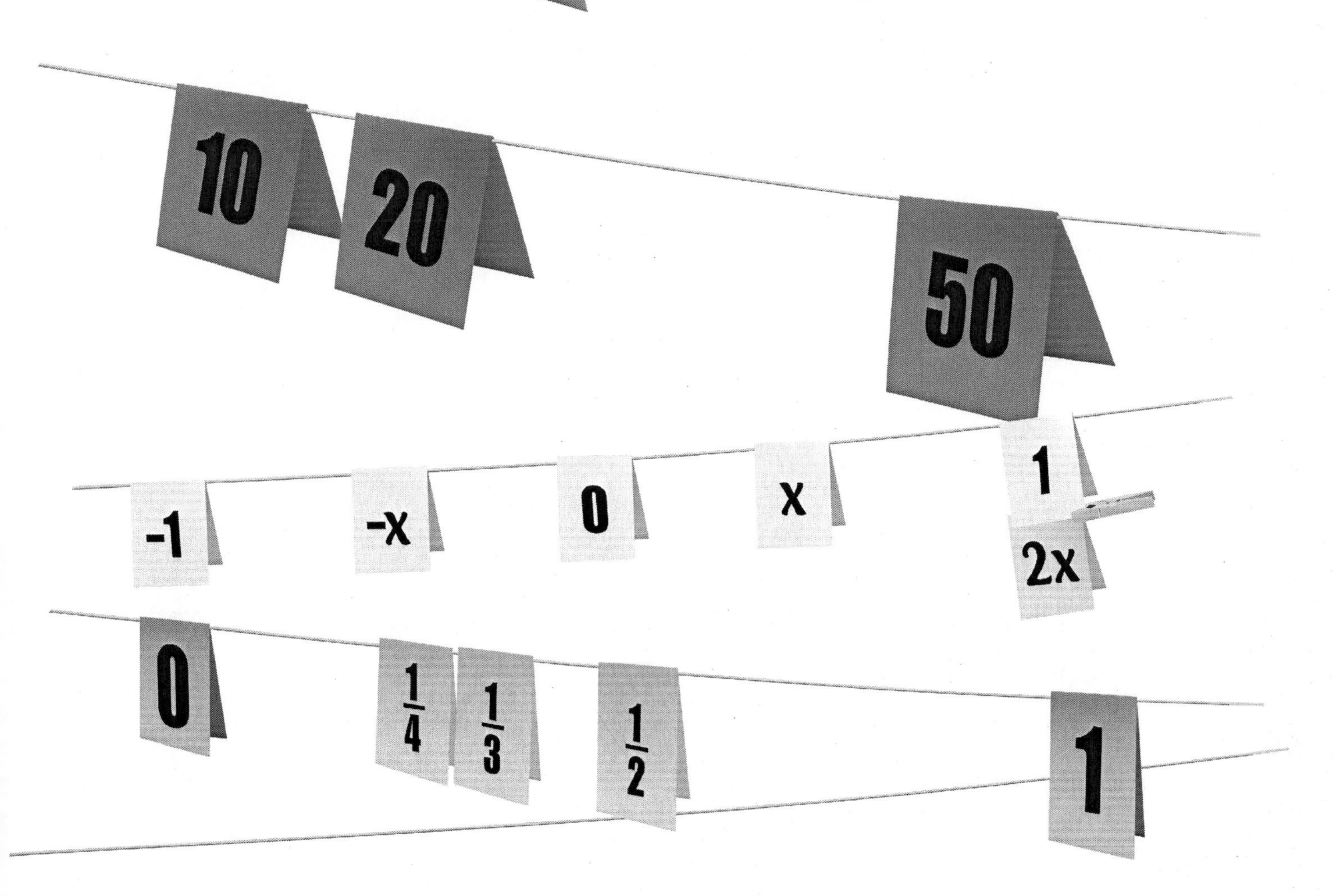

Chris Shore, M.Ed.

Publishing Credits

Corinne Burton, M.A.Ed., *Publisher*
Conni Medina, M.A.Ed., *Managing Editor*
Diana Kenney, M.A.Ed., NBCT, *Content Director*
Kyleena Harper, *Assistant Editor*
Fabiola Sepulveda, *Graphic Designer*

Image Credits

Photos © Walter Mladina.

Standards

Shell Education
A division of Teacher Created Materials
5301 Oceanus Drive
Huntington Beach, CA 92649-1030
www.tcmpub.com/shell-education

ISBN 978-1-4938-8514-5

Foreword 5

Acknowledgments 7

Introduction: A Universal Example 9

PART I: Teaching with the Clothesline

Chapter 1: The Master Number Sense Maker 19
Chapter 2: Setting Up Your Clothesline 25
Chapter3: Making the Most of Your Clothesline 33

PART II: Clothesline Lessons for Elementary School

Chapter 4: Whole Numbers for Elementary School
Lesson 1: Counting Numbers 41
Lesson 2: Place Value 50
Lesson 3: Single-Digit Addition 57
Lesson 4: Single-Digit Multiplication 62
Chapter 5: Fractions for Elementary School
Lesson 5: Fractions: Almost Half 65
Lesson 6: Comparing Fractions with Common Denominators 74
Lesson 7: Comparing Fractions with Common Numerators 78
Lesson 8: Equivalent Fractions 84
Lesson 9: Adding Fractions 91
Chapter 6: Algebraic Reasoning for Elementary School
Lesson 10: Order of Operations 103
Lesson 11: Variables 111
Lesson 12: Expressions 117
Lesson 13: One-Step Equations 123

PART III: Clothesline Lessons for Middle and High School

Chapter 7: Arithmetic for Middle School
Lesson 14: Multiplying Fractions 137
Lesson 15: Converting Fractions, Decimals, and Percentages 148
Lesson 16: Operations with Integers 154
Chapter 8: Algebra
Lesson 17: Evaluating Expressions 161
Lesson 18: Variable on Both Sides 166
Lesson 19: Two-Step Equations 172
Lesson 20: Rules of Exponents 179
Lesson 21: Rational Expressions 191

Chapter 9: Geometry
Lesson 22: Ratio and Proportion 199
Lesson 23: Linear Pairs 203
Lesson 24: Vertical Angles 209
Lesson 25: Transversals 213
Lesson 26: Special Right Triangles 220
Lesson 27: Polygon Angle Properties 226
Chapter 10: Statistics
Lesson 28: Average 239
Lesson 29: Mean, Median, Mode, and Range 244
Lesson 30: Five-Number Summary 250
Lesson 31: Standard Deviation 255
Chapter 11: Functions
Lesson 32: Slope 263
Lesson 33: Slope-Intercept Form 269
Lesson 34: Exponential Growth and Decay 274
Lesson 35: Exponential Decay 282
Lesson 36: Three-Equation Models 289
Lesson 37: Logarithms 297
Lesson 38: Radians and Degrees 304
PART IV: Appendixes
Appendix A: References Cited 312
Appendix B: Additional Resources 313
Appendix C: Reproducibles 317
Appendix D: Digital Resources 336

Imagine almost missing an international flight because you and a friend were engrossed in something that was rockin' your world. In October 2015, Chris Shore and I were those two friends, and we almost missed our flight to Vancouver. Picture two grown men surrounded by hundreds of strangers inside Los Angeles International Airport. We were excitedly talking and deeply focused on my math notebook as we used an open number line to solve a multistep math equation with variables on both sides. At one point, we had this magical eureka moment, and we celebrated our collaborative findings with a vigorous chest bump and extremely loud, "Oh yeah!"

Rest assured, we made our flight. More importantly, thanks to the open number line that day, our number sense received a serious boost. *Clothesline Math: The Master Number Sense Maker* by Chris Shore is your ticket for embarking on a similar journey that will boost your number sense and the number sense of your students.

Two months before our airport eureka chest bump, Chris came to Tustin Unified School District (TUSD) and provided secondary math teachers and instructional coaches with professional development. As a math instructional coach at TUSD, I was beyond excited to learn from him. He introduced TUSD to the open dynamic number line known as Clothesline Math. Shore immediately blew our math minds and challenged our number sense with lessons ranging from numerical relationships of fractions to mean absolute deviation. I immediately thought, "This is awesome! I wish I had this in my classroom when teaching middle-school math!" Looking across the room at all the other math teachers, I could see their faces filled with both excitement and apprehension. On one hand, teachers were thrilled to receive this amazing and transformative math gift. On the other hand, whenever teachers explore something so radical and dynamic like Clothesline Math, they desire and need a roadmap to guide their teaching and student learning. *Clothesline Math: The Master Number Sense Maker* is your roadmap to classroom-tested lessons and teacher moves for successfully facilitating Clothesline Math in your classroom.

This book begins with Chris doing a wonderful job citing research and best practices to explain why it is so valuable for teachers to provide their students with opportunities to strengthen their number sense. Number sense has a direct impact on the mathematical content students are expected to learn each year of their schooling. As Shore points out, "Students don't flunk current content; they flunk prior content." He then goes on to explain why teaching number sense and prerequisite skills are essential for students to be more successful in mathematics at their current grade level. An open number line, such as the Clothesline, is both a vehicle and a destination for students and teachers to boost their number sense and prerequisite skills.

After quickly convincing us why number sense lessons need to be a frequent part of the teaching and learning of mathematics, Chris takes us on a ride through various K–12 Clothesline Math lessons, such as numerical relationships, rational numbers, algebraic expressions, logarithms, and trigonometric relationships. The lessons are not "just one more thing" you should be doing. The lessons and resources in this book are framed in a way to enhance the content you're supposed to teach while empowering you to strengthen your craft of teaching mathematics at the same time.

I've seen how Chris has continued to both gather and share Clothesline Math wisdom and strategies with teachers around the country at conferences and other professional development. Most of the wisdom and strategies Chris shares in the book come from his own classroom experience as a math teacher. If you have ever met and had a conversation with him, you know Chris has the gift of intently listening and remembering every detail of conversations. His gift of listening for details shows in this book, as he has meticulously captured student and teacher conversations, showcasing them through vignettes. The vignettes capture the essence of a Clothesline lesson, benefiting teachers who are either new to Clothesline Math or are already familiar with its transformative powers. These vignettes model the necessary and essential mathematical discourse that go along with any Clothesline Math lesson, providing the roadmap teachers desire. Chris says it best: "Clothesline Math is intended for discourse, not lecture."

Chris has tested and experienced these lessons with students; he has listened intently to their comments, questions, and ideas. As you'll discover, Chris uses the perspective and voice of students to demonstrate how powerful and accessible Clothesline Math lessons are for all types of math students. Through these conversations, you'll quickly see that the clothesline is a student-centered tool for building number sense. It's our job as educators to identify when we need to get into the passenger seat and let students drive the learning. But don't worry—this book is still your roadmap as it provides card sets, organizational tips, conversations, and key questions for each lesson to help guide the conversations and learning that will transpire in your classroom.

Before you know it, you will find yourself engrossed in *Clothesline Math: The Master Number Sense Maker,* exploring the numerical relationships you teach and skills students need to be successful in learning mathematics. I highly encourage you to find a math friend and explore Clothesline Math lessons together. Get immersed in your notebooks, celebrate your findings (maybe with a chest bump), and, as Chris Shore would say, "Go change the world, one math lesson at a time." Just don't miss your flight.

—Andrew Stadel
Math Teacher, Instructional Coach, and Consultant

Thank you to...

Kelli Wise, for introducing me to the potential of Clothesline Math. That one session of yours at the MaTHink conference changed my teaching profoundly. This book would never have been written without your initial inspiration.

Tim McCaffery, for sharing that original Clothesline experience and giving the Clothesline its nickname, The Master Number Sense Maker.

Andrew Stadel and Daniel Luevanos, for being my co-conspirators. Our exchange of ideas has truly sharpened the edges on this instructional tool. You have been fellow evangelists of this message. May David Lee, Angus, and John Frusciante rock on together forever. Kristen Acosta, Stacy Zagurski and your colleagues at Merlinda Elementary School for your insights into elementary's implementation of Clothesline Math, and to the multitude of other clothesline enthusiasts for blazing a trail through this new territory.

My kindred spirits in the Math Twitter Blogosphere, for making me a phenomenally better teacher. So many of your ideas, like Annie Fetter's Notice and Wonder, have made their way into my teaching, and thus into the examples in this book. #MTBoS rocks!

Diana Kenney, for bringing this book to life so quickly. Conni Medina and all the people at Shell Education who believed so enthusiastically in this project from the very beginning. My editor, Kyleena Harper, for walking me through my first professional publication process, knowing when to nudge and when to kick. Your leadership, organization, and vision, Walter Mladina's photography, and Fabiola Sepulveda's design navigated the uncommon challenges arising from the unique structure of the Clothesline to create a book of which we can all be proud.

My many heroes and mentors, Dr. Tim Kanold, Dr. Uri Treisman, Dr. William Schmidt, Dr. Ilana Horn, Dr. Harris Schultz, Dr. Tom Bennett, Dr. Maggie Smith, Dr. John Star, Dr. Rick Dufour, Dr. Patrick Campbell, Dr. Juli Dixon, Dr. William Doll, Dr. Titu Andreescu, and David Foster, for the solid theoretical foundation on which I have constructed my world view of math education.

My students, for being the reason that I love teaching so much, for playfully receiving the Clothesline in class, and for teaching me far more than I ever taught you.

My children, Rylie and Preston, for growing up as the guinea pigs of your math teacher dad. I am so proud of the special adults that you have become. In your own unique way... change the world. My wife, Casie, for being my #1 fan and for teaching me to always teach to the hearts of children as well as to their minds. I love you dearly. As always, thanks for the adventure.

The Lord, for the talents, relationships, knowledge, and opportunities that you have so blessed me with in number and magnitude. May my use of them serve you well.

Dedication

Dedicated to my mother, Jo Lancaster

A Universal Example

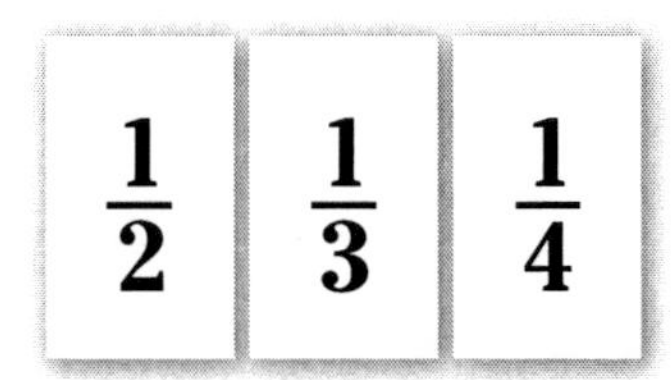

Objective: Accurately place and space three unit fractions on an open number line.

Teacher: **Good day, class. Today, I am introducing you to the clothesline. This number line will help us better understand the math we are studying. Today, though, we will start with a topic we already know in order to show you how the Clothesline works.**

First, we call it the Clothesline because it is literally a clothesline with clothespins that we will use to show that certain values are equal.

I will give you numbers written on folded cards that look like tents. This shape makes it really easy to hang our numbers on the clothesline.

The teacher shows students the folded cards, as shown in Figure I.1.

Figure I.1 Folded Clothesline Cards

You will notice there are no "benchmarks"—those numbers and tick marks you usually see on a number line. That's because this is an open number line, so you get to choose how to label it. When it is your turn to place numbers on the number line, I have benchmarks printed and ready for you on the table at the front of the room.

The teacher gestures to the pile of benchmark cards at the front of the room, as shown in Figure I.2.

Figure I.2 Sample Benchmark Cards

If you want to use a benchmark that I don't have up here, I have provided blank cards that you may use to create your own.

The teacher shows blank cards sitting next to the benchmark cards, as shown in Figure I.3.

Figure I.3 Blank Cards

I'm going to give this first set of numbers to one group that will come to the front of the class and place them on the clothesline. While that group is up here, the rest of you will draw your responses on your lapboards with your elbow partners. The person whose birthday is closest to today will go first. Whose birthday is closest?

Steve raises his hand.

Steve. Here is our first set. Class, we are placing $\frac{1}{2}$, $\frac{1}{3}$, and $\frac{1}{4}$ on the number line. Please draw a number line on your lapboards with your partners. You are to show any benchmarks you need. Steve, you and your group may use any benchmarks I have provided or none at all. Remember, if you need a number that I did not prepare, you may write it on a blank card.

After some rich conversation at the clothesline, students place the cards in the wrong order (see Figure I.4) and return to their seats.

Figure I.4 First Clothesline Placement

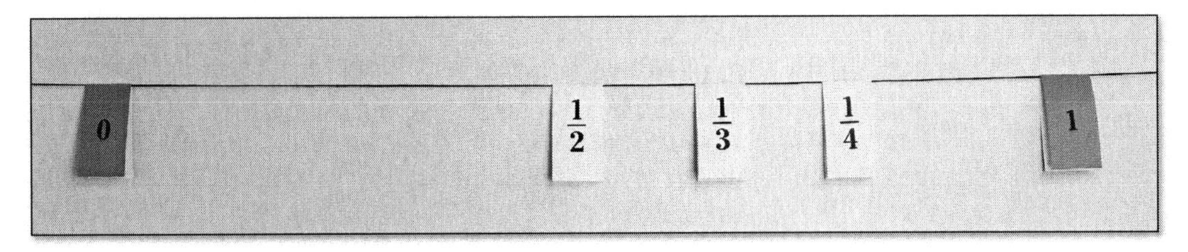

Class, let's see who agrees with this response. Please declare your lapboards.

Student pairs raise their lapboards overhead to display their answers. About half the class agrees with the improper placement of the three values. The other students have them properly placed.

Steve, I see that your group used the benchmarks 0 and 1. So, are you claiming that all three fractions are less than 1, but greater than 0?

Steve: Yes.

Teacher: Why did you place the fractions in this order?

Steve: Because we saw the numbers 2, 3, and 4, so we just counted up.

Teacher: Most of the class agrees with your response here. Does anyone agree with the order of the numbers but for a different reason?

Jackson raises his hand.

Yes, Jackson?

Jackson: I did it in that order because you gave them to us in that order.

Teacher: You disagreed, Kendra. Why?

Kendra: Because $\frac{1}{4}$ is smaller than $\frac{1}{3}$ which is smaller than $\frac{1}{2}$.

Teacher: How do you know?

Kendra: If I cut one pizza into two slices, another pizza into three slices, another into four slices, and then take one slice from each, half of a pizza will be bigger than a fourth of a pizza, and a third will be in the middle.

Teacher: Assuming the pizzas are the same size to begin with?

Kendra: Yes.

Teacher: **You have now heard three different explanations. Let's thumb vote: thumbs-up if you agree with the current placement; thumbs-down if you do not. Are these fractions in the correct order?**

The whole class votes thumbs-down.

Steve, your group voted against your own response. If you would like to change it, please feel free to come up and do so.

Steve's group returns to the front of the classroom and rearranges the cards (see Figure I.5).

Figure I.5 Second Clothesline Placement

So, the first thing we want to do with the clothesline is *place* the cards. That means we want to get the values in the correct numerical order. We will talk about spacing in a minute, but do you all agree that these three are now in the right order? All in favor, say "aye."

Class: Aye!

Teacher: **I noticed that your group equally spaced the numbers this time. Can someone from the group tell me why?**

Akira: Well, we realized $\frac{1}{2}$ had to be in the middle. And $\frac{1}{4}$ is half of a half. Then, $\frac{1}{3}$ goes in between those two.

Teacher: **Okay. So after we *place*, we want to *space*. That means we want to slide the values on the number line so they are in their proper locations on the number line. Do you agree these values are properly spaced? Thumb vote.**

Most, but not all, vote no.

Teacher: **Arturo, why not?**

Arturo: $\frac{1}{3}$ needs to be closer to $\frac{1}{4}$.

Teacher: **Okay, I'm going to tap $\frac{1}{3}$ in the direction in which you point as a class. When the value is in the right place, clap.**

Students point to the left as the teacher moves the cards, as shown in Figure I.6. Several are mumbling, "Keep going." There is finally a collective clap.

Figure I.6 Third Clothesline Placement

Who wants to explain why we stopped here? Cindy.

Cindy: I converted the fractions to decimals.

Teacher: How did that help you?

Cindy: You get 0.25, 0.33, and 0.50. I know that 33 is closer to 25 than 50, so I kind of estimated where 0.33 would go.

Teacher: Thank you. Did anyone do it a different way? Skylar.

Skylar: Common denominators give you $\frac{3}{12}$, $\frac{4}{12}$, and $\frac{6}{12}$. The number 4 is closer to 3 than 6. So, I estimated by counting 3, 4, 5, 6.

Teacher: Thank you. Any other ways to do this? Jesse.

Jesse: On our board, I used my pencil to measure $\frac{1}{3}$ and then checked whether that length fit into 1 three times.

Teacher: So, you are using something we call finger reasoning. Everyone, close one eye and place your fingers in front of you so you can remotely measure from 0 to $\frac{1}{3}$.

Students close their eyes and hold up their fingers.

Now, see whether three times that distance is equivalent to 1.

Students use their fingers to remotely measure the spacing of the fractions on the clothesline, as shown in Figure I.7.

Figure I.7 Sample Finger Measuring

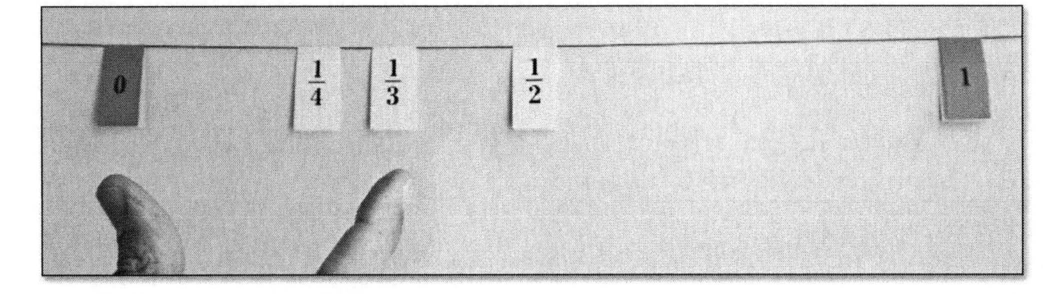

If you now agree with both the placing and the spacing of our three values, clap twice.

The class gives a resounding two claps. The cards stay as they are, as shown in Figure I.8.

Figure I.8 Final Clothesline Placement

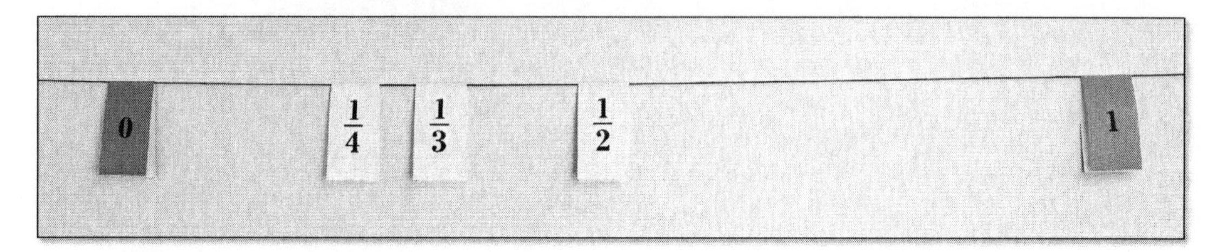

Now class, it is time to record our discussions, deductions, and decisions on your activity sheets. On the three blanks, write $\frac{1}{2}$, $\frac{1}{3}$, and $\frac{1}{4}$. Then, record them on the number line provided with the benchmarks. Be sure to be just as diligent about placing and spacing on your paper as we were on our clothesline. Well done, math crew.

Why These Numbers?

The key to a good clothesline lesson is in the selection of the proper values to pose to the class. For example, I could have chosen any three fractions, but $\frac{1}{2}$, $\frac{1}{3}$, and $\frac{1}{4}$ are the most common. In addition, the consecutive denominators of 2, 3, and 4 keep calculations simple and bring out the misconceptions that a larger denominator implies a larger fraction.

The Key Questions

- Why did you include 0 and 1 as benchmarks?
- Why did you not include any benchmarks?
- Do we need benchmarks?
- Why did you place them in that order?
- How do you know $\frac{1}{4}$ is smaller than $\frac{1}{3}$?
- Did anyone come to the same conclusion, but in a different way?
- Why do you disagree?

For the record, this example occurred during an Algebra 2 class. A room full of juniors in high school struggled with accurately placing the three simplest fractions on a number line. With these three fractions, I anticipated some students would make the error of placing them in ascending order of the denominator, as some indeed did. I did not expect students to simply copy the order in which I offered the three values. This reveals just how trained students are to copy from and regurgitate back to the teacher.

The strategies of classmates, particularly those of decimal conversions and common denominators, helped many students sort their thinking on fractions. The proportional reasoning encouraged by the finger reasoning was beneficial to all and will be further enhanced as we revisit number sense on the clothesline throughout the school year.

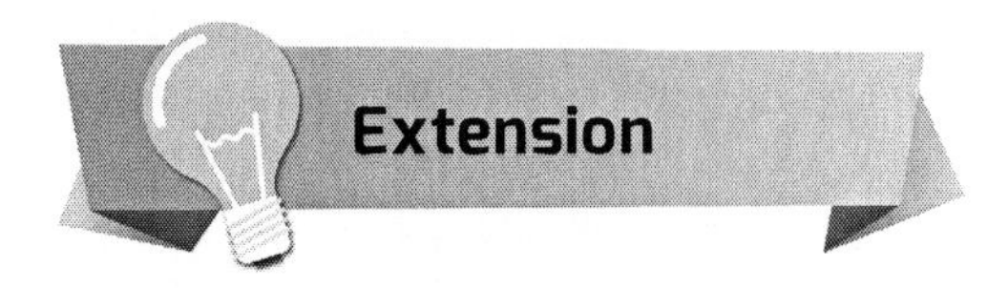

Place a fraction on the number line that is between $\frac{1}{3}$ and $\frac{1}{2}$.

Figure I.9 Final Clothesline Placement

The Structure of This Book

What you just read was a typical introduction to Clothesline Math.

The majority of this book is filled with lessons like these. The values or expressions to be placed on the number line are posed first, followed by a mock classroom discourse that leads to the correct placement on the clothesline, which is to be recorded on the *Clothesline Math* activity sheet (pages 315–316) at the end of each lesson. A teacher debrief finishes each model lesson with the following format:

Why These Numbers: This is an explanation of the importance behind the choice of numbers for each lesson.

Key Questions: Clothesline Math is intended for discourse, not lecture. Since good classroom discussion is led by questions, not statements, a bank of prompts is offered for each lesson.

Analysis: This is the reflection on the lesson vignette, including discussion of student thinking and teacher decisions during the lesson.

Extension: This is an additional activity suggestion to be used for differentiation, assessment, or independent assignment.

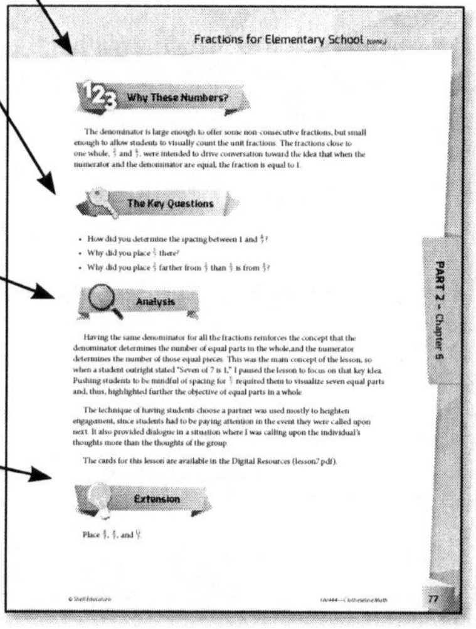
Fractions for Elementary School (cont.)

Why These Numbers?

The denominator is large enough to offer some non-consecutive fractions, but small enough to allow students to visually count the unit fractions. The fractions close to one whole, [illegible] and [illegible], were intended to drive conversation toward the idea that when the numerator and the denominator are equal, the fraction is equal to 1.

The Key Questions

- How did you determine the spacing between 1 and [illegible]?
- Why did you place [illegible] there?
- Why did you place [illegible] farther from [illegible] than [illegible] is from [illegible]?

Analysis

Having the same denominator for all the fractions reinforces the concept that the denominator determines the number of equal parts in the whole,and the numerator determines the number of those equal pieces. This was the main concept of the lesson, so when a student outright stated "Seven of 7 is 1," I paused the lesson to focus on that key idea. Pushing students to be mindful of spacing for [illegible] required them to visualize seven equal parts and, thus, highlighted further the objective of equal parts in a whole.

The technique of having students choose a partner was used mostly to heighten engagement, since students had to be paying attention in the event they were called upon next. It also provided dialogue in a situation where I was calling upon the individual's thoughts more than the thoughts of the group.

The cards for this lesson are available in the Digital Resources (lesson7.pdf).

Extension

Place [illegible], [illegible], and [illegible].

PART 2 - Chapter 5

© Shell Education 100444—Clothesline Math 77

This text also includes justification for teaching number sense at all grade levels, the simple materials and setup needed to conduct clothesline lessons in your classroom, and techniques for effectively facilitating the discourse spawned by our favorite number line tool. With these examples and tips, I hope to empower your teaching with Clothesline Math.

Teaching with the Clothesline

The Master Number Sense Maker

In February 2015, I attended the annual MaTHink Conference in Riverside, California. I sat in on a session that a friend of mine, Kelli Wise, was presenting on Clothesline Math. Although I had seen the clothesline used for fractions and decimals, I was intrigued by her description of demonstrating algebra on an open number line. As promised, Wise placed a variable on the clothesline, and my mind was blown. As she continued through her examples, the clothesline touted its prowess of promoting conceptual understanding of algebra, revealing students' misconceptions of variables, strengthening basic number sense, enhancing critical thinking, and generating mathematical discourse. I was in awe of this display of instructional muscle. Wise concluded her session by offering tips on choosing values to place on the number line and on asking the right questions to lead discourse, which I had already learned from her skillful presentation.

At the end of the session, my friend sitting next to me, Tim McCaffrey, claimed, "The clothesline is the master number sense maker." I wholeheartedly agreed and returned to my classroom excited to implement this new tool with my own Algebra students. The Clothesline proved to be even more powerful in the classroom since students didn't know all the answers as did the room full of math teachers at the conference. I started to see more potential for the clothesline as I experimented with it in my geometry classes, developing new applications and improving my own skills for its facilitation.

I began to demonstrate it for other teachers as part of my math coaching duties, as well as at district trainings and math conferences, and the popularity of clothesline grew. To assist in its expanding use, I created the website www.clotheslinemath.com and started the Twitter hashtag #clotheslinemath. With more teachers at various grade levels in various math courses sharing ways of using the Clothesline, enthusiasm for it swelled. I started receiving thank you emails and photographs of Clothesline lessons from across the country. Then, I went to the annual conference for the National Council of Teachers of Mathematics in 2017 and saw five different sessions besides my own in which the Clothesline was being presented. I realized that my love for the "Master Number Sense Maker" was shared by many.

The Need for Number Sense

The clothesline is more than a trending activity, however. It serves a crucial, yet overlooked, purpose in mathematics education—the need to teach number sense.

Several longitudinal studies show that students who lack number sense in primary grades are nearly guaranteed to fail their high school math courses, and students who have strong number sense are nearly guaranteed to excel in their high school math courses (Jordan 2010; Duncan 2007). Number sense has a similar influence on students' success in math as reading has on their school success overall. In fact, "mathematics difficulties and disabilities have their roots in weak number sense" (Jordan 2010, 2).

So, what is number sense exactly? In essence, it is a "child's fluidity and flexibility with numbers" (Gersten and Chard 1999, 3). In other words, being fluent in math is much the same as being fluent in a language. To be fluent in a language, one must know the fundamentals of the language, but one must also have the ability to respond to a question or take part in a conversation with sentences that have not been simply memorized and regurgitated. Just as people need to be fluid and flexible with their words, students must be able to respond in the same fluid and flexible way when encountering unique math problems. Their ability to do this relies heavily on their number sense.

Number sense is acquired through exploration and play. Much of the number sense children arrive with in kindergarten is acquired informally in the home in much the same way their language skills are acquired. Students who are more engaged with words at home enter school with larger vocabularies and stronger language structures. Thus, they are better prepared to learn how to read and write. Students who are engaged with numbers and their properties at home enter school better prepared to learn math, too. However, students do not obtain language readiness by studying pre-K vocabulary lists. They informally interact with their language, usually through conversation with adults. Correspondingly, students do not obtain number sense by memorizing lists of math facts. Instead, they interact informally with numbers. Oftentimes this comes through games, such as card and board games, in which students are required to count, add, and strategize. Duncan (2007) describes that "play-based, as opposed to 'drill-and-practice,' curricula designed with the developmental needs of children in mind can foster the development of academic and attention skills in ways that are engaging and fun" (8). The clothesline straddles that boundary between developing informal play and the formal needs of classroom management. This is enormously helpful when teaching students who do not have adequate number sense to learn the content at their current grade level.

Teaching Number Sense and Prerequisite Skills

Why, then, does the building of number sense get so little attention in the mathematics classroom? There are two valid reasons. The first is the lack of time. Teachers are already hard pressed to cover current content standards. The second is the lack of resources to help teach number sense. Worksheets for procedural practice fall short of building number sense, take too much time, and often have already failed students in their prior experiences. Therefore, teachers need a simple tool that can quickly strengthen students' prior knowledge while simultaneously moving them forward with grade-level content. The open number line serves as such an instrument that can be used at the beginning of a unit, as a warm-up, on assignments, or as part of assessments.

Besides a lack of number sense, there is another hindrance to learning math: the lack of prerequisite skills. This is well-known to most math teachers. At all grade levels, one of the most common complaints made by teachers is that students are not prepared for the current grade level or math course. Calculus teachers claim students don't flunk calculus, they flunk algebra. Students don't flunk algebra, they flunk middle-school math. If they can't multiply negative numbers, it's because they can't multiply positive numbers. Elementary school students can't add two-digit numbers because they can't add single-digit numbers. Nearly every teacher agrees: Students don't flunk current content; they flunk prior content.

So, if we are all in consensus that number sense is a critical ability too many students lack and that many students are missing several of the skills required to be successful, then the obvious response from all those teaching math should be to teach number sense and the prerequisite skills. The Clothesline is prepared for the task.

The Open Number Line

As shown in Figure 1.1, an open number line is one that does not have benchmarks already established. In other words, the numbers 0, 1, 2, 3, etc. that you would see on a traditional number line are missing.

Figure 1.1 Open versus Traditional Number Line

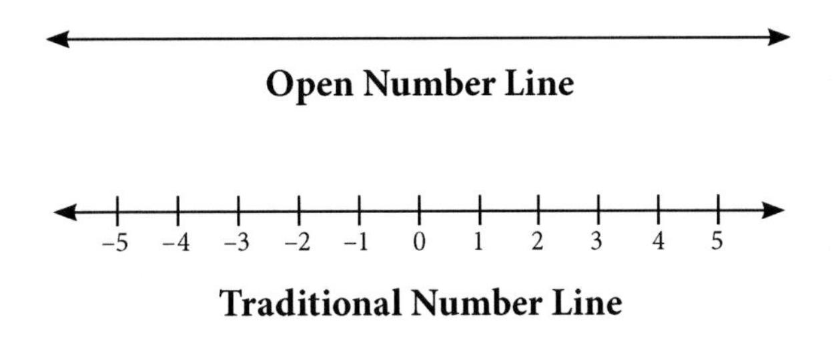

A lack of benchmark numbers requires students to establish a scale. Demanding this of students serves two purposes. The first is to help students build understanding of cardinality, a sense of magnitude, and proportional reasoning. The second is that the new twenty-first century standards insist that high school students be able to "choose and interpret the scale and the origin in graphs and data displays" (CCSS HSN.Q.A.1) and students of all levels be capable at "considering the units involved; attending to the meaning of quantities, not just how to compute them" (CCSS MP 2). There are standards that explicitly claim the use of number lines throughout the elementary and middle school grades, such as "represent a fraction $\frac{a}{b}$ on a number line diagram by marking off lengths $\frac{1}{b}$ from 0," (CCSS 3.NF.A.2.A) and "Use ratio and rate reasoning to solve real-world and mathematical problems, e.g., by reasoning about…double number line diagrams" (CCSS 6/RPA.3).

The open number line is a means and an end. It is a tool to develop number sense and reteach prerequisite skills, but having students proficiently use it is a goal in and of itself.

NCTM's Eight Mathematics Teaching Practices

The National Council of Teachers of Mathematics (2014) laid out "eight effective teaching practices" in their publication *Principles to Actions: Ensuring Mathematical Success for All.* This well-regarded document claims that effective mathematics teachers do the following:

1. Establish mathematics goals to focus learning.
2. Implement tasks that promote reasoning and problem solving.
3. Use and connect mathematical representations.
4. Facilitate meaningful mathematical discourse.
5. Pose purposeful questions.
6. Build procedural fluency from conceptual understanding.
7. Support productive struggle in learning mathematics.
8. Elicit and use evidence of student thinking.

The Clothesline helps implement all of these, particularly facilitating math discourse, building conceptual understanding, and eliciting evidence of student thinking. In other words, the clothesline can aid you in becoming a more effective twenty-first century math teacher. As you read the lessons in this book, notice how the card sets are chosen and placed to drive students toward cognitive dissonance. I joke to my pupils that I purposely try to get their brain cells to fight. This intellectual challenge is what brings conceptual understanding to the forefront, so students gain procedural fluency more easily.

The lesson vignettes in this book also serve as extensive models of how to facilitate mathematical discourse. They demonstrate solid teacher questioning and guidance of the discussion as well as typical statements made by students. The entire conversation, of course, is driven by the evidence of students' reasoning found in their responses on the Clothesline. The ultimate goal of these lessons is to develop the mathematical understanding of the students. NCTM calls for teachers to take students beyond simple answer getting. The Clothesline answers that call.

The Eight Standards of Mathematical Practice

The new twenty-first century standards not only have framed what students should know but also how they should think. These "habits of mind" are most commonly known as Standards of Mathematical Practice. However, in some states they are also known as Mathematical Processes. They are all very similar and nearly identically described. This noble goal of teaching students to think is the hallmark of the new standards, and thus is the hardest part of the new standards to teach students. They are not teacher moves like NCTM's Practices; they are student habits to be instilled at every grade level. They are exquisitely stated as the following:

1. Make sense of problems and persevere in solving them.
2. Reason abstractly and quantitatively.
3. Construct viable arguments and critique the reasoning of others.
4. Model with mathematics.
5. Use appropriate tools strategically.
6. Attend to precision.
7. Look for and make use of structure.
8. Look for and express regularity in repeated reasoning.

The Clothesline can be used to teach and reinforce any one of these. However, I think its strengths lie in supporting Practices 1, 2, 3, 6, and 7:

- **Make sense of problems and persevere in solving them**—Sense making is one of the key features of the clothesline. Taking symbols and numbers from paper to the number line forces students to grapple with the relationships and properties of these values until they are understood, not just memorized.
- **Reason abstractly and quantitatively**—Reasoning quantitatively is a given...it's a number line after all. Of course it's going to get students to think about numbers! But instead of students just seeing a value written as 0.003, placing it very close to zero on the clothesline helps them understand the relative minuteness of this quantity.

  Additionally, reasoning abstractly is the holy grail of an education. While concrete reasoning involves things we can see and touch, abstract reasoning involves things that we cannot see and touch. It means we think about ideas. For example, $\frac{1}{2^1}$ and $\frac{1}{2^2}$ are easy to think about, because I have experience cutting something in half, and I have taken half of a half before. But what about $\frac{1}{2^{100}}$? I have never repeatedly cut something in half 100 times. Therefore, placing this fraction on the number line requires abstract reasoning. Many students place it well to the right of the number line, claiming that since the exponent is so big the value is close to infinity. Others place it well to the left because they figure if something gets smaller 100 times, it must approach negative infinity. Imagining cutting something in half 100 times (without actually doing the cutting) and realizing the size of the quantity is approaching 0 takes some serious abstract thinking.

- **Construct viable arguments and critique the reasoning of others**—This is the fun part of the clothesline. Getting students to debate ideas in math leads them to a deeper understanding of mathematics, but even more so, it trains them to think like mathematicians. Having the confidence to defend their own conjectures and challenge others is a phenomenal gift that Clothesline Math can give our students.
- **Attend to precision**—Attending to precision is given particular importance when using proportional reasoning to get the spacing of the cards correct and in the promotion of properly using academic vocabulary.
- **Look for and make use of structure**—This process is often required to accurately respond to Clothesline prompts. Several structures were being wrestled with by the students in the Introduction example: the denominator establishes the total parts of the same whole, fewer parts of the same whole implies larger parts, and the parts are equivalent. These are some of the underlying structures of unit fractions that become more evident on a number line than when simply writing them with pencil and paper.

My final thoughts on the relationship between teaching mathematical practices and processes and the Clothesline deal with dual objectives. Every lesson should have some type of instructional objective. A twenty-first century math lesson should always have two: one content objective and one practice/process objective. I do this in my classroom on a regular basis. For example, "We will *persevere* in solving problems involving *trigonometric ratios*" is a dual objective I have used in my geometry class. While the content standard is focused on applying trigonometry, the practice standard is on persevering. I state this at the beginning and revisit the idea of not quitting throughout the lesson as students work on higher level trigonometry problems.

Since Clothesline activities are rarely used as stand-alone lessons, I have not stated dual objectives here in this book; I have stated only the content objective of the lesson. Besides, nearly all of the math tasks I have used in my own classes can be used for teaching several mathematical practices/processes, even though we only want to focus students on one for any given lesson. For these reasons, I leave it up to you as to which practice/process you want to emphasize in any given clothesline lesson. I strongly encourage you, though, to make it a point to instill these habits of mind on a daily basis.

Setting Up Your Clothesline

Now that I have made the case for why the Clothesline is such a powerful tool for teaching mathematics, let's dive into what you will need to make it happen in your classroom. The following sections will outline the items you will need as well as how to utilize each element in your lessons.

The Materials

The Clothesline gets its name from being, well, literally a clothesline. You can purchase a clothesline and a package of clothespins for less than $5. You will also need a few other materials for attaching the clothesline to the wall and for students to participate at their desks. In this chapter, I will discuss several options for you to choose from, but the materials needed to conduct an amazing Clothesline lesson are minimal. For example, the materials list for my own classroom is as follows:

- 2 nylon clotheslines
- 12 clothespins (assorted colors)
- 4 magnetic hooks (2 per clothesline to attach to the whiteboard)
- clothesline cards (number cards, benchmark cards, and blank cards)
- 5 oversized envelopes (to store the cards)
- lapboards (1 board per student pair or group)
- *Clothesline Math* activity sheet (pages 315–316)

Let me show you the setup I use in my classroom as well as examples I have seen from other classrooms.

The Front of the Class

In my classroom, I use magnetic hooks on my whiteboard (see Figure 2.1). The clotheslines are attached and hang nicely to the side, so they are always ready to go. The added bonus of the whiteboard is that I can write notes and examples above and below the clothesline.

Figure 2.1 Sample Clothesline on Whiteboard

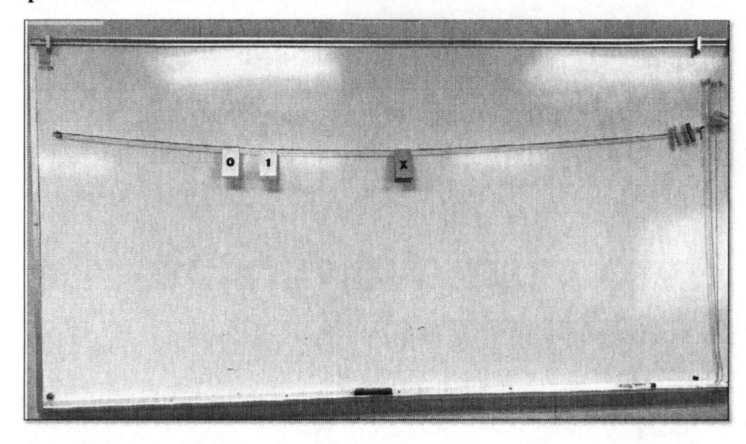

This is not the only setup, however. The first time I saw the Clothesline demonstrated, it was suspended across the corner of a room (see Figure 2.2), so that is how I initially did it. I have seen clotheslines tied between individual chairs (see Figure 2.3), stacks of chairs, and even homemade PVC posts. I once made a makeshift clothesline with a computer printer cable and two music stands. I've even seen a hi-tech version on an interactive whiteboard.

Figure 2.2 Corner Clothesline

Figure 2.3 Chair Clothesline

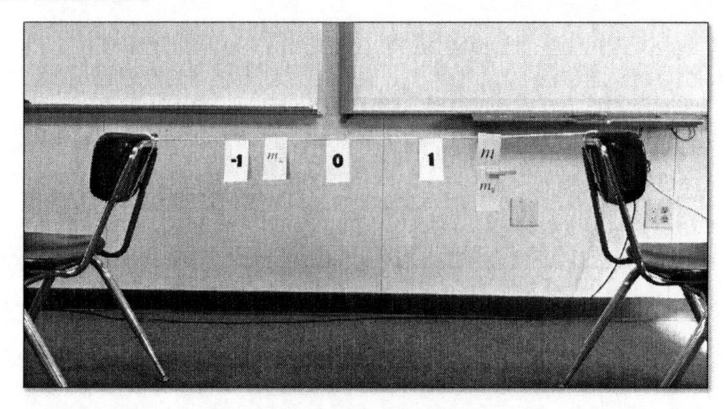

The easiest setup, of course, is a clothesline pinned or taped to a wall (see Figure 2.4).

Figure 2.4 Wall Clothesline

Choose what works best for you and your classroom. All that matters is that you get the open number line in front of students.

The Cards

Clothesline cards are best folded in half so they hang nicely on the clothesline. This also allows them to easily move back and forth along the line. I typically cut three cards from one sheet of paper. Cards that have longer expressions may only fit two on a single sheet. While you can handwrite them, common number cards for the lessons in this book have been included in Appendix C (pages 317–335). All the cards needed for each lesson are available as individual PDFs in the Digital Resources. You can also find a multitude of prepared cards online at www.clotheslinemath.com.

> Digital resources to support the lessons in this book are available online. A complete list of available documents is provided on page 336. To access the digital resources, go to this website: http://www.tcmpub.com/download-files. Enter this code: 34583417. Follow the on-screen directions.

While the open number line offers few to no benchmarks to start, benchmarks should be available to students. Benchmark cards need to be printed and available, usually sitting on a desk or table near the front of the room. As shown in Figure 2.5, benchmark cards typically have whole numbers from 0 to 10 for elementary school and integers from -3 to 5 for secondary.

Figure 2.5 Integer Benchmark Cards

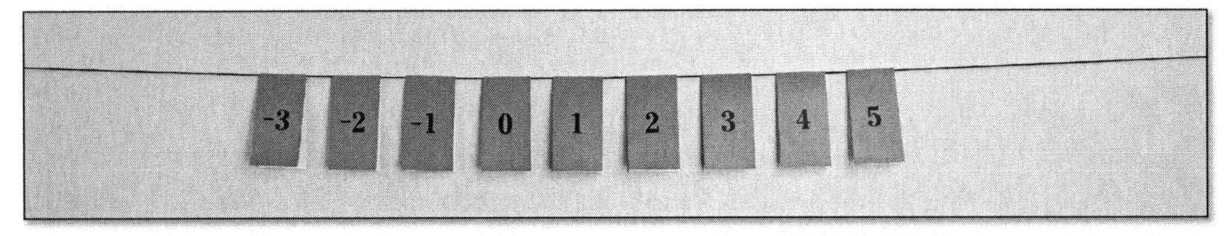

For courses that teach angle measure, you will also want benchmarks for the most common degree measures of 0°, 30°, 45°, 60°, 90°, 180° (see Figure 2.6).

Figure 2.6 Degree Benchmark Cards

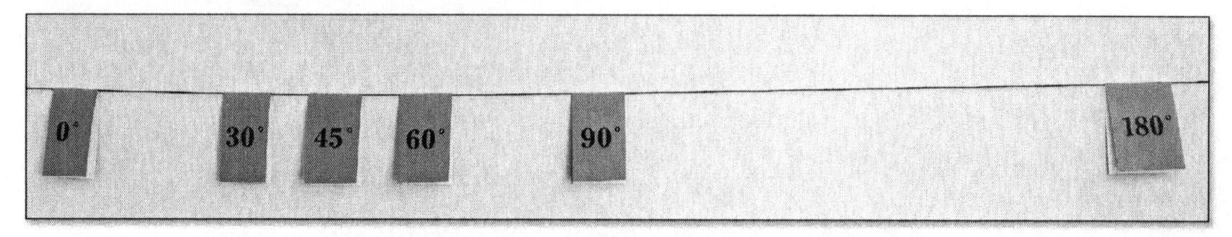

You should also have several blank cards on a table nearby for students to use.

Like the fraction example in the Introduction, a lesson's card set is usually comprised of three values. However, some lessons may vary. I store my card sets in oversized envelopes that I keep pinned to the front wall just under the board where the clothesline is displayed. I also color-code my cards, though this is not at all necessary. As shown in Figure 2.7, I print the cards on colored paper as follows:

- **White:** benchmarks
- **Green:** numbers
- **Red:** single variables and statistic symbols
- **Orange:** algebraic expressions
- **Yellow:** angle measures and drawings

Figure 2.7 Color-Coded Card Sets

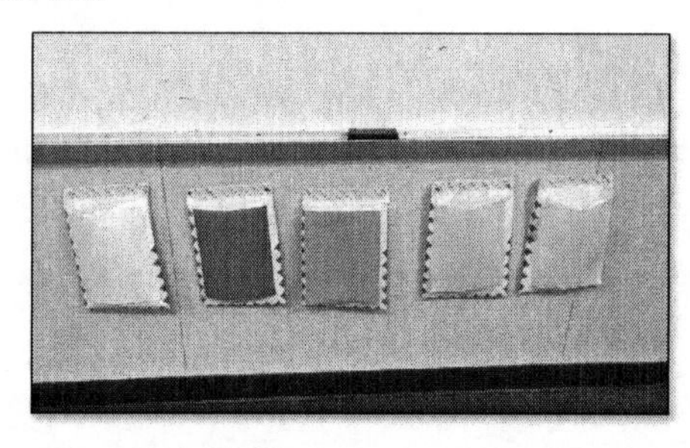

For readability purposes of the photos in this book, all number cards will be on white paper, and benchmarks will be on colored paper.

Drawings are often needed for Clothesline lessons on geometry and functions. While I often have these drawings on paper, I also display drawings by enlarging them on the whiteboard or interactive whiteboard for the sake of easier viewing (see Figure 2.8).

Figure 2.8 Sample Drawings on Whiteboard

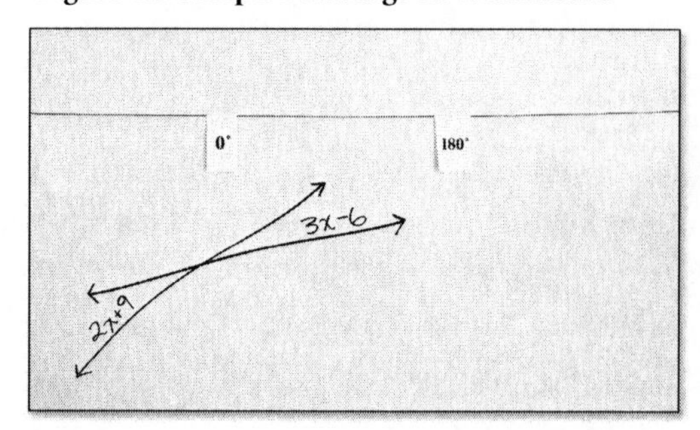

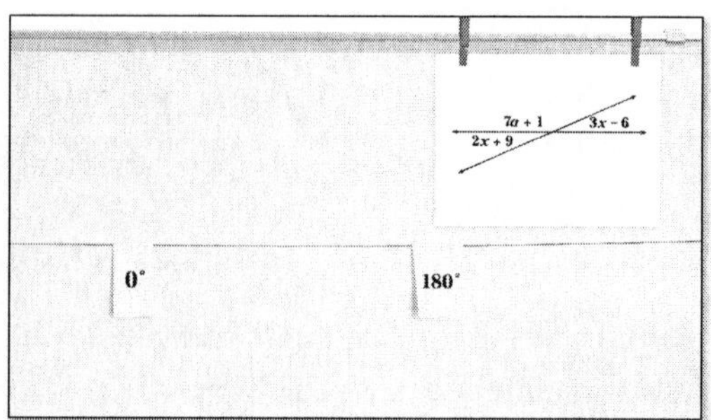

The Rest of the Class

It is very important that, while one group is at the clothesline, the rest of the class has something to do at their desks to engage them in the activity. I learned this the hard way.

One day, I was telling a fellow teacher how excited I was to try the clothesline for the first time. He asked if he could come watch. My first thought was, "Did you not hear me say that I was doing this for the first time?" As a fellow teacher, he should have known that we teachers don't want people watching us until we have the lesson down pat. However, I was a math coach for the department and needed to promote transparency and risk taking, so I had no recourse but to invite my buddy along.

I introduced the Clothesline to the class and sent a group of students up to place the cards. The conversation within the group was amazingly rich. But, it was also long. Do you want to guess what happens when one group of students stands with their backs to the class for an extended period of time and you haven't given the rest of the class something to do? You've got it…the animals escaped from the zoo. And, my colleague watched the whole thing implode.

We debriefed afterward and both agreed the group at the front of the class proved how engaging and instructional the Clothesline is, but the other students needed something to do at their desks to engage them as well. The next day, he came back to observe another Clothesline activity again. This time, we had students pair up and use lapboards (see Figure 2.9) to give their responses to the prompt being discussed at the front of the room. This not only helped classroom management and engaged all students in the mathematics of the activity, it also greatly enhanced classroom discourse. We were able to discuss who agreed and disagreed with the card placement on the clotheslines, plus students had the opportunity for the same rich discussion with their partners that the group at the clothesline had.

Figure 2.9 Sample Lapboard

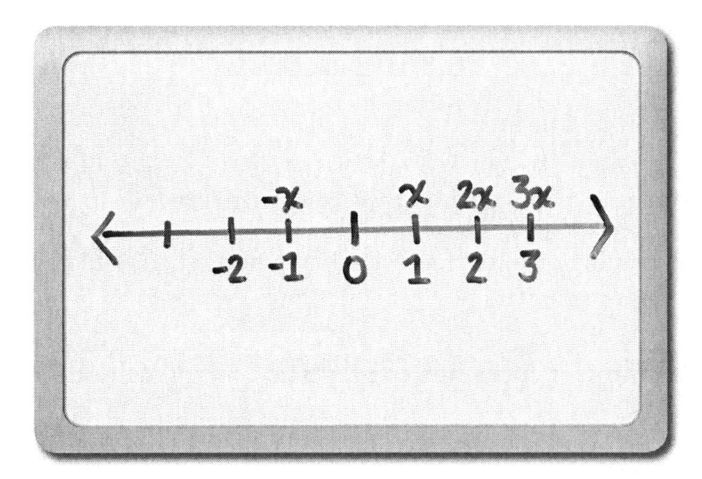

I find lapboards to be the easiest tool to engage students at their desks but not the cheapest. If you do not have a set of lapboards available and your school does not have the funds to purchase a class set for you, you can make them by inserting blank paper or graph paper into sheet protectors.

The Clothespins

When students place equal values on the same location of the clothesline for the first time, they inevitably set one atop the other, so that one of the values is no longer visible. I joke with them at this point, "Hey, where did the other one go?" When they reveal the other value underneath, I ask, "Why did you do that?" Once they accurately claim the values are equal, I grab a clothespin and pin the values together vertically so both are visible (see Figure 2.10). The clothespins are used to show equivalency. Sometimes, there may be more than one pair or more than two values that are equivalent to one another.

Figure 2.10 Equivalent Cards Pinned

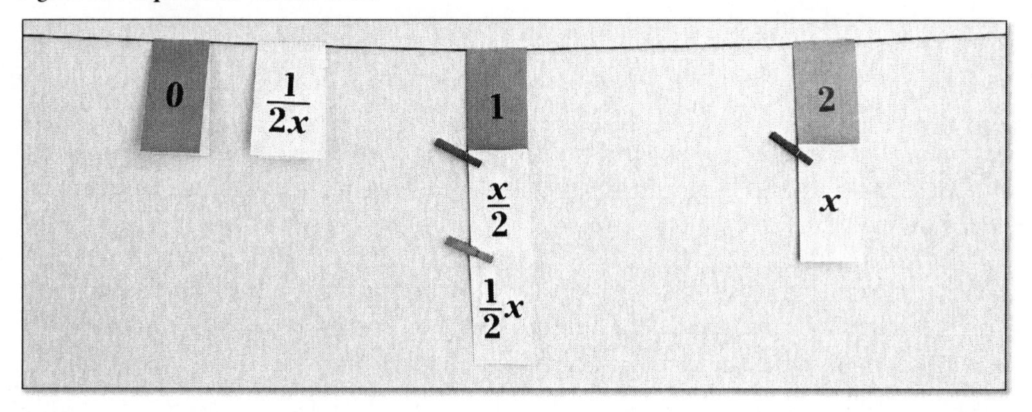

Sometimes, equivalency becomes the focal point of the lesson. That is the time to make the move to multiple clotheslines.

The Double and Triple Clothesline

A double clothesline is an effective setup when you have situations with different units of measure, like with rates, or when teaching multiple forms of a value, such as percentages and decimals. For example, a double clothesline is a terrific tool for presenting the concept of unit rate.

Consider this scenario, shown to me by Kristen Bennett of Tustin USD: Jody buys a 5-ounce serving of frozen yogurt for $2. Her older brother, Timmy, buys the 10-ounce serving. How much does Timmy's yogurt cost? What is the price of yogurt per ounce?

In this example, pinning a 5 and 2 together on one clothesline would be confusing and inaccurate. We are not claiming that 5 is equal to 2! We are claiming there is a relationship between ounces and dollars. As shown in Figure 2.11, the double clothesline allows us to label these units in a way that illuminates the unit rate concept of dollars per ounce.

Figure 2.11 Sample Double Clothesline—Price versus Ounces

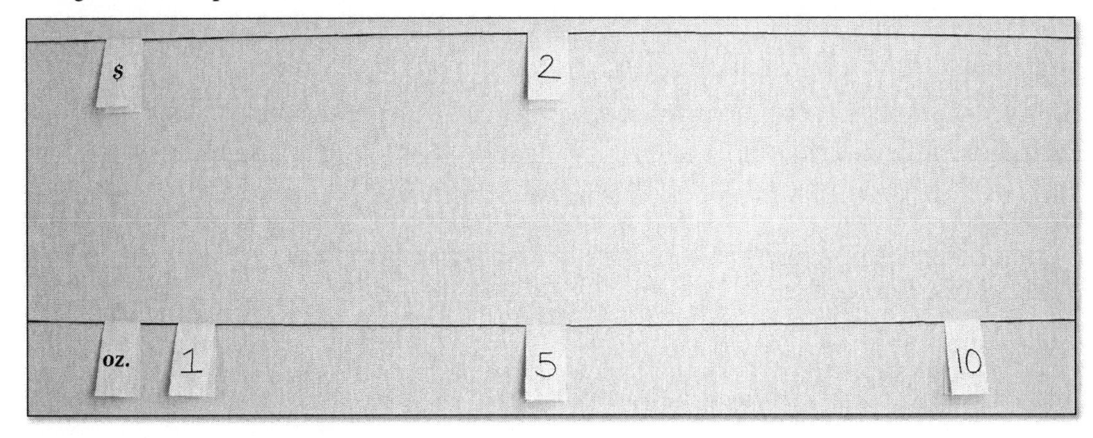

This way of comparing different units or representations with multiple clotheslines spans all grade levels, as you can see in Figure 2.12.

Figure 2.12 Sample Double Clothesline—Number versus Pictoral

I have even seen a triple clothesline tie together the various forms of numbers (see Figure 2.13)

Figure 2.13 Sample Triple Clothesline

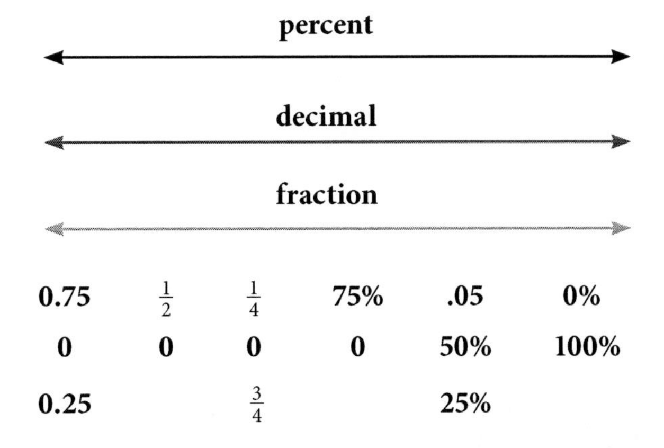

The Activity Sheet

Students can easily record the results of the Clothesline lessons on notebook paper. However, the *Clothesline Math* activity sheet (pages 315–316) is available in Appendix C. Its use is discussed in Chapter 3.

Name: ______________________ Date: ______________

Clothesline Math

Directions: For each set, record the given values, expressions, or drawings. After the discussion of their placement on the clothesline, record them on the number line.

1. __________, __________, __________

2. __________, __________, __________

3. __________, __________, __________

© Shell Education 100444—Clothesline Math 315

Making the Most of Your Clothesline

Now, you have what is needed to bring Clothesline Math to your classroom. The next step is to review how you can best facilitate it.

Choosing Values for the Card Sets

The key to an effective Clothesline lesson is choosing the appropriate values for the card set. This set of values needs to create some kind of cognitive dissonance that will assist you in illuminating the mathematical objective. For example, math teachers know that students have a very difficult time discerning between $(-3)^2$ and -3^2, yet this is a critical concept for evaluating expressions. One of the best ways to clear up any misconception is to take it to the Clothesline. Therefore, the card set for this lesson would be the expressions $(-3)^2$, $-(3)^2$, and -3^2. The last two terms are equivalent since both equal negative 9, but the first equals positive 9.

Choosing values in which some are equal and others are not, while still looking very similar, adds power to a lesson's ability to clarify the mathematical concept that is confusing students. Card sets for the Clothesline lessons in this book are available in the Digital Resources. In the event you want to create your own card sets, take your time choosing values that you will use, making sure to choose values that best drive student attention toward the point of proper understanding.

Questioning with Purpose

One of NCTM's Eight Teaching Practices—purposeful questioning—is an especially important teacher move during Clothesline lessons. You need to do more than simply state, "That's wrong. You should have placed it here." Good questioning drives class discussions to the core mathematical objectives. Therefore, two things are offered in this book to support your purposeful questioning. The first are the vignettes, which model questioning by the teacher. You will notice the teacher often responds to student answers with another question that drives discussion closer to the lesson objective. Sometimes, the purpose of a question is just to have the next student restate the previous student's response to keep the class actively engaged. Other times, the question is intended to reveal multiple solutions or strategies.

The second tool to support purposeful questioning is a set of suggested questions for each lesson. Eventually, you will get comfortable generating your own questions. In the meantime, these will help you get started and will show just how far you can push student thinking with the Clothesline. To help in this endeavor, let us explore a proven framework for leading classroom discourse in mathematics.

Facilitating Classroom Discourse (NCTM's 5 Practices)

Without classroom discourse, much of the potential of Clothesline Math will not be achieved. In the *5 Practices for Orchestrating Productive Mathematics Discussions* (2011), Dr. Peg Smith and Dr. Mary Kay Stein cleanly lay out a framework that will be utilized many times in our Clothesline lessons. Here is a brief summary of these simple, yet crucial, practices and how they apply to teaching with Clothesline Math as well as the corresponding Teaching Practices.

1. **Anticipate:** Choose values to place on the number line that illuminate the math concept you are targeting. Then, think about how students might respond. This is done during the planning of a lesson. (#1 Math Goals, #2 Tasks, and #6 Procedural from Conceptual)
2. **Monitor:** While one group is at the clothesline and the rest of the class works in pairs with a number line at their desks, walk the room to observe what students are producing and revealing. (#7 Productive Struggle)
3. **Select:** Choose student samples you know will drive the upcoming discussion. You want more than just the "right answer." Students learn just as much, if not more, from mistakes as they do from correct examples. When available, choose an incorrect example that may highlight a key understanding of the math goal. Of course, we don't want to leave students with incorrect responses, so choosing an accurate example is important, too. Also, identifying a model of an unusually clever or high-level response is always a treasure. During those times when the class unanimously produces correct answers, selecting various strategies can then be the focus of discussion. (#8 Evidence of Student Thinking)
4. **Sequence:** This is when the meat of class discussion occurs. After selecting student examples to drive discourse, sequence them in such a way that the math goal is progressively revealed. This is often best accomplished by displaying erroneous answers or simpler strategies first, more straightforward strategies second, and any advanced examples last. (#4 Math Discourse and #5 Purposeful Questioning)
5. **Connect:** Few students tie all the pieces of a lesson together on their own. It will be your job to do that for them. They need you to show how the examples and discussions are connected to both the current lesson objective as well as any previously learned math concepts. (#3 Multiple Representations)

The Monitoring-Selecting-Sequencing phase of NCTM's 5-Practice protocol comes into play once the teacher sends students to the clothesline at the front of the room. While students are at the clothesline, the teacher is not only observing the responses there but also the responses of students still seated. While monitoring of student thinking is occurring, the teacher makes a mental note of particular responses to be shared in the upcoming discourse. Once class discussion begins, the order in which students are called should be very intentional. Indeed, the teacher needs to remain open to improvisation to support unexpected ideas, yet there is also a particular stream of mathematical consciousness that needs to be revealed.

Typically, once the leading group sits down, the teacher will ask for an explanation of their very public responses. Then, the teacher will call for any disagreement. The classic thumb vote technique (thumbs-up for agreement, thumbs-down for disagreement) is my personal go-to as you will see throughout the vignettes in this book, but you should use whatever technique is your go-to for facilitating student engagement. Before the vote is cast, however, I already know who I'm going to call first (sequencing) and next, while leaving room for others to chime in.

Sending Students to the Clothesline

Students need to engage with the clothesline, not watch the teacher engage with it. Therefore, they must be sent to the front of the room to work out their thinking publicly. It is often best to send students in pairs or groups, but there are times when sending students up individually can enhance the dynamics of the lesson as well. Often, one lesson or card set will require several successive groups of students to be called to the clothesline. There are several methods for selecting students to work at the front of the classroom.

Group Chain

My favorite selection method is the group chain, in which the next group to go up is chosen by the previous group selected. This keeps students engaged with the mathematical story being developed at the front of the room. It may sound contrary to the select and sequence process that I just professed, but it is not. While there is some randomness and ownership in choosing students to go to the clothesline, the selection and sequencing is dependent upon the student work occurring at the desks.

Student Chain

A student chain is similar to a group chain in that the next student is chosen by the previous one—with one caveat. I start by handing the first value to the first student, who places it on the clothesline. The next student is given a new value and a choice. The student either places the card on the clothesline and sits back down, or changes something on the clothesline that is thought to be wrong and hands the card I gave him or her to another student. If the card is placed, I choose the next student with my selection sequence in mind. Student chains are best used when value sets are large.

Random Selection

Of course, there is always the method of randomly selecting a group or student with a random number generator, equity sticks, or a deck of cards. Similar to the group and student chains, however, while one group is up front at the clothesline, the teacher is working the room in preparation for the class discussion.

Gradual Reel In

Gradual release is a technique that gained a great deal of popularity during the era of No Child Left Behind. It is encapsulated by the phrase "I do, we do, you do." With this method of instruction, the teacher models the skill to be emulated, then has the class mimic the procedure while making any needed corrections. Then, students practice independently. This technique is effective once students are at the procedural fluency phase of learning a math goal, but when we precede that with the gaining of conceptual understanding of the mathematics underlying the procedure, we want to reverse this order of gradual release to a sequence that I refer to as gradual reel in. In other words, "You do, we do, I do." This is a preferred model when students are working on mathematical tasks that often involve critical thinking. With tasks, we want students thinking on their own or in small groups first (you do). Then, the teacher leads a discussion about the student thinking (we do). Finally, the teacher summarizes and highlights the key objectives of the lesson (I do). You will see this structure in almost every Clothesline lesson in this book.

Using the Activity Sheet

Delay the recording of anything until the closure of the discussion on a particular card set. As you can see from the fraction example in the Introduction, there will be many changes made on the number line during the activities. You do not want students having to erase multiple times. That is why we use lapboards or other such manipulatable tools first. Once mathematical conversations are done, then students can write the original values or expressions from the card set, the final display of the number line(s), and notes of key ideas. In other words, the *Clothesline Math* activity sheet is a record of our discussions, deductions, and decisions. This practice of taking notes at the end of the lesson is contrary to most students' experiences. I often tell students, "When I do most of the talking, we take notes at the beginning of class. When you do most of the talking, we take notes at the end of class."

Using Clothesline Math

There are several different ways to utilize Clothesline Math within your daily math lessons. From a warm-up, to an assignment, to a full lesson, there is a way for the Clothesline to fit within any time constraint you may have. The following examples detail different options based on classroom need.

Clothesline as a Warm-Up on Prerequisite Skills

The most common use of Clothesline Math is for a brief warm-up activity that ties a mathematical skill to the lesson objective of the day. These skills are often considered prerequisite to the current lesson. For example, a Clothesline warm-up on equivalent fractions would be very useful before a lesson on adding fractions using common denominators.

Clothesline as a Warm-Up for Current Content

A Clothesline warm-up activity for the day can also focus on a concept currently being learned. For example, if a task being presented involves adding fractions, the warm-up may entail placing two fractions and their sums on the open number line.

Clothesline as an Introduction of Current Content

While most Clothesline activities are short (approximately 10 minutes) and, thus, serve as terrific lead-ins to a lesson, some are longer and serve as investigative introductions to important concepts. For example, before we teach students the algorithm of adding fractions with like denominators, we can present these fractions on the clothesline (see Figure 3.1) to generate student conjectures regarding the operation.

Figure 3.1 Sample Fraction Cards

$\frac{1}{5}$ $\frac{1}{5}+\frac{1}{5}$ $\frac{1}{5}+\frac{2}{5}$

Since $\frac{1}{5}+\frac{1}{5}=\frac{2}{5}$, $\frac{1}{5}+\frac{2}{5}=\frac{3}{5}$, and $\frac{2}{5}+\frac{3}{5}=\frac{5}{5}=1$, students see a pattern to keep the denominator and add the numerator. The procedure for adding fractions is then generated from their own understanding to make adding fractions easier than using number lines.

Clothesline as an Assignment or Assessment

While the clothesline is a manipulatable tool, placing an open number line prompt on paper (see Figure 3.2) can be just as valuable to generate discourse during a group assignment or when revealing conceptual understanding on individual quizzes and tests. NCTM emphasizes the importance of engaging students in high-level thinking. To do that, NCTM (2014) encourages the use of "mathematics through multiple entry points, including the use of different representations and tools" (17), such as the use of the Clothesline. By having students work on open number lines on the Clothesline Math activity sheet, they are able to "foster the solving of problems through varied solution strategies" (17) by applying what they have learned on the clothesline in the classroom to the number lines in front of them. Students are also able to write important algorithms, expressions, and more next to their number lines to help reinforce the connection between the visual learning and the application. Group assignments using the Clothesline also help teachers "move away from past practices, such as tracking that separated students, and instead develop productive practices that support learning for all" (NCTM 2014, 65).

Figure 3.2 Open Number Line Prompt

Place *w*, RT, TS, and RS on the number line.

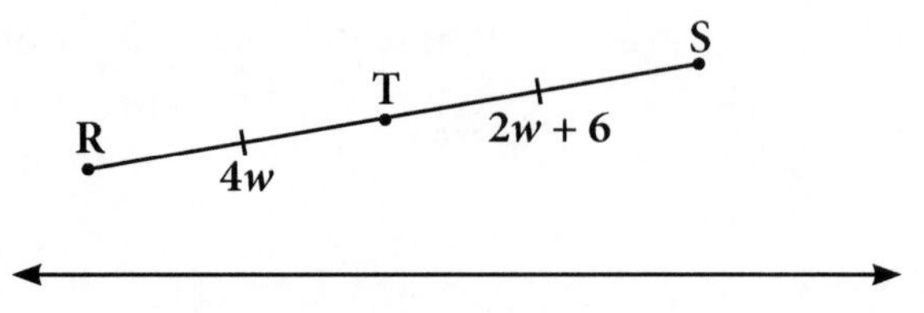

Given that $\frac{a}{b} = \frac{c}{d}$, $a \neq c$, and the position of *a* and *b* on the number line below, show a possible palcement of *c* and *d*.

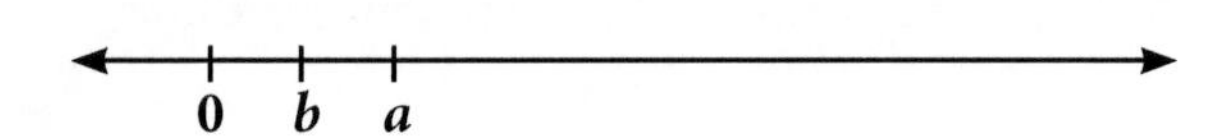

Clothesline as a Review

A pervasive practice in American mathematics education is to review for a test with a long problem set offered at the end of a textbook chapter. While this may serve well for independent practice, it does not take advantage of the learning potential available through teacher and peer interaction. For example, when reviewing a unit on the rules of exponents, an extended sample set of cards (see Figure 3.3) for the following values may be used.

Figure 3.3 Sample Set of Cards

$\sqrt[3]{8}$	$25^{\frac{1}{2}}$	$\frac{3}{2}$	$\sqrt{42}$	$8^{-\frac{1}{2}}$	$4^{\frac{3}{2}}$	73%
$(-2)^{\circ}$	-2.5	$-\sqrt{3}$	-0.08	$16^{\frac{1}{4}} + 32^{\frac{3}{5}}$	5^{-2}	

With this example, a student chain selecting process works very well. It demands that each student carefully pay attention and analyze values on the clothesline. It also allows for various concepts and skills to be viewed and compared simultaneously. The class discussion, which is absent from independent practice, helps clarify and solidify key math principles.

Now that you are better equipped to conduct a Clothesline Math activity, find a lesson in this book that is appropriate for your grade level. Have fun with it. Enjoy discovering your students' thinking. And change the world, one math lesson at a time.

Clothesline Lessons for Elementary School

Whole Numbers for Elementary School

Lesson 1: Counting Numbers

Objective: Place the counting numbers 1 through 5 in order on the number line.

Teacher: Hello, everybody.

Class: Hello!

Teacher: You sound excited to learn math today. Come join me on the rug in front of my display.

Students gather on the floor in front of the diagram (see Figure 4.1).

Figure 4.1 Number Line

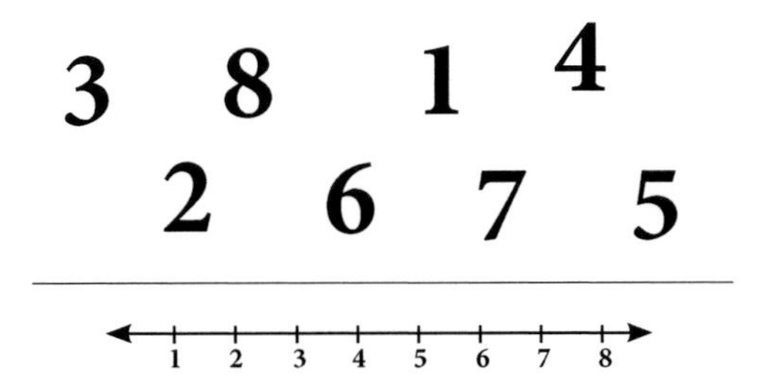

Let me see your hands in your laps and your eyes looking at the board. Can anyone tell me what you notice about this display?

Dione raises her hand.

Dione.

Dione: The top and bottom both have a 2.

Teacher: Thank you. Skye?

Skye: There is a 7 in the top part and another 7 in the bottom part.

Teacher: Thank you, Skye. Yolanda?

Yolanda: The top has all the same numbers as the bottom.

Teacher: What numbers do you see?

Laura raises her hand.

Laura.

Laura: One all the way up to 8.

Teacher: What do you notice is different?

Ollie raises her hand.

Ollie.

Ollie: The top numbers are bigger than the bottom.

Teacher: Bigger in what way?

Ollie: The bottom numbers are teeny, the top numbers are bigger this way and this way.

Ollie spreads her hands vertically and horizontally.

Teacher: I see what you mean. Just because it is written bigger, does that mean this top 8 is bigger than the bottom 8? Class?

Class: No!

Teacher: That's right. They are still the same amount. Do you see anything else that is different?

James raises his hand.

James.

James: The top numbers are random and the bottom ones are in order?

Teacher: Random is an interesting word. What do you mean by that?

James: They are all splattered.

Teacher: Class, are the numbers easier to count when they are splattered or in order?

Class: In order!

Teacher: Thank you. What else do you notice about the ones that are in order?

Marla raises her hand.

Teacher: Marla.

Marla: They have a line.

Teacher: Gina?

Gina: And the line has marks and arrows?

Teacher: I also notice those marks are evenly spaced. What might the arrows mean? Joshua?

Joshua: The arrows mean you can keep counting.

Teacher: Thank you, Joshua. Is there anything that you wonder about this diagram?

Ben raises his hand.

Ben.

Ben: I wonder how high we can count and how long the line can go.

Joshua: I can draw the longest line!

Teacher: That would be interesting to see. Thank you, Joshua, but not now. Sit back down, please.

Sunny raises her hand.

Yes, Sunny?

Sunny: I wonder what numbers are that way?

Sunny points to the left.

Teacher: Well, we have zero over there and some numbers we call negative numbers that you learn about when you are a little bit older. Well class, you have noticed some very good things about this tool that we call a number line. We call it a number line because it is a line with numbers on it. The numbers are equally spaced, in order, and the line can go on forever if you want it to. We have been learning our counting numbers and today we get to do an extra cool thing and place those numbers on a number line. Let's get our chairs and place them on the other side of the room in front of that piece of string you see.

Students collect their chairs and form two rows in front of the clothesline.

Teacher: Hands in your laps, eyes back on me, please. This is our number line today.

The teacher points to the clothesline hanging on the board with no number cards on it yet.

Sunny: But I don't see any numbers.

Teacher: Not yet, because you are the ones that are going to put the numbers on it. Here are some things to think about with the number line. Just like we read from the left to the right, we count on the number line from left to right. That means smaller numbers are on the left; bigger numbers on the right. The numbers also like to be evenly spaced from each other to show that each are the same amount from the previous number and to the next number. So, smallest to biggest and evenly spaced. Are we ready to put some numbers up here?

Class: Yes!

Teacher: I shuffled these cards that each have a number 1 through 5 on them. You are to draw a card and place it on the number line. Karim, you go first.

Karim selects 4.

You drew 4. Please put it on our number line.

Karim places the card on the clothesline (see Figure 4.2).

Figure 4.2 Karim's Placement

4

Thank you, Karim. Everyone, I will point, and you say the number.

The teacher points to 4 on the number line.

Class: Four!

Teacher: Good. Robynn, you draw next.

Robynn draws 2.

Place it on the number line, and if you want to move anything that is already there, you may.

Robynn places 2 on the clothesline (see Figure 4.3).

Figure 4.3 Robynn's Placement

Teacher: Are you thinking 2 is less than 4 since you placed it to the left of 4?

Robynn: Uh huh.

Teacher: Thank you, Robynn. I point, and you say the number.

The teacher points at 2.

Class: Two!

The teacher points at 4.

Class: Four!

Teacher: Let's do that backward.

The teacher points to 4.

Class: Four!

The teacher points to 2.

Class: Two!

Teacher: Good. Eva, it's your turn.

Eva pulls 1 and places it on the clothesline (see Figure 4.4).

Figure 4.4 Eva's Placement

Why do you have those three cards evenly spaced?

Eva: You said to do that.

Teacher: You are right, I did. And you followed those directions well. I need to be clearer. I meant the numbers need to be evenly spaced, not the cards. For example, is there a number that goes between 2 and 4?

Eva: Three?

Teacher: Correct. So, we will need to leave enough room between 2 and 4 so we can put 3 there when it gets drawn. Do you want to fix that for me?

Eva adjusts the spacing of the cards (see Figure 4.5).

Figure 4.5 Eva's Card Adjustment

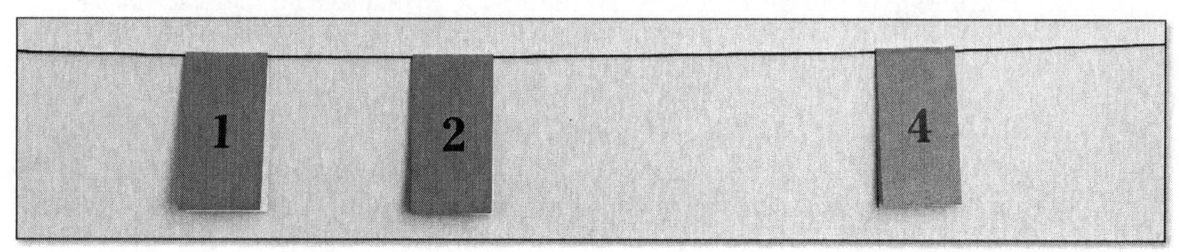

Thank you, Eva. That will be very helpful when we draw our last two cards. I'll point; you count.

The teacher points to 1.

Class: One!

The teacher points to 2.

Class: Two!

The teacher points to 4.

Class: Four!

The teacher points to 2.

Class: Two!

The teacher points to 4.

Class: Four!

The teacher points to 1.

Class: One!

Teacher: Ha, I challenged you with that, but you were paying attention! Great job. Hakeem, your turn.

Hakeem draws 5.

You drew 5. Where does that go?

Hakeem places his card on the number line (see Figure 4.6).

Figure 4.6 Hakeem's Placement

Teacher: Are these cards in the right order so far?

Class: Yes!

Teacher: I point; you count.

The teacher points to the cards, but skips 3.

Class: One! Two! Four! Five!

Teacher: Backward now!

The teacher points to the cards in the opposite order, but still skips 3.

Class: Five! Four! Two! One!

Teacher: I wonder what this last card is.

Class: Three! Three! Three!

Teacher: Wow, how do you know that? Can you see through the card?

Hakeem: No, it's the only one left!

Teacher: Oh, well then I will give it to…

Paz: Me! Me!

Teacher: Paz.

Paz places 3 on the clothesline (see Figure 4.7).

Figure 4.7 Paz's Placement

Paz, thank you. I'll point; you count.

The teacher points to the cards in numerical order.

Class: One! Two! Three! Four! Five!

The teacher points to 3.

Class: Three!

The teacher points to 2.

Class: Two!

The teacher points to 5.

Class: Five!

Teacher: **Outstanding. Now, I am going to give each group a picture card. I want you to talk about the card and decide which number it should be pinned to. Once you decide, send one person up to the board form your group to pin your picture.**

Students discuss amongst themselves which picture should attach to which number. One by one, students walk to the board and pin their pictures (see Figure 4.8).

Figure 4.8 Addition of Pictures

Teacher: **Now that we have our first five counting numbers on the clothesline, let's draw the line on our paper and put these numbers in order there, too. Then, create your own picture under each one to show the amount the number means. Very good math thinking today, everyone.**

Why These Numbers?

These numbers were an obvious choice for introducing young students to the number line. The list was kept to five so as to not tax the attention span of students or clutter the clothesline.

The Key Questions

- Are you thinking 2 is less than 4 since you placed 2 on the left side of 4?
- Is there a number between 2 and 4?
- Where does that number go?
- How do you know which number is still left?

Analysis

Prior to this lesson, students were showing proficiency with counting in order to 10. The purpose of this lesson was to help them map each number and its name to its corresponding symbolic representation. This is why I had students read the numbers aloud after each number was placed.

The spacing directive was to not only introduce students to the structure of the number line, but also to reinforce the idea that counting involves equivalent, incremental increases. Those are concepts we discuss with students. They will not learn that concept informally unless we insist they pay attention to the spacing of the values.

The random order of the numbers was also intentional. Again, students were already proficient at counting from 1; otherwise, this component of the lesson would have been premature. The randomness of the draws required some higher-order thinking about the ordering, spacing, and vacant values.

The cards for this lesson are available in the Digital Resources (lesson1.pdf).

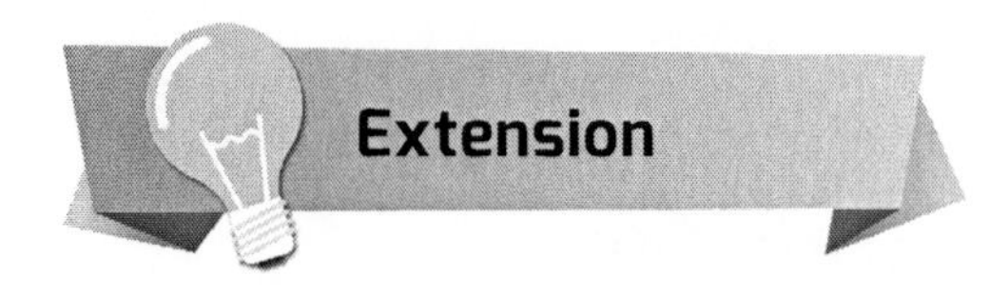

Extension

Pull one end of the clothesline off the wall, causing the numbers to fall on the floor. Call students up one at a time to place numbers back on the clothesline. This time, they must pick them off the floor in numerical order.

Lesson 2: Place Value

1 10 100

Objective: Use place value to properly place the values of 1, 10, and 100 on the number line.

Teacher: **Good day, class. We have been studying place value to 999. Today, we are going to put various values up to three digits on the open number line.**

Our first group is going to put the following values on the clothesline: 1, 10, and 100. The rest of you will work with your partners to draw a number line on your lapboards.

The first group places the values in the correct order but spaces them improperly (see Figure 4.9).

Figure 4.9 First Group's Placement

Class, declare your answers.

Student pairs show their lapboards, revealing that most of the class has similar answers. The values of 1, 10, and 100 are in proper numerical order. Many students show 1 and 10 closer together than 10 and 100, but none of them show proportional spacing. The value of 0 as a benchmark is not in anyone's response.

We all agree these numbers are placed in the correct order. One, then 10, then 100 as we read them left to right. However, everyone seems to disagree with the spacing. How can we decide how far we should space these from one another? Blanca.

Blanca: Ten needs to be closer to 1.

Teacher: **So, you want me to move 10?**

Blanca: No. I meant 10 has to be closer to 1 than to 100.

Teacher: **Good use of precise language. It's time to thumb vote. Give a thumbs-up if you agree with Blanca or a thumbs-down if you disagree.**

Everyone gives a thumbs-up.

Teacher: All of you agree. Jill, why?

Jill: Because, 10 is only a little bit bigger than 1, but 100 is a lot bigger than 1.

Teacher: How can we tell how much bigger?

Kirk raises his hand.

Kirk.

Kirk: Ten is only 9 more than 1, but 100 is 90 more than 10.

Teacher: How did you figure that out?

Kirk: I subtracted 10 – 1 and 100 – 90.

Teacher: How did you subtract 100 – 90?

Kirk: I started at 100 and counted down 10.

Teacher: Let's thumb vote on Kirk's strategy.

Everyone gives a thumbs-up again.

Okay, you all agree that Kirk's subtraction strategy is correct. Did anyone subtract 100 – 90 differently?

Saul raises his hand.

Saul.

Saul: Since 1 is 9 away from 10, then 10 is 90 away from 100, since you just add a zero to all of them.

Teacher: I know what you mean by "adding a zero," but let's be careful that people don't think you mean to add zero, which will just equal the same thing. What are we actually doing to 1, 9, and 10 to get your extra zero, or to equal 10, 90, and 100?

Saul: Multiplying by 10?

Teacher: Yes, and that is exactly what place value is all about. Each place is 10 times the place to the right. So, let's go back to our original numbers—10 is 10 times 1, and 100 is 10 times 10. I am going to challenge you to apply that idea of 10 times greater. Let's set 0 to the left (see Figure 4.10).

Figure 4.10 Teacher Placement

Teacher: **Would it now be easier to space 10 or 1? If you think 10, hold up 10 fingers; if you think 1, hold up 1 finger.**

Students hold up 1 or 10 fingers, with most of the class holding up 10 fingers.

Since the majority of you voted for 10, let's try that first. I am going to take 1 off the clothesline for a moment (see Figure 4.11).

Figure 4.11 Teacher Adjustment

You tell me how to adjust. Point left or right and clap when we have it.

Students point to the left and watch the teacher slowly move the card toward zero. A few students clap, which prompts more students to clap.

That clap was not all together, so that makes me think you are not sure. Let's do some finger reasoning and see whether 100 is truly equal to ten 10s.

Josylynn: It should be a little more to the left.

Teacher: **Okay, Josylynn. I will adjust it until you tell me when to stop.**

The teacher slowly moves 10 toward 0.

Josylynn: Stop.

Teacher: **All in favor, clap twice.**

The class claps twice. The 10 card is in the correct place, as shown in Figure 4.12.

Figure 4.12 Correct Placement of 10

Teacher: **Now, I challenge you to apply that same reasoning for placing 1. Where should 1 go so 10 is 10 times that distance from 0? On your boards, go.**

Students work on their lapboards.

3, 2, 1. Declare.

The class holds up their lapboards to show similar spacing among students.

I see much more agreement now. Let me place 1 where I think you want it on the clothesline, and you check with your finger reasoning.

Patricia: More to the left!

Teacher: **Clap when I get there.**

Students simultaneously clap when the teacher moves the card to the left.

Wow, you are much more confident in your answer now. Everyone, does 100 look 10 times greater than 10?

Class: Yes!

Teacher: **Does 10 look 10 times greater than 1?**

Class: Yes!

Teacher: **Does 100 look 100 times greater than 1?**

Class: Yes!

Teacher: **Great. Copy the spacing you see on the clothesline** (see Figure 4.13) **onto your number lines.**

Figure 4.13 Correct Spacing of Cards

Teacher: How does this spacing relate to the idea of place value? Partner on the left, please tell the partner on the right how the idea of spacing 1, 10, and 100 applies to what we have been studying about place value.

Students discuss place value with each other at their desks.

Do I have a volunteer who would like to share what their partner told them?

Jamie raises his hand.

Jamie.

Jamie: We are not sure what you mean.

Teacher: Thank you for asking me to clarify. On the number line, 10 is 10 times 1. One hundred is 10 times 10. How does that apply to place value when we write a number like 111, for example?

Jamie: Oh, each number is 10 times more than the one next to it.

Teacher: Reading which way?

Jamie: The first numbers are bigger than the next numbers.

Teacher: I think I understand what you are saying. Let me see whether I can rephrase it. Each place is 10 times the value of the place to its right. If you agree with Jamie, show me a thumbs-up.

Students show the teacher their thumbs-up.

Tyson, can you please summarize this idea of place value for us again?

Tyson: Each place is 10 times more than the next one.

Teacher: Which way is next?

Tyson: Each place is 10 times more than the next one to the right.

Teacher: All in favor, shout "aye."

Class: Aye!

Teacher: Well then, since you are so convinced now, let's record our discussions, deductions, and decisions on your *Clothesline Math* activity sheet. On the three blanks, write 1, 10, and 100. Then, record the numbers on the number line provided with the benchmarks. Be sure to space the values to show that 10 is 10 times greater than 1, and that 100 is 10 times greater than 10. Also, make note of how this applies to place value.

Why These Numbers?

The key to this lesson is having students see that place value is based on 10, so the simplest numbers for teaching the fundamentals of place value were chosen—namely, 1, 10, and 100.

The Key Questions

- Are these numbers placed correctly? In other words, are the numbers in correct numerical order?
- Are these values spaced correctly?
- How did you determine the spacing?
- Does 100 look 10 times greater than 10?
- Does 10 look 10 times greater than 1?
- Does 100 look 100 times greater than 1?
- How does the spacing of these three numbers on the number line apply to what we know about place value?
- Do you agree or disagree?

Analysis

I anticipated that most students would place the values in correct numerical order on the number line, but I was wondering whether any would place them in the reverse order since that is how we read them. In other words, with the number 111, the 100 is written left of the 10, even though the 100 is placed to the right of 10 on the number line. No one made that error, so I was able to immediately address the spacing issue. All students placed 10 closer to 1 than to 100, but no one demonstrated an understanding of a factor of 10. I knew the zero would be imperative to accomplish this, but I refrained from placing it on the number line until I saw their initial responses.

I wanted to generate the conversation around their strategies by selecting a subtraction strategy while I monitored their work and how they chose to sequence the first clothesline in the discussion. I was looking for a multiplication strategy to sequence second, but that did not arise with any students. Therefore, I chose to pose it as a challenge.

The finger reasoning was key to emphasize the fact that each place value is literally 10 times greater than the place value to the right. I made sure to have a long enough number line because the span of 0 to 100 takes up a great deal of space. The visual of the spacing from 0 to 1, 1 to 10, and 10 to 100 is such a powerful teaching tool. It is critical to connect that visual to the idea of place value, as in the given example of 111, because the spacing is lost.

I could have gone further and asked students what each 1 in 111 truly represents (e.g., the first 1 represents 100). However, I chose to keep the focus on the factor of 10 and save that line of questioning for another lesson.

The cards for this lesson are available in the Digital Resources (lesson2.pdf).

Extension

Have students place multiples of 10, such as 20, 50, and 80.

Lesson 3: Single-Digit Addition

3+6 1+4 7+8

6 7 8 9 10 15

Objective: Display two single-digit numbers and their sum on a number line.

Teacher: **Hi there, my young mathematicians.**

Class: Hi!

Teacher: **We have learned about adding two numbers in several different ways. Today, we are going to show it on a number line. This is known as an open number line because it is missing the numbers. That is going to allow you to imagine where some of the numbers should be. I am going to show you an example first and then start calling some of you to the clothesline, which is our real-life number line up on the board** (see Figure 4.14).

Figure 4.14 Clothesline with Benchmark Cards

You can see I have 0, 1, and 4 placed here. Zero is there to help me figure out where everything else should be placed. It's kind of like your house is the place you use to figure out where to walk around your neighborhood. I'm using 10 to tell us whether any of our sums will have two digits. The 1 and 4 are the numbers I am going to add together. In fact, I have that written already on the board as well (see Figure 4.15):

Figure 4.15 Teacher's Prompt

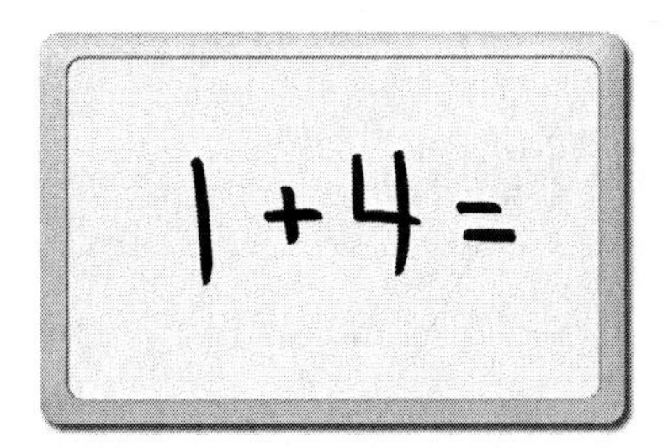

Teacher: **When I add numbers on the number line, I am really adding their distance from 0. Watch how I use my hands to measure. I see how far it is from 0 to 4 with my hands. Then, I move this distance so that it actually starts now at the 1 instead of the 0. Look where my other hand is now.**

The teacher moves his hand to show where 5 would be on the number line.

It looks like it is around how much?

Class: Five!

Teacher: **Yes, 5 because a distance of 1 plus a distance of 4 is a distance of 5. But that is a mouthful of words, so we just say 1 plus 4 equals 5.**

The teacher pins the equation 1 + 4 to 5, as shown in Figure 4.16.

Figure 4.16 Teacher Adds Expression Card

If you are trying to measure from where you are sitting, you can use finger reasoning. Close one eye and hold your index finger and thumb up to it. Record this on the first number line on your *Clothesline Math* activity sheet. Are you ready to try it on your own?

The class is silent.

Don't be scared, you can handle it. Watch. Our next addition problem is on the board (see Figure 4.17):

Figure 4.17 New Addition Problem

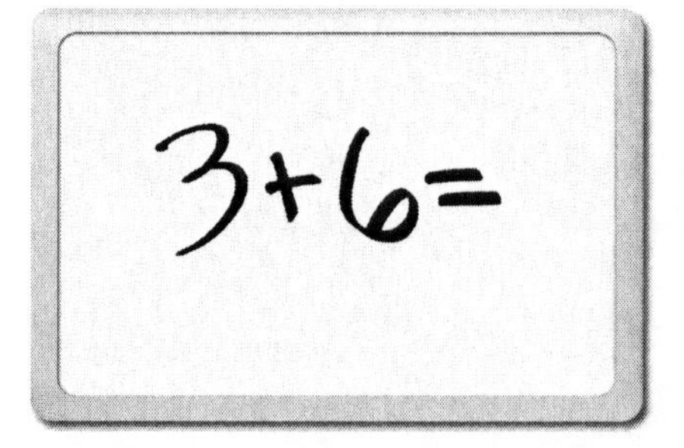

Let's draw a number line on our lapboards and mark 3 and 6. Now, measure 6 and add it to 3. Where should this sum go?

Students work on this equation on their boards.

Teacher: Now that you have it on your lapboards, I will take a brave volunteer group for the first one.

Davida raises her hand.

Davida, have your group place this sum on our clothesline.

Davida's group adds 9 to the clothesline and then pins the expression to it, as shown in Figure 4.18.

Figure 4.18 Davida's Placement

Class, use finger reasoning to check this group's work while Davida explains what they did.

Davida: We measured 6 and then moved our hands over so we could add it to 3.

Teacher: All in favor, clap twice.

Students clap twice.

I wonder whether this is still 9 if I switch the order of the numbers. Let's try 6 + 3. That means measure the 3 and add it to the 6. Does that get us to the same place? Are we at 9? Give a thumbs-up if you get 9 that way or a thumbs-down if you get something different.

Students all give a thumbs-up.

It looks like we all agree the order doesn't matter when we add. Please record this on the second number line on your activity sheet.

Students write the new order on their second number line.

One more. Davida, please choose the next group to go.

Davida selects Johnny.

Johnny, your group has the equation on the board (see Figure 4.19):

Figure 4.19 New Prompt

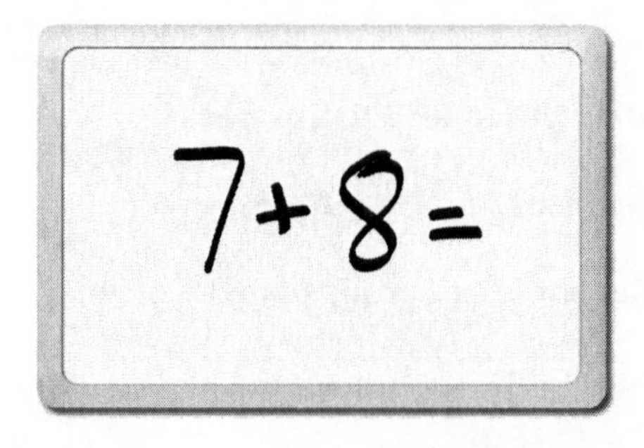

Teacher: Everyone else, work on your individual boards. Draw the number line and place 0, 7, 8, and 10. Then, place 7 + 8.

Johnny's group places the benchmark cards on the line and pins the expression to 15, as shown in Figure 4.20.

Figure 4.20 Johnny's Placement

Before you declare your boards, compare your answers with others in your group, and choose one board to hold up.

Students discuss their boards with each other.

Declare.

Students hold up one board per group, showing their work matches with the clothesline.

Johnny, the class seems to agree with you. How did you know it was 15?

Johnny: We did the hand thingy. We counted 8 from 7.

Teacher: Very good work. Did anyone get 15 a different way?

Sabrina raises her hand.

Sabrina.

Sabrina: I added 3 to get to 10. Out of 8, that means I had 5 left.

Teacher: Thank you both for sharing your thinking. Let's record our final number line diagram on our activity sheets.

The values were chosen so the first two sums were less than 10, but the third sum would yield a two-digit quantity.

- What does this distance look like on the number line?
- If I switch the order of the addition, do I get the same sum?
- Where should this sum go?
- How did you know the sum was 15?

There are distinct differences between addition on the open number line and the traditional number line that students are usually familiar with (see Figure 4.21).

Figure 4.21 Traditional Number Line

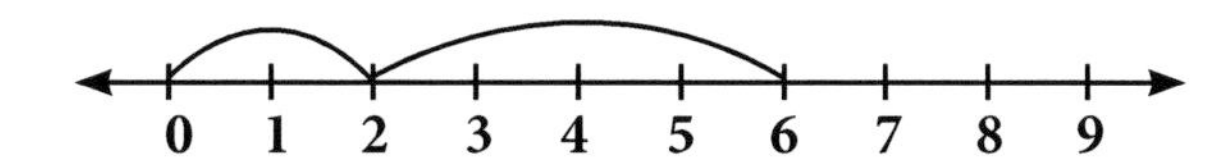

Traditional number lines allow students to count tick marks. On an open number line, students are encouraged to measure distance instead, which is why benchmarks are intentionally left off for a Clothesline activity. Both number line experiences are healthy, as they require the brain to view adding in different ways.

The cards for this lesson are available in the Digital Resources (lesson3.pdf).

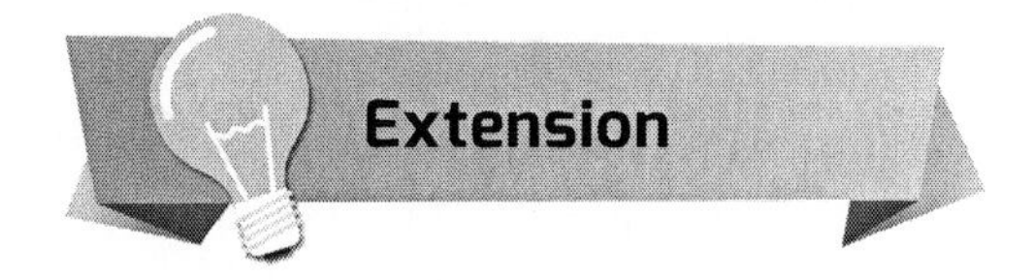

Offer the corresponding subtraction expressions. For example, given 3 + 6 = 9, ask students to show what 9 – 6 looks like on the number line.

Lesson 4: Single-Digit Multiplication

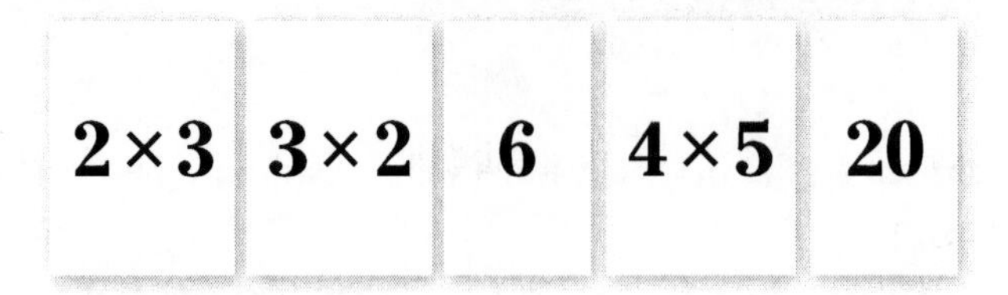

Objective: Show products of two single-digit numbers on a number line, showing understanding of groupings.

Teacher: **Good day, students. We have been learning how to multiply by grouping. Today, we are going to see how this looks on a number line. Take a look at what I have for you here** (see Figure 4.22)**.**

Figure 4.22 Benchmark Cards

I am going to ask a group to show us where 3 × 2 should be placed without using any other benchmarks. Let's go with someone who has the same last initial as me. Leon, bring your group to the clothesline. The rest of you, work with your partner on your lapboards.

Leon's group places 6 on the clothesline and pins the equation to it, as shown in Figure 4.23.

Figure 4.23 Leon's Placement

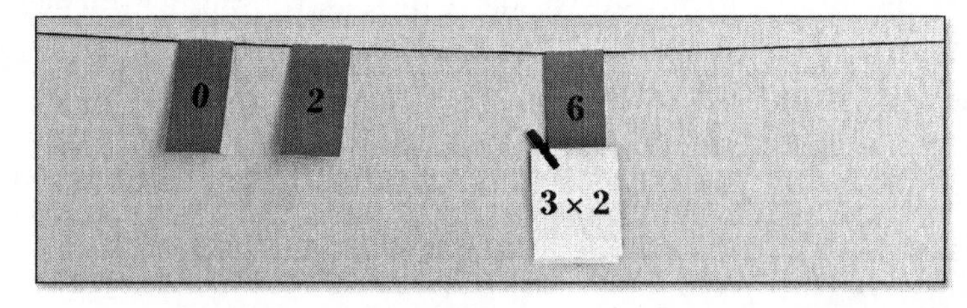

Leon, can you share how your answer relates to multiplication by grouping?

Leon: We just used our hands to measure how far between 0 and 2 and moved until we had 3 of that much.

Teacher: **Thank you, you get to pick the next group.**

Leon selects Ansel.

Teacher: Ansel, your group gets to apply the commutative property for us today. Does 2 × 3 also equal 6? Show us whether grouping will yield a product of 6 as well.

Ansel's group pins 2 × 3 to 6 as well, as shown in Figure 4.24.

Figure 4.24 Ansel's Placement

Ansel, what did your group decide?

Ansel: Yes, it gives you the same answer.

Teacher: How do you know?

Ansel: Two times that much to 3 equals the same as 6.

Teacher: So 2 times 3 literally means we want 2 groups of 3 or that we want 3 things 2 times. Thank you, Ansel. Who is going to do our last example of 4 × 5?

Callie raises her hand.

Callie, it's time for your group to head to the clothesline.

Callie's group places 5 and 20 on the number line and pins the equation to 20, as shown in Figure 4.25.

Figure 4.25 Callie's Placement

Callie, what do you have for us?

Callie: Same as the others. We put 5 there because we noticed you started with the second number and then said how many times do you do that much.

Teacher: Let's do some finger reasoning, everyone. Is Callie's group accurate? Do 4 groups of the distance between 0 and 5 equal the distance from 0 to 20?

Students measure the distance using finger reasoning.

Teacher: If you think the equation is placed correctly, give a thumbs-up. If you think it needs to be moved, point in the direction you think it needs to go.

Students all give a thumbs-up.

Great job, everyone. Callie, your group is correct with your reasoning. Everyone, write this on your activity sheet.

Why These Numbers?

These numbers (2, 3, 4, and 5) are four of the smallest values we can multiply and still have a product that can fit on a number line without a great deal of scaling.

The Key Questions

- Can you explain how your answer relates to multiplication by grouping?
- Do four groups of the distance between 0 and 5 equal the distance from 0 to 20?
- By finger reasoning, is this group accurate?

Analysis

The Clothesline is not just a memorization tool; it is a tool for understanding. Therefore, we do not want to waste time seeing whether students properly place the product on the number line through rote memorization of the multiplication table. The clothesline allows students to see how grouping (e.g., how many of that distance) is the structure underlying the multiplication table. This way the discussion is in the contextual frame of explaining how multiplication by grouping works.

The cards for this lesson are available in the Digital Resources (lesson4.pdf).

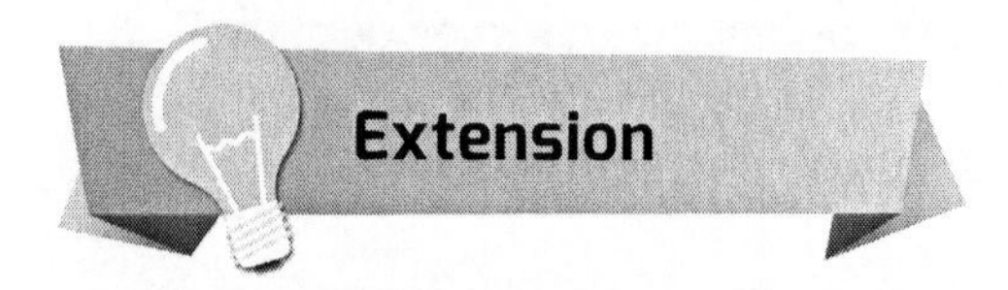

Extension

Scale up by multiplying 7×8. Benchmarks for units of 10 would be helpful for this extension activity.

Chapter 5

Fractions for Elementary School

Lesson 5: Fractions: Almost Half

Objective: Estimate the relative value of fractions that are approximately one-half.

Teacher: **Good morning, class. We have been learning a great deal about fractions lately, so I wanted to use fractions today to build our brains. We know that by thinking really hard, our brains get stronger, just like when you exercise and your muscles get stronger. To get our brains stronger today, we are going to think really hard about fractions on the number line. Are you ready to build your brains?**

Class: Yeah!

Teacher: **I want to start by giving this fraction, $\frac{3}{7}$, to the person whose birthday is closest to today.**

Kimble: Mine's tomorrow!

Teacher: **Is anyone's today? No? Well then, Kimble, happy early birthday. You seem excited to place this on our clothesline. Before you get started, I want to know from the whole class: What benchmarks will help Kimble know where the fraction should go?**

Students sit silently.

In other words, do you want me to place any other numbers first?

Several students raise their hands.

I see plenty of hands. Kimble, hold on to your card for a second so you can tell me whether you agree with what the class says. Darren?

PART II – Chapter 5

Darren: Zero. Zero on a number line always helps.

Teacher: Okay. Any others? Helen?

Helen: Five.

Teacher: Why 5?

Helen: It's my favorite number.

Teacher: I like 5, also. Are there other numbers that would help even more so? Sammie?

Sammie: Three and 7.

Teacher: Why?

Sammi: Because there is a 3 and a 7 in the fraction.

Teacher: Any other suggestions. Gali?

Gali: One.

Teacher: Why?

Gali: Because lots of fractions are less than 1.

Teacher: Thank you for your suggestions. Kimble, which of the numbers mentioned will be most helpful to you?

Kimble: Zero and 1.

Teacher: Why?

Kimble: Because this number is less than 1.

Teacher: Kimble, I will give you three numbers to place on the clothesline. Those numbers will be 0, 1, and $\frac{3}{7}$. Go ahead.

Kimble walks to the front of the class.

Everyone else, while Kimble places the cards, here is what I would like you to do. The first partner in your pairs will draw the number line. Then, the second partner will mark 0 and 1. After that, the first partner will mark $\frac{3}{7}$. Finally, the second partner will hold the board up when I ask. You may use any of the numbers you heard or use others you are thinking of. Use whatever strategies help you the most.

While the teacher speaks, Kimble places his three cards on the clothesline (see Figure 5.1).

Figure 5.1 Kimble's Placement

Teacher: Kimble, please explain.

Kimble: I imagined the whole thing was chopped into seven parts, and I picked the third one of those.

Teacher: What do we call one of those seven parts?

Kimble gives a shrug.

Blake, can you help out Kimble? What do we call each of those seven parts?

Blake: Sevenths?

Teacher: Kimble, are you okay if we call those sevenths during our discussion?

Kimble: Yeah.

Teacher: Class, let's see what you claimed on your boards. Second partner, please politely hold the board up so I can see it. Xavier and Danny, you did not use sevenths. Please tell us what you were thinking.

Xavier: We chopped 1 into threes because that is the top number, and then we counted out seven of those, so we had to put a 2 and 3 on the line (see Figure 5.2).

Figure 5.2 Xavier and Danny's Placement

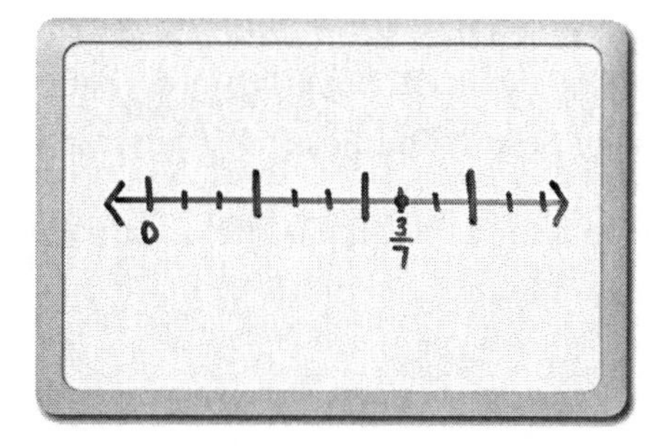

Teacher: Danny, what do each of your fraction markings mean if you have three in every one?

Danny: Sevenths?

PART II – Chapter 5

Teacher: **Sevenths is what Kimble is using because he portioned 1 into seven parts. You portioned 1 into three parts, so what is each of those parts called?**

Danny: Thirds.

Teacher: **Yes. Class, you get to help us decide. Should we use thirds or sevenths? All those who say sevenths, raise your hand. All those who say thirds, give a thumbs-up.**

Most students raise their hands, but some give a thumps-up. The teacher notices Xavier and Danny discussing the options instead of doing either.

Xavier, you two were talking it over?

Xavier: We want to change ours.

Teacher: **Why?**

Xavier: The bottom number…

Teacher: **The what?**

Xavier: The bottom…the denominator is the number of sections.

Teacher: **What does the numerator tell you?**

Xavier: How many of those sections you want.

Teacher: **So, would Kimble's placement of $\frac{3}{7}$ be correct?**

Xavier: Yes.

Teacher: **Yazmin, I didn't see sevenths or thirds on your board.**

Yazmin: We did it differently.

Teacher: **You built your brain differently. I am curious how you two did it then.**

Yazmin: We doubled 3 to get 6. Since 2 times 3 is close to 7 but not exactly 7, we thought 3 would be close to $\frac{1}{2}$, but not half exactly.

Teacher: **Wow, that is some interesting thinking. Did you get the same placement for $\frac{3}{7}$?**

Yazmin: Yes.

Teacher: **Since you mentioned $\frac{1}{2}$, I think I will add that benchmark to our line** (see Figure 5.3).

Figure 5.3 Teacher Placement

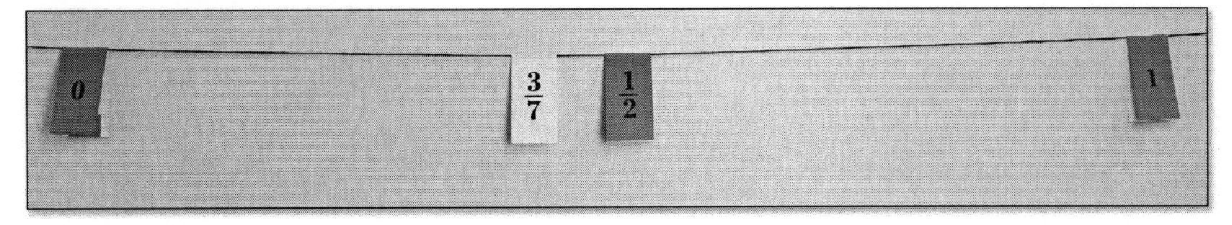

Teacher: **Class, if you are in favor of Kimble's answer, clap twice.**

Students all clap twice.

Kimble, you get to choose who is going to place our next fraction of $\frac{6}{13}$.

Kimble: Karla.

Teacher: **Karla, Kimble will take your place with your partner, and you will go to the clothesline to place $\frac{6}{13}$. The rest of you may make any adjustments to your lapboards that you wish; although, I suggest you keep 0, 1, $\frac{1}{2}$, and $\frac{3}{7}$. Go.**

Karla pins $\frac{6}{13}$ to $\frac{3}{7}$, as shown in Figure 5.4.

Figure 5.4 Karla's Placement

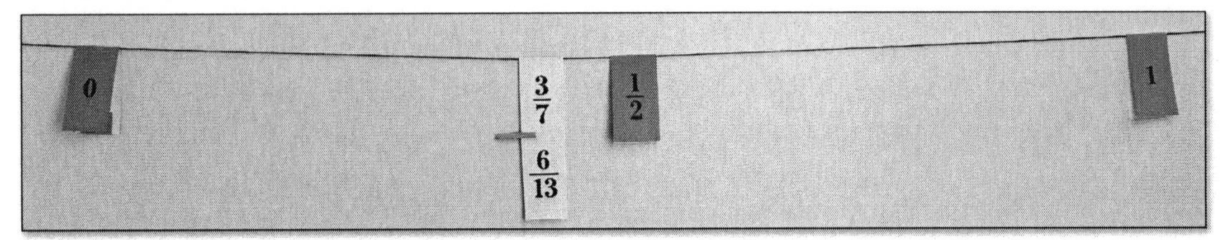

Karla, can you explain?

Karla: I did the same thing that Yazmin said. The 6 doubled is less than 13, so just the 6 is less than half.

Teacher: **You are showing that it is equal to $\frac{3}{7}$.**

Karla: They are both just a little less than half.

Teacher: **By the same amount?**

Karla stares at the teacher silently.

Let's see what the class thinks. Declare your boards.

Students hold up their boards, and most of the number lines match Karla's.

Everyone agrees with you that it is a little less than half. Some also agree that it is equal to $\frac{3}{7}$. Pablo, you have $\frac{6}{13}$ being less than $\frac{3}{7}$, and Rhona has it being greater than $\frac{3}{7}$. Class, take a look at their boards (see Figure 5.5).

Figure 5.5 Pablo and Rhona's Lapboards

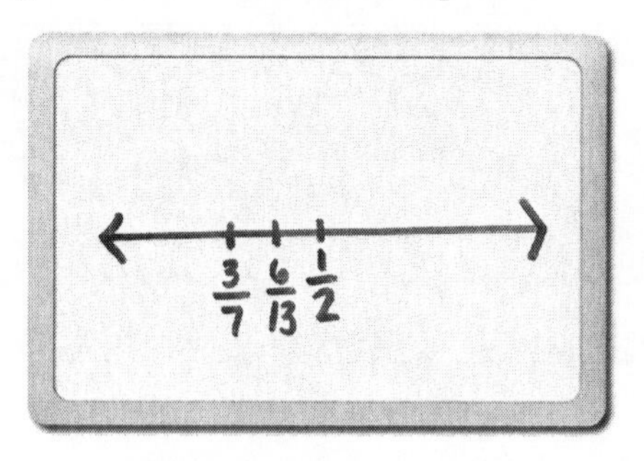

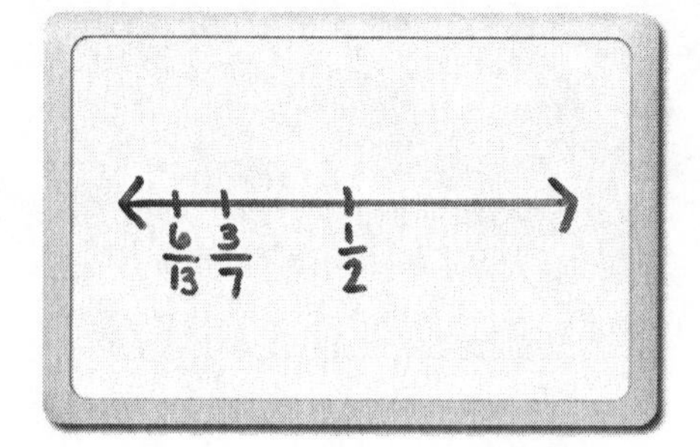

Teacher: Pablo, why less than?

Pablo: Well, $\frac{1}{13}$ is really small, it would be closer to 0 than a seventh.

Teacher: Thank you. Rhona, why greater?

Rhona: Thirteen is bigger than 7, so it is greater.

Teacher: Does anyone have a different thought?

The class is silent.

May I think out loud and build my own brain?

Class: Yes!

Teacher: I agree with Rhona that 13 is bigger than 7, but it doesn't mean my fraction will be bigger. In fact, Pablo is correct in saying the greater number of pieces into which we divide something, the smaller each piece is. If I take both of those ideas, I'm thinking $\frac{6}{13}$ is a smaller piece, or distance, away from half. What do you all think? Thumbs-up if you agree with me, or thumbs-down if you don't.

The entire class shows a thumbs-up.

Ah, you're just saying that because I'm the teacher.

Carlos: But even teachers can build a bigger brain. You even told us that.

Teacher: Yes, we can. So, let me adjust our fraction to show that thinking (see Figure 5.6).

Figure 5.6 Teacher Adjustment

If you agree with my logic, then use it to place $\frac{15}{29}$. I will take a volunteer on this one.

Several hands shoot into the air.

Teacher: Oh, so many willing mathematicians. Barkely, Karla will take your place with your partner.

Barkely places the card on the clothesline, as shown in Figure 5.7.

Figure 5.7 Barkely's Placement

Whoa, Barkley. You changed things on us and went on the other side of half.

Barkely: Because 15 doubled is a little more than 29, so 15 is a little more than half of 29.

Teacher: Declare your boards.

Students' boards show the same results as Barkely.

Everyone agrees. Does anyone want to tell us whether $\frac{15}{29}$ is closer to $\frac{1}{2}$ than $\frac{6}{13}$ or further away? Ahmad.

Ahmad: Closer, because 29 pieces are a lot smaller than 13 pieces or 7 pieces, and we are only half a piece away on all of them.

Teacher: Half a piece? What do you mean?

Ahmad: Fifteen doubled is 1 piece away from 29, so 15 will only be half a piece away from half.

Teacher: All in favor, say "aye."

Class: Aye!

Teacher: I will tap it over a bit closer then (see Figure 5.8).

Figure 5.8 Teacher Card Adjustment

All those whose brains grew so big and fast today that it hurts, say, "ouch!"

Class: Ouch!

> **Teacher: I'm glad we were able to make your brains stronger with fractions today. Copy this diagram, and write the mathematical ideas we discussed. Great mental workout, team.**

Why These Numbers?

These three fractions were all chosen because they are all close to $\frac{1}{2}$—two were just below and one was just above. I chose $\frac{3}{7}$ because it allowed a low entry point to the activity so students could imagine sevenths. The next two fractions had large enough denominators to make counting too cumbersome, so it forced more complex strategies for estimating.

The Key Questions

- What kind of benchmarks will help you place the number?
- Why do you think $\frac{6}{13}$ is less than $\frac{3}{7}$? Why do you think it is greater?
- Are these two fractions less than $\frac{1}{2}$ by the same amount?
- What does the denominator tell you?
- What does the numerator tell you?
- Does a greater denominator imply a larger or smaller part of the whole?
- What do we call one of those 7 parts?

Analysis

The benchmarks could have been provided or simply left to the first student to choose. Having the class discuss which benchmarks would be most helpful not only challenged students to be reflective, but also revealed a great deal about the lack of understanding of fractions in some students. The Clothesline offers terrific opportunities like this to dig deep for student misconceptions. Prompting students for benchmarks really opened the door of insight into student thinking.

The hot spot of the lesson was when students knew $\frac{6}{13}$ and $\frac{3}{7}$ were both less than $\frac{1}{2}$ and relatively close, but they struggled in determining which one was closer to half. The strategy discussion proved fruitful because students became far more confident and adept at determining whether $\frac{15}{29}$ was even closer.

Math teachers are called upon in NCTM's *Principles to Actions* (2014) to use evidence of student thinking to drive instruction. The student work shown by Pablo and Rhona on the lapboards regarding the placement of $\frac{6}{13}$ was gold. It begged for an explanation that all students would have to grapple with. The silence of students when asked to discern whether a larger denominator implies a larger or smaller part to the whole was more evidence that spoke volumes about student understanding.

The technique of having partners take turns drawing the number line, placing the values, and displaying the board is simply an engagement technique to prevent the "markers hogs" from dominating the partner participation.

The cards for this lesson are available in the Digital Resources (lesson5.pdf).

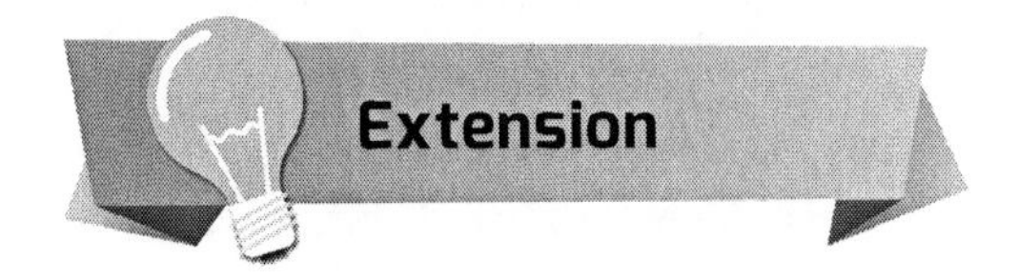

Extension

Estimate fractions close to one whole: $\frac{5}{6}$, $\frac{9}{10}$, and $\frac{11}{12}$.

Lesson 6: Comparing Fractions with Common Denominators

$\frac{5}{7}$ $\frac{3}{7}$ $\frac{6}{7}$

Objective: Estimate the relative value of fractions that all have the same denominator, understanding the denominator determines the number of equal parts in the whole.

Teacher: **Good day, all. We have been learning about fractions, so today I brought something interesting to make you think deeply about them. Remember that we call the top number of a fraction the numerator and the bottom number the denominator. Today, we are going to compare fractions that have the same denominator. In fact, today's fractions are all going to have a denominator of 7. I'm going to call on one person who is then going to call on any partner in the class who has not been chosen yet. The two of you will place the fraction that I give you on the clothesline together. The rest of the class will place the fraction on a number line on your lapboards. Are you ready for our first fraction?**

Class: Yes!

Teacher: **I am giving Chase the fraction $\frac{6}{7}$. Chase, choose a partner and head to the clothesline.**

Chase: Jason.

Teacher: **Jason and Chase at the clothesline with $\frac{6}{7}$. Everyone else, work on your boards.**

Chase and Jason place $\frac{6}{7}$ on the clothesline together (see Figure 5.9).

Figure 5.9 Chase and Jason's Placement

Chase, what did you two decide?

Chase: Seven out of 7 is 1, so 6 out of 7 is less than 1.

Teacher: What you just said is important for our discussion today: "Seven out of 7 is 1." The denominator determines how many equal parts are in the whole. Thank you. Would anyone like to challenge that?

No one responds to the offer to challenge.

Declare your boards.

Students hold up their boards, showing almost everyone placed $\frac{6}{7}$ in the same location.

Everyone agrees with you on the placement, but some people have a little different spacing. Tricia?

Tricia: Well, if I think about 7 pieces, and 6 is 1 piece less than 7…

Teacher: Do you mean that $\frac{6}{7}$ is $\frac{1}{7}$ less than 1?

Tricia: Yes, but I think that $\frac{1}{7}$ is bigger than what the boys are showing.

Teacher: Please adjust it while everyone watches.

Tricia moves $\frac{6}{7}$ slightly to the left, as shown in Figure 5.10.

Figure 5.10 Tricia's Placement

All in favor, clap twice.

Students give two uniform claps.

Even Chase and Jason clapped for you, so I think we showed some good teamwork on this one. Tricia, while you are up there, why don't you pick a partner?

Tricia: Donna.

Teacher: I will give you our next fraction, $\frac{5}{7}$.

Tricia and Donna place $\frac{5}{7}$ to the left of $\frac{6}{7}$, as shown in Figure 5.11.

Figure 5.11 Tricia and Donna's Placement

Teacher: Donna, why there?

Donna: We need one less seventh, so we measured between $\frac{6}{7}$ and 1 and just gave it that much more space.

Teacher: Thank you. Now, for our last fraction of $\frac{3}{7}$. This one is going to Chloe.

Chloe: Anna, come with me.

Teacher: Please make sure you have all three fractions on your boards while Chloe and Anna think about this one.

Chloe and Anna place $\frac{3}{7}$ on the clothesline much farther to the left than the other two fraction cards (see Figure 5.12).

Figure 5.12 Chloe and Anna's Placement

Chloe, why way over there?

Chloe: We just kept counting down. $\frac{6}{7}$, $\frac{5}{7}$, $\frac{4}{7}$, then $\frac{3}{7}$ goes here.

Teacher: All in favor, say "aye."

Class: Aye!

Teacher: Any challengers? Let me see your boards.

Students hold up their boards to show the same spacing on their number lines as well.

Everyone is showing the same location, so let's record this on our activity sheet. Be as precise as possible when estimating the sevenths.

Why These Numbers?

The denominator is large enough to offer some nonconsecutive fractions, but small enough to allow students to visually count the unit fractions. The fractions close to one whole, $\frac{5}{7}$ and $\frac{6}{7}$, were intended to drive conversation toward the idea that when the numerator and the denominator are equal, the fraction is equal to 1.

The Key Questions

- How did you determine the spacing between 1 and $\frac{6}{7}$?
- Why did you place $\frac{5}{7}$ there?
- Why did you place $\frac{3}{7}$ farther from $\frac{5}{7}$ than $\frac{5}{7}$ is from $\frac{6}{7}$?

Analysis

Having the same denominator for all the fractions reinforces the concept that the denominator determines the number of equal parts in the whole, and the numerator determines the number of those equal pieces. This was the main concept of the lesson, so when a student outright stated "Seven out of 7 is 1," I paused the lesson to focus on that key idea. Pushing students to be mindful of spacing for $\frac{6}{7}$ required them to visualize seven equal parts and, thus, highlighted further the objective of equal parts in a whole.

The technique of having students choose a partner was used mostly to heighten engagement, since students had to be paying attention in the event they were called upon next. It also provided dialogue in a situation where I was calling upon the individual's thoughts more than the thoughts of the group.

The cards for this lesson are available in the Digital Resources (lesson6.pdf).

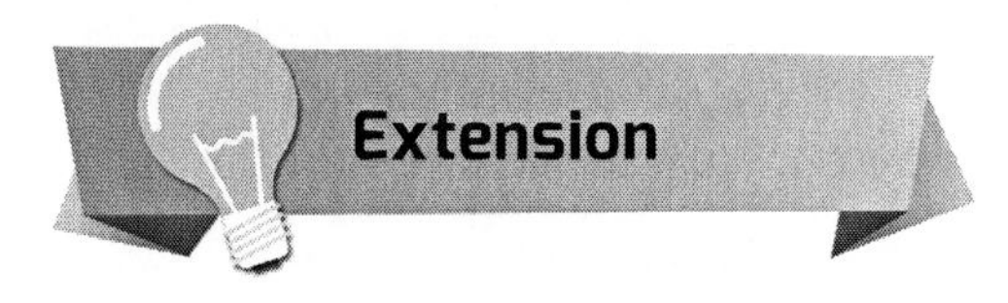

Extension

Place $\frac{8}{7}$, $\frac{9}{7}$, and $\frac{12}{7}$.

Lesson 7: Comparing Fractions with Common Numerators

$\frac{4}{1}$ $\frac{4}{2}$ $\frac{4}{3}$ $\frac{4}{4}$ $\frac{4}{5}$ $\frac{4}{6}$ $\frac{2}{3}$

Objective: Estimate the relative value of fractions that all have the same numerator, understanding the numerator indicates the number of equal parts present.

Teacher: **Good morning, kiddos.**

Class: Good morning!

Teacher: **Thank you for such a warm greeting. You must be ready to do some math. Recently, we placed fractions that had the same denominator on an open number line. Today, we are going to place fractions that have the same numerator. That means all the top numbers of the fractions will be the same. Today, that number will be 4. There will be six fractions all with a numerator of 4. We are going to do the student train today, so one person at a time will come to the clothesline. Everyone else needs to pay close attention because you may be asked to change something on the number line. Phil, you are first. Where would you place $\frac{4}{1}$?**

Phil pins his number card to a benchmark card already on the line, as shown in Figure 5.13.

Figure 5.13 Phil's Placement

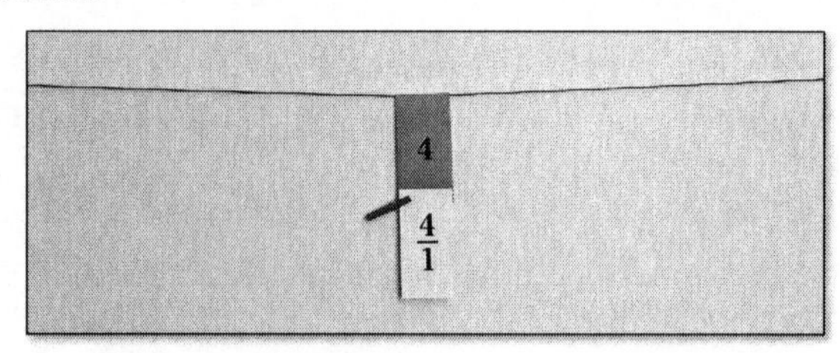

Lily, I am now going to give you $\frac{4}{2}$. If you agree with everything that is on the clothesline, place this one and have a seat. If you disagree with something, change it and then pass this card to anyone else in the class.

Lily chooses a 2 benchmark card from the table and places it on the clothesline, then pins $\frac{4}{2}$ (see Figure 5.14).

Figure 5.14 Lily's Placement

Teacher: We are going to keep rolling with our student train. Raul, here is $\frac{4}{3}$. Decide whether you want to place it or change something that is there.

Raul skips using a benchmark card and hangs the fraction card directly on the clothesline (see Figure 5.15).

Figure 5.15 Raul's Placement

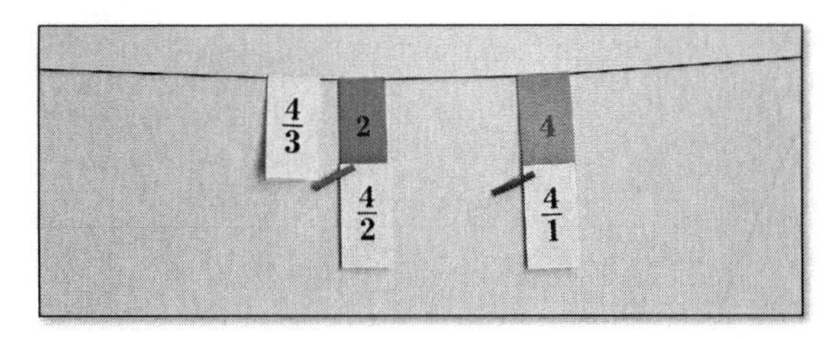

I think we need to organize our clothesline a bit, but first I want to see whether anyone notices a pattern as the denominator grows under the same numerator. What is happening to the value of our fraction? Ollie?

Ollie: The numbers are getting smaller.

Teacher: I'm noticing that same pattern, so I'm going to shift everything over to the right and put a 0 benchmark on the clothesline to help us. I will put the 4 benchmark all the way to our right, the 0 benchmark all the way to our left, and the 2 benchmark in the middle. Then we have…oops. Raul, I'm sorry, I remember you put this card less than 2, but I don't remember where exactly.

Raul: It's one out of three left over, so it goes bigger than 1 but less than 2.

Teacher: Okay, how about here?

The teacher places the card in between benchmarks 1 and 2.

Raul: That's good.

Teacher: Thank you.

Betty raises her hand.

Teacher: **Yes, Betty.**

Betty: If you take 4 and divide it into three equal parts, you can see where it goes, also.

The teacher moves the card slightly.

Teacher: **So, is this good?**

Betty: You are still a little bit off.

Teacher: **Is that bugging you?**

Betty: A lot.

Teacher: **Then you get the next card. You can place $\frac{4}{4}$ on the line, or you can fix $\frac{4}{3}$ and pass this on to someone else in the class.**

Betty hands $\frac{4}{4}$ to Francis before heading to the clothesline. While at the clothesline, she moves $\frac{4}{3}$ closer to 1 than 2, as shown in 5.16.

Figure 5.16 Betty's Placement

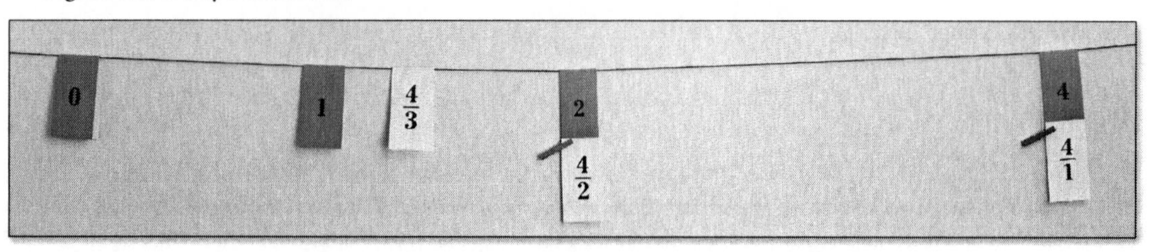

Francis, Betty just gave you $\frac{4}{4}$. Place it, or correct and pass.

Francis places $\frac{4}{4}$ on the clothesline, as shown in Figure 5.17.

Figure 5.17 Francis's Placement

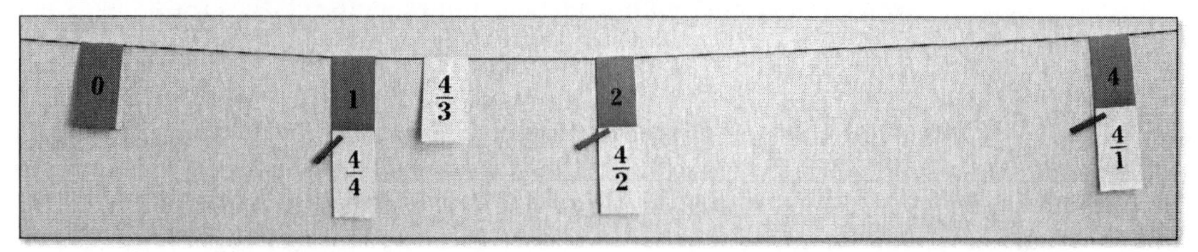

Thank you, Francis. Li, you get $\frac{4}{5}$. Please place the card, or pass it to correct something on the clothesline.

Li adds $\frac{4}{5}$ to the clothesline to the right of 1, as shown in Figure 5.18.

Figure 5.18 Li's Placement

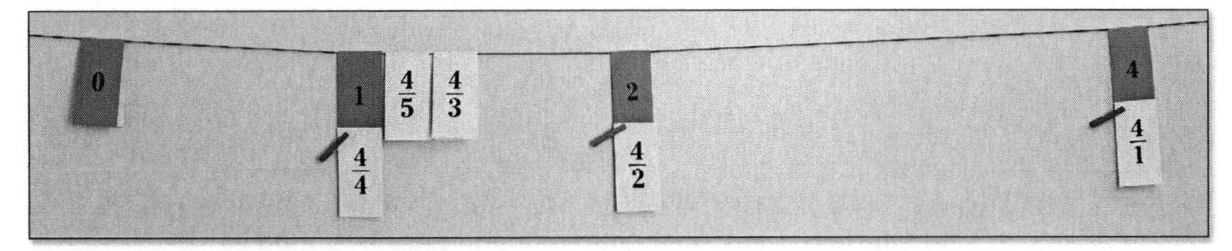

Teacher: **Thank you, Li. Luke, you get our last card, which is $\frac{4}{6}$.**

Luke moves $\frac{4}{5}$ to the left of 1, as shown in Figure 5.19.

Figure 5.19 Luke's Placement

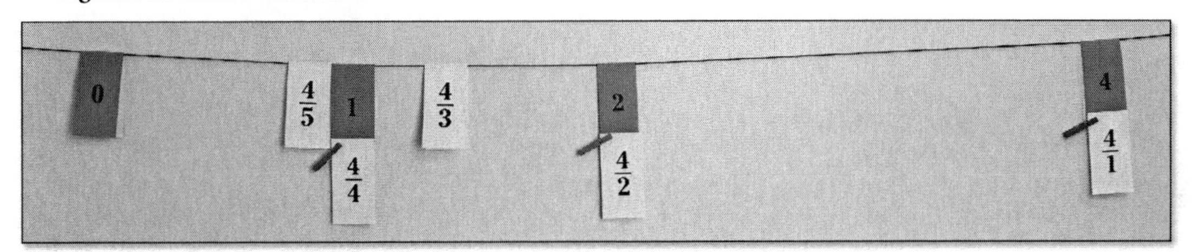

Luke, you chose to correct where Li placed $\frac{4}{5}$. Can you explain why?

Luke: I think Li was counting fourths and using the numerator. I used the denominator to put it on the line because $\frac{4}{5}$ is less than 1, not more.

Teacher: **Li, does that make sense to you?**

Li nods.

Why is that less than one?

Li: Because it is 4 out of 5 parts, not 5 out of 4 parts.

Teacher: **Great. The last fraction goes to Flo. Where does $\frac{4}{6}$ go?**

Flo places $\frac{4}{6}$ on the clothesline. Then, she grabs $\frac{2}{3}$ and pins it to the bottom of $\frac{4}{6}$ to show the simplified version of the original fraction (see Figure 5.20).

Figure 5.20 Flo's Placement

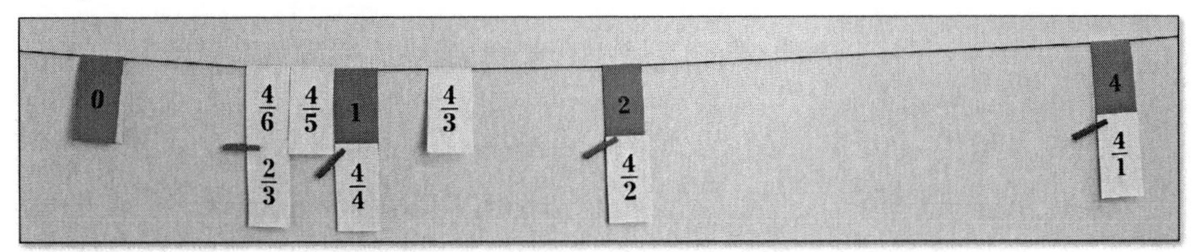

Any challengers? No? If you like what you see, applaud.

There is loud applause from students.

PART II – Chapter 5

Teacher: **That was a terrific discussion. We got the chance to learn from both correct and incorrect answers. Please record our very intelligent discussion on your activity sheet.**

There is a pause in discussion while students work on their activity sheets.

Now that we have all our values together, let's review them one at a time to see what we can learn from fractions that have the same numerator. I am going to point to a fraction and call your name. When I call your name, read the fraction aloud. For example, I will point to $\frac{4}{1}$ and say 4 wholes.

The teacher points to $\frac{4}{2}$.

Lily.

Lily: Four-halves.

Teacher: **We can see 1, 2, 3, and 4 halves.**

The teacher points to $\frac{4}{3}$.

Raul.

Raul: Four-thirds.

Teacher: **Correct. We have three parts, and we want 1, 2, 3, and 4 of those thirds.**

The teacher points to $\frac{4}{4}$.

Francis.

Francis: Four-fourths.

Teacher: **One, 2, 3, and 4 fourths. Yes.**

The teacher points to $\frac{4}{5}$.

Luke.

Luke: Four-fifths.

Teacher: **One contains 5-fifths, and we want 1, 2, 3, and 4-fifths.**

The teacher points to $\frac{4}{6}$.

Flo.

Flo: Four-sixths.

Teacher: **Yes—we want 1, 2, 3, and 4-sixths. Great work, class. Let's record our decisions on the activity sheet. Be sure to note that the numerator communicates how many equal parts we have.**

Why These Numbers?

By starting with wholes and incrementing the denominator by one, students see how the parts get smaller but the number of parts counted remains the same.

The Key Questions

- As the denominator grows under the same numerator, what happens to the value of our fraction?
- What does the numerator communicate to us?

Analysis

The benchmarks were intentionally not included at the beginning of the activity. The pattern of receiving smaller numbers as the denominator increases under the same numerator was allowed to develop first, which then made the benchmarks necessary. This led to a terrific conversation regarding $\frac{4}{3}$ and to the precious moment when a student confessed to being annoyed at the inaccuracy of its placement.

Fractions equivalent to 1 are a key concept in future operations with fractions. So, $\frac{4}{4}$ was an important value to include in this lesson. Students hit the common bump of $\frac{4}{5}$ by placing the fraction at $\frac{5}{4}$ instead.

Knowing how to read the fractions (e.g., four-thirds) leads to better understanding of the value the fraction represents. Therefore, I methodically made the connection for students between the written form of the fraction and the verbal pronunciation and meaning of the fractions on the number line.

The cards for this lesson are available in the Digital Resources (lesson7.pdf).

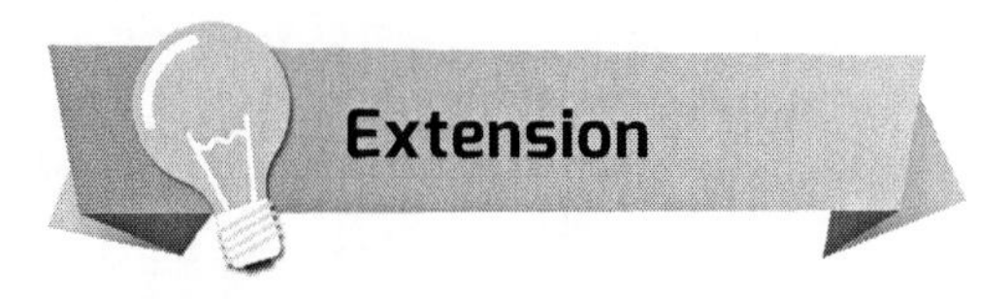

Extension

Place a variety of fractions, such as $\frac{5}{6}$, $\frac{3}{3}$, $\frac{2}{1}$, and $\frac{2}{5}$.

Lesson 8: Equivalent Fractions

$\frac{1}{2}$ $\frac{2}{4}$ $\frac{3}{6}$ $\frac{4}{8}$

Objective: Find equivalent fractions for one-half.

Teacher: Good morning, everybody. We have been having so much fun learning about fractions, I thought it would be good to challenge you with a new fraction concept today—equivalent fractions. Equivalent means "equal," so we will write different fractions to represent the same amount. For example, I am going to challenge you to find equivalent fractions for $\frac{1}{2}$. Let's start with our open number line. Take a look at where I am placing $\frac{1}{2}$ on the clothesline.

The teacher places 0 and 1 benchmark cards along with $\frac{1}{2}$ (see Figure 5.21).

Figure 5.21 Teacher Placement

Since this is a challenge today, I will take volunteers. Does anyone want to put another fraction on the clothesline that is equivalent to $\frac{1}{2}$?

Louisa raises her hand.

Louisa.

Louisa places $\frac{2}{3}$ on the clothesline (see Figure 5.22).

Figure 5.22 Louisa's Placement

Teacher: **Louisa, can you explain why you chose $\frac{2}{3}$?**

Louisa: For $\frac{1}{2}$, 1 is more than 2, so I added 1 to the top and bottom. The answer came out with the top still being more than the bottom, so they have to be equal.

Teacher: **Louisa, thank you for such a complete explanation. You did a lot of math in your head. Let's analyze where you placed $\frac{2}{3}$. Everyone, please use finger reasoning to see whether Louisa placed her fraction at the second of three equal parts. If you think Louisa placed $\frac{2}{3}$ correctly on the number line, applaud.**

There is loud applause from the class.

I also agree, Louisa, but I remember showing equivalency differently in past clothesline activities. Don't we pin equal numbers together?

Louisa: Oh yeah, should I do that?

Teacher: **Let's think about this together and decide. Are you okay with that?**

Louisa: Yes.

Teacher: **Alright. Your $\frac{2}{3}$ is in the right place, but it can only be equal to $\frac{1}{2}$ if they are in the same spot on the number line. Since they are not in the same location, what does that mean for the two fractions?**

Louisa: They are not equal.

Teacher: **That is correct. Your answer helped us learn something about finding equivalent fractions. We now know not to add to the numerator and the denominator. That is very important, so thank you. Let's give Louisa a round of applause for helping us reach this conclusion.**

Students clap. Patty keeps clapping.

Patty, you seem excited to participate. Here is a blank card; write your fraction on it, and place it on the clothesline.

Patty writes $\frac{3}{4}$ and places it on the clothesline (see Figure 5.23).

Figure 5.23 Patty's Placement

Teacher: **How did you decide on $\frac{3}{4}$, Patty?**

Patty: We can't add 1 to each number in the fraction, so I tried adding 2.

Teacher: **Okay everyone, let's do finger reasoning on Patty's $\frac{3}{4}$. Is it located at the third of four equal portions? If so, please applaud.**

The class applauds loudly again.

Patty, I also agree on your placement of $\frac{3}{4}$. I am still noticing your number is not pinned to $\frac{1}{2}$, so it is not equivalent. However, as with Louisa, we now know that adding 2 to the numerator and denominator won't work either. That was very helpful, so thank you. Who would like to go next?

Campbell raises his hand and waves it around.

Wow, we have another enthusiastic volunteer. Campbell, your turn.

Campbell writes $\frac{2}{2}$ on a blank card and pins it to the 1 benchmark (see Figure 5.24).

Figure 5.24 Campbell's Placement

This is something different. Campbell, what are you thinking for this one?

Dev: Well, 2 is equal to 2, and anything over itself is 1, so it is equivalent to 1.

Teacher: **You are correct that $\frac{2}{2}$ and 1 are equivalent. So, that helps us better understand the idea of an equivalent fraction. However, it is still not equivalent to $\frac{1}{2}$. Let me do something on the clothesline I think might help. I'm going to take away these three numbers and place pins instead.**

The teacher pins clothespins to the clothesline, as shown in Figure 5.25.

Figure 5.25 Clothespins

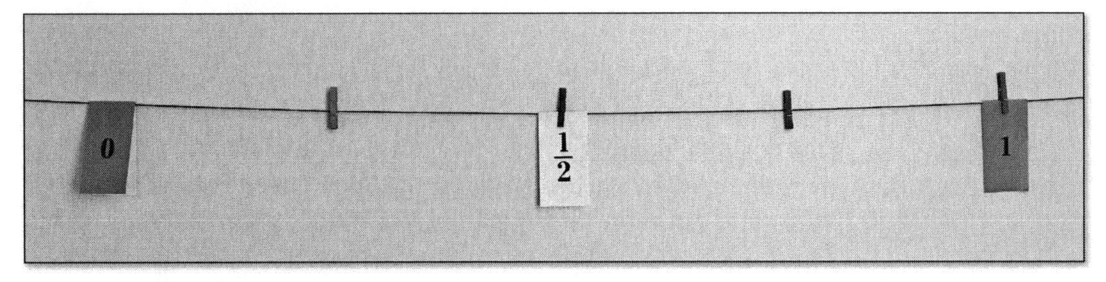

Teacher: **What fractions are these clothespins marking on our number line?**

Class: Fourths!

Teacher: **Yes! My challenge to you now is to determine how many fourths are equivalent to $\frac{1}{2}$. Think for a minute. With your lapboard partner, draw this number line. Then, write the fraction that represents the number of fourths that is equivalent to $\frac{1}{2}$. Declare your boards.**

Several students show boards that have $\frac{2}{4}$ written below $\frac{1}{2}$.

So, the hint seemed to help. Porter, please show us your answer on the clothesline.

Porter writes $\frac{2}{4}$ on a blank card and pins it to $\frac{1}{2}$ (see Figure 5.26).

Figure 5.26 Porter's Placement

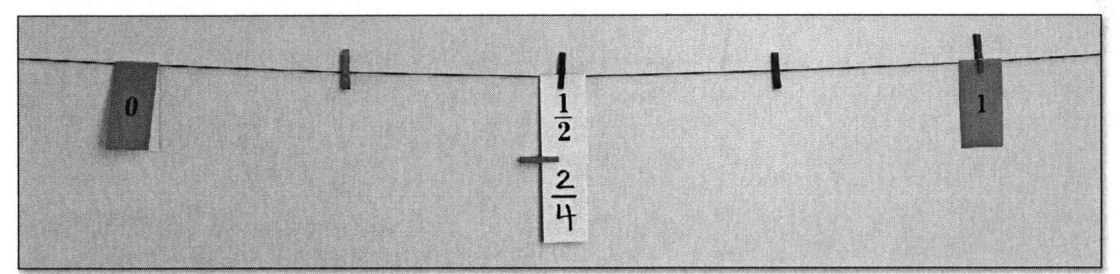

Terrific. Porter, while you are here, can you think of any other fractions we can portion our whole into?

Porter: What do you mean?

Teacher: **We went from halves to fourths. Can we break it into more equal pieces?**

Porter: Eighths?

Teacher: **Why eighths?**

Porter: We cut the halves in half, so can't we just cut everything in half again?

PART II – Chapter 5

Teacher: **By everything, do you mean the fourths?**

Porter: Yeah.

Teacher: **Sounds good. Everyone, create eighths on your lapboards and determine the number of eighths equivalent to $\frac{1}{2}$ and $\frac{2}{4}$. While everyone is working with their partners, do I have a pair of partners who want to show their answer on the clothesline?**

Gwynn and Penn raise their hands.

Gwynn and Penn.

Gwynn and Penn write $\frac{4}{8}$ on a blank card and pin it to the clothesline, as shown in Figure 5.27).

Figure 5.27 Gwynn and Penn's Placement

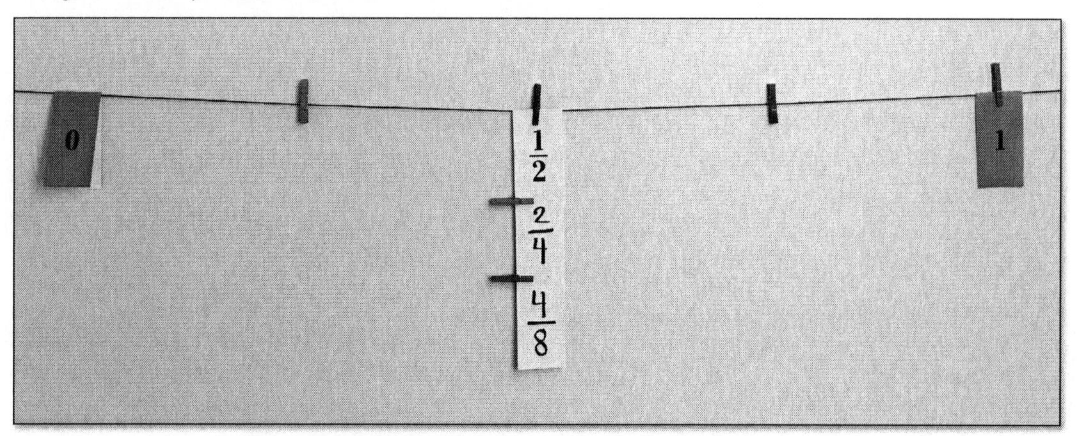

So, we have divided 1 into halves, the halves into fourths, and the fourths into eighths. I am now going to remove the pins showing fourths and ask: Is it possible to break the halves into something other than two pieces each side?

Alva raises her hand.

Alva.

Alva: Three pieces on each side would be sixths.

Teacher: **If we had sixths, what would be the equivalent fraction to $\frac{1}{2}$? Everyone, work with a partner on your board. Alva, would you like to place this one?**

The teacher pins six clothespins to the clothesline. Then, Alva writes $\frac{3}{6}$ on a blank card and pins it to the line (see Figure 5.28).

Figure 5.28 Alva's Placement

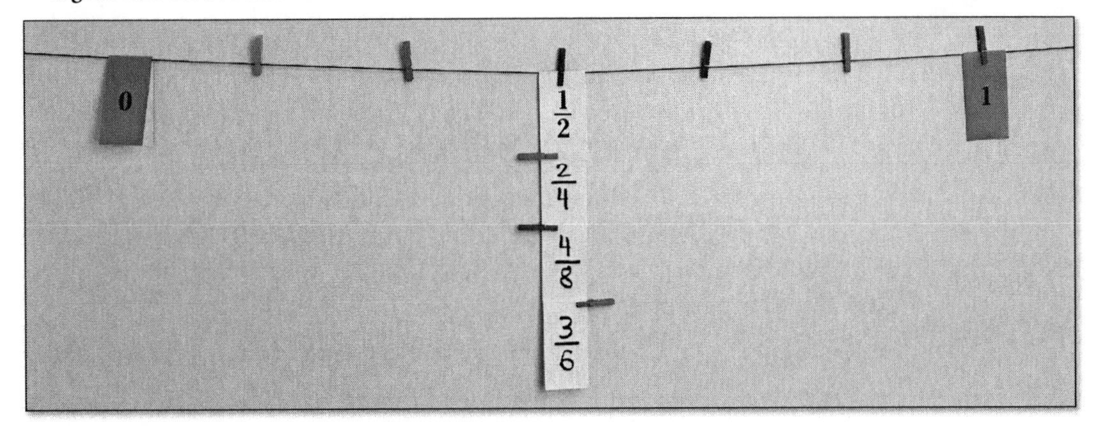

Teacher: Well done! This was a terrific conversation, but that was a lot of work to find equivalent fractions. It was good work, which is why I called it a challenge. Now, I would like to show you how to find the equivalent fractions much easier and faster.

Francis: You always show us the longer, harder way first.

Teacher: That is why you are such great, young mathematicians. Instead of adding the same number to the numerator and denominator of a fraction, we should multiply the same number to each. Take a look at the examples I'm writing on the board (see Figure 5.29).

Figure 5.29 Teacher's Examples

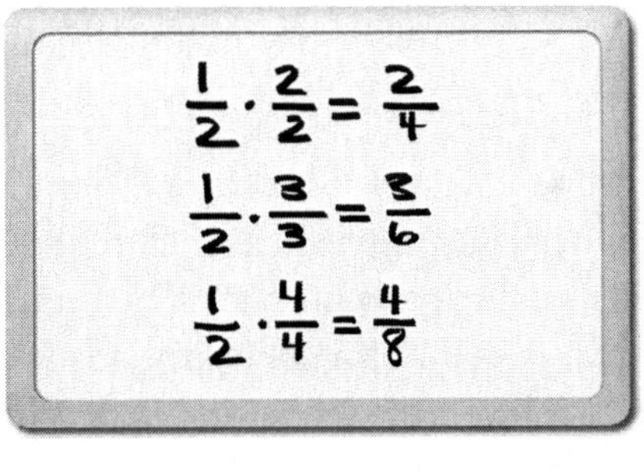

$$\frac{1}{2} \cdot \frac{2}{2} = \frac{2}{4}$$

$$\frac{1}{2} \cdot \frac{3}{3} = \frac{3}{6}$$

$$\frac{1}{2} \cdot \frac{4}{4} = \frac{4}{8}$$

Notice in each of these examples, we are multiplying by a fraction that is equivalent to 1. Remember Campbell's example of $\frac{2}{2}$ being equivalent to 1? When we multiply by 1, we always get an equivalent number. The key to finding an equivalent fraction is to multiply by a fraction equivalent to 1. We will conclude our lesson by drawing our number line diagram with our three fractions equivalent to $\frac{1}{2}$. After you've finished the number line, write these three examples of our algorithm next to it.

Why These Numbers?

One-half is the simplest of all fractions and made the visual, computations, and algorithm the cleanest.

The Key Questions

- What was your strategy in determining an equivalent fraction?
- Why do you think that value is equivalent to $\frac{1}{2}$?
- How many sixths are equivalent to $\frac{1}{2}$?
- What other fractions could we use in this situation?

Analysis

Had I simply told students how to find an equivalent fraction, I would not have uncovered their errant thinking that led them to answers like $\frac{2}{3}$, $\frac{3}{4}$, and $\frac{2}{2}$. When teaching this lesson, I figured by the third time of telling students to find a fraction that is equivalent, they would get it. When they didn't, I altered the course of the lesson somewhat. Using the clothespins was improvisation. In the past, I would have written the other fractions cards and posted them to the number line. The idea of using the clothespins seemed much cleaner. We can clearly see how many fourths, sixths, etc., are equivalent to $\frac{1}{2}$.

The conceptual understanding should precede the procedural fluency. So, as the irritated students proclaimed, I showed the longer method first before showing the shorter method. The Clothesline is effective for learning mathematics but not necessarily the most efficient for using it; just as algorithms are great for using mathematics but not learning it. With that said, students don't make a natural leap from the number line to the algorithm, so I had to show students the connection before concluding the lesson and giving them some needed rote practice. As a note, it will probably take much longer than a one-hour class period for elementary students to make the leap from the number line to the algorithm. However, I included this leap to show the full range of this lesson. The inclusion of the connection should be determined by the readiness of each individual class and when one decides to use this lesson during the course of instruction.

The cards for this lesson are available in the Digital Resources (lesson8.pdf).

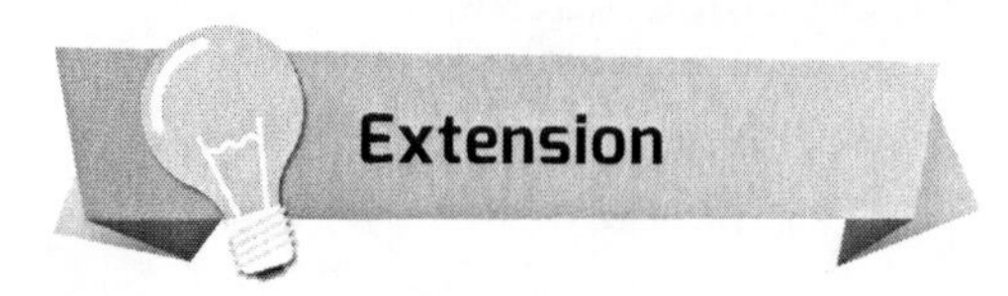

Extension

Use the Clothesline in a similar fashion to find equivalent fractions for $\frac{2}{3}$.

Lesson 9: Adding Fractions

$\frac{3}{12}$ $\frac{8}{12}$ $\frac{2}{3}$ $\frac{1}{2}$ $\frac{1}{4}$ $\frac{1}{2}+\frac{1}{4}$ $\frac{1}{4}+\frac{2}{3}$ $\frac{11}{12}$

$\frac{3}{6}$ $\frac{4}{6}$ $\frac{1}{2}+\frac{2}{3}$ $\frac{7}{6}$ $\frac{3}{4}$ $\frac{2}{4}$ $\frac{1}{5}$ $\frac{3}{5}$ $\frac{1}{5}+\frac{3}{5}$

Objective: Add two fractions with unlike denominators.

Teacher: **Hi, everyone. We know how to add fractions as long as the denominators are the same. We call these common denominators. Today, we are going to explore fractions that have unlike denominators and see whether we can still apply our method for adding fractions. We are going to warm-up with what we know first. Work on your whiteboards, please. Take a look at what I have written on the board** (see Figure 5.30).

Figure 5.30 Teacher's Equation

$\frac{1}{5}+\frac{3}{5}=$

Declare your boards.

Students show their boards but are split between two answers.

The two answers that I see are $\frac{4}{5}$ and $\frac{4}{10}$. The clothesline will help us settle the debate. For the warm-up, I will take a volunteer group.

Rami raises his hand.

Thank you, Rami. Please have your group place $\frac{1}{5}$, $\frac{3}{5}$, and $\frac{1}{5}+\frac{3}{5}$ on the clothesline while everyone else does it on their lapboard partner.

Rami's group places the three cards on the number line, as shown in Figure 5.31.

Figure 5.31 Rami's Placement

Teacher: Rami, can you explain?

Rami: We put 0 and 1 on the number line because all our cards are less than 1. Then, we chopped 1 into five parts and marked one of those parts and three of those parts. You can see that if you add how long $\frac{1}{5}$ is to how long $\frac{3}{5}$ is, then you end at $\frac{4}{5}$.

Teacher: Terrific explanation. And, notice how $\frac{4}{10}$, which is less than $\frac{1}{2}$, would not make sense for the sum of these numbers. To be sure everyone sees the fifths like you do, I'm going to put some clothespins at the unmarked fifths.

The teacher adds clothespins to the line, as shown in Figure 5.32.

Figure 5.32 Added Clothespins

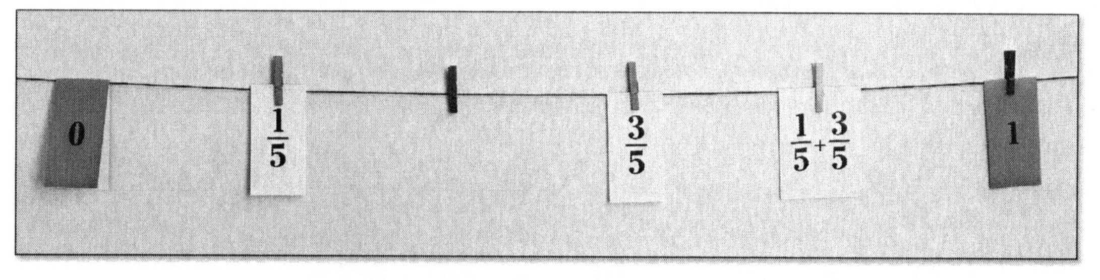

While I point, you say the number.

The teacher points to each successive fifth by pointing to the clothespins.

Class: One-fifth. Two-fifths. Three-fifths. Four-fifths. Five-fifths.

Teacher: Excellent. We are ready for the lesson now regarding fractions that have different denominators. I am placing $\frac{1}{4}$ and $\frac{1}{2}$ on the number line. I would like to have their sum here as well. Wendy, it is your group's turn. Please place $\frac{1}{4}$ + $\frac{1}{2}$ on the clothesline while everyone else works with their elbow partners on lapboards.

Wendy and her group pin the equation to the line and pin $\frac{3}{4}$ to it, as shown in Figure 5.33.

PART II – Chapter 5

Figure 5.33 Wendy's Placement

Teacher: Wendy, why did you choose $\frac{3}{4}$?

Wendy: We took how far it is to $\frac{1}{2}$ and measured that far from $\frac{1}{4}$.

Teacher: Everyone who agrees, clap twice.

Students clap twice.

I agree it sounds logical, but we learned that when adding fractions, we keep the denominators and add the numerators. Here, we only kept one denominator, and the sum of the numerators is 2 rather than 3. So, what's up with that?

The class is silent.

Are you all confused about it, too?

Several students nod.

Well, that makes me feel better. But, there has to be an answer because math always makes sense.

Wendy: Is our group's answer wrong?

Teacher: We all agreed the logic of adding the distances was correct. However, I think we need another method to use when the denominators are different. Heather, do you have an idea?

Heather: I don't know.

Teacher: Throw it out there. What are you thinking?

Heather: That we make $\frac{1}{2}$ be $\frac{2}{4}$?

Teacher: Let's see. I will pin $\frac{2}{4}$ to $\frac{1}{2}$.

The teacher pins $\frac{2}{4}$, as shown in Figure 5.34.

Figure 5.34 Teacher's Placement

Wendy: Oh yeah…it does work!

Teacher: Heather?

Heather: Cool. It works.

Teacher: Keep the denominators?

Heather: Yes, 4.

Teacher: Add the numerators?

Heather: Yes, that would match the 3.

Teacher: Let's draw this on a number line on the *Clothesline Math* activity sheet. Also, write next to your number line what we call this procedure in mathematics: the algorithm. Everyone, say "algorithm."

Class: Algorithm!

Teacher: Here is what the algorithm looks like for adding these fractions that have different denominators (see Figure 5.35).

Figure 5.35 Adding Fractions

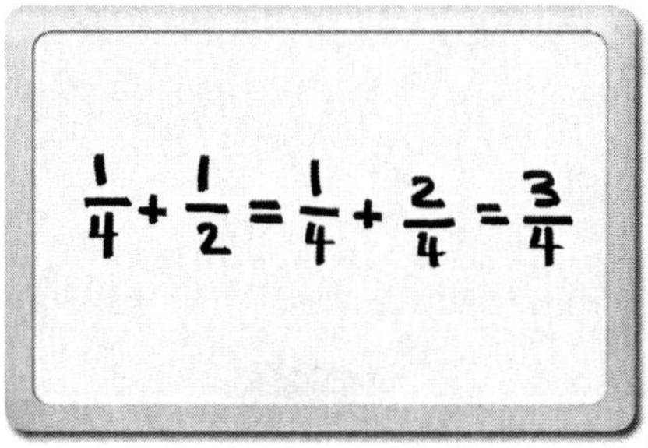

$$\frac{1}{4}+\frac{1}{2}=\frac{1}{4}+\frac{2}{4}=\frac{3}{4}$$

I wonder, does that algorithm apply to this next situation?

The teacher changes the cards on the clothesline, as shown in Figure 5.36.

Figure 5.36 New Situation

Teacher: What if I want to add $\frac{1}{2} + \frac{2}{3}$? Can anyone tell me how adding these two fractions is different from adding the previous fractions?

Edmund raises his hand.

Edmund.

Edmund: The 2 does not go into the 3.

Teacher: So 4 is a multiple of 2, but 3 is not a multiple of 2?

Edmund: Correct.

Teacher: Let's take this to the clothesline and see whether the algorithm still works. Wendy, who is taking this?

Wendy chooses Micah.

Micah, have your group head to the clothesline. Everyone else, erase your boards, and draw a new number line.

Micah's group places the equation on the right side of 1, as shown in Figure 5.37.

Figure 5.37 Micah's Placement

Interesting that you put the sum greater than 1.

Micah: Is that wrong?

Teacher: I am going to ask the class to decide by following your explanation. Go ahead.

Micah: Like last time, we used our hands to measure 0 to $\frac{1}{2}$ and slid it over until we lined up with $\frac{2}{3}$. That puts us past 1, but we aren't sure exactly what it would equal.

Teacher: I would think you just change one of the denominators like we did last time.

Micah: Yes, but you can't change 2 into 3 or 3 into 2.

Teacher: Do we just change stuff? Remember, this is math, not magic.

The class laughs, including Micah.

Micah: No, we can't just change stuff.

Teacher: You are correct, though—just like we discussed in the previous problem, 4 is a multiple of 2. This means 2 times another whole number will equal 4. However, there is no whole number that we can multiply 2 by to equal 3. But, is there a multiple of 3 that is also a multiple of 2? In other words, 2 times a whole number will equal the 3 times a different whole number.

Danika raises her hand.

Danika.

Danika: Six.

Teacher: That would mean we would see sixths. Will you help us see the sixths?

Danika: Sure.

Teacher: Place this clothespin at $\frac{1}{6}$.

Danika places the clothespin and turns around to return to her seat.

Stay here for a just a moment while I point this out to everyone. Can you all see equal sixths in $\frac{1}{2}$? In other words, can we cut $\frac{1}{2}$ into equal parts this size?

Class: Yes!

Teacher: How many sixths will be in $\frac{1}{2}$?

Class: Three!

Teacher: Danika, please use another clothespin to show there are three sixths in $\frac{1}{2}$.

Danika adds one more clothespin to the line, so there are now two clothespins on the number line.

I see only two sixths. Can you please pin the other sixths?

Danika pins the final sixths with the clothespins, as shown in Figure 5.38.

Figure 5.38 Danika's Placement

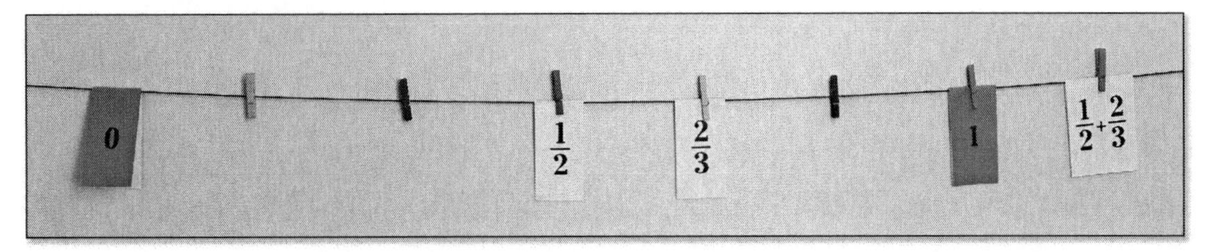

Teacher: **If we do this right, then $\frac{1}{2}$ and $\frac{2}{3}$ would also be located at one of these sixths. Sure enough, we see clothespins at both of them. These sixths are what we call what?**

Class: The least common multiple.

Teacher: **Sweet.**

Joanie raises her hand.

Yes, Joanie?

Joanie: We can rewrite them with common denominators.

Teacher: **Come on up and show us.**

Joanie uses blank cards to rewrite the fractions and pins them, as shown in Figure 5.39.

Figure 5.39 Joanie's Placement

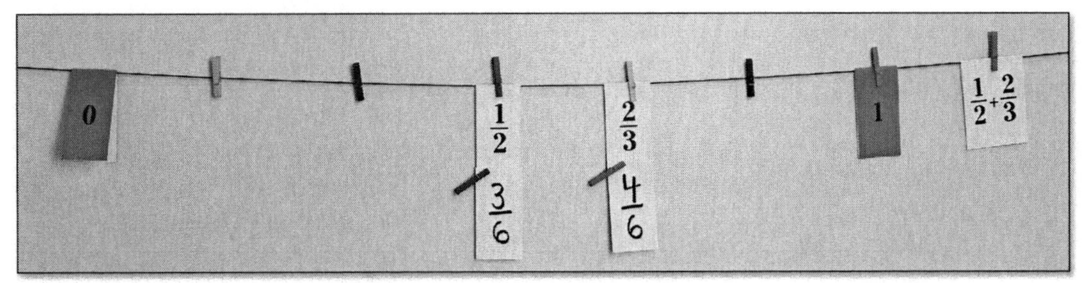

Class, all those who agree with what Joanie has shown here, say "aye."

Class: Aye!

Teacher: **Nice work showing us the equivalent fractions. I'm going to now take a volunteer to pin the sum of $\frac{1}{2}$ and $\frac{2}{3}$ written as some number of sixths.**

Monica raises her hand.

Monica, go ahead.

Monica writes $\frac{7}{6}$ and pins it to the expression card (see Figure 5.40).

Figure 5.40 Monica's Placement

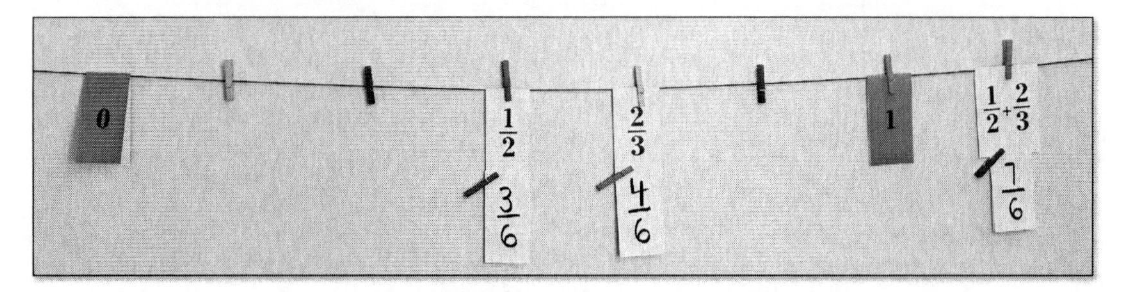

Teacher: Is Monica properly applying our method for adding with common denominators? Thumb vote time—thumbs-up if you agree, and thumbs-down if you disagree.

The entire class votes thumbs-up.

Monica, the whole class agrees. Let's record this number line and the algorithm on our activity sheet. I will write the algorithm on the board (see Figure 5.41).

Figure 5.41 Teacher's Algorithm

$$\frac{1}{2}+\frac{2}{3}=\frac{3}{6}+\frac{4}{6}=\frac{7}{6}$$

For our last sum, $\frac{1}{4}+\frac{2}{3}$, I want a volunteer group to use clothespins to show the common denominator of these fractions.

Adam raises his hand to volunteer.

Adam's group has volunteered, thank you.

Adam's group places 12 clothespins, as shown in Figure 5.42.

Figure 5.42 Adam's Placement

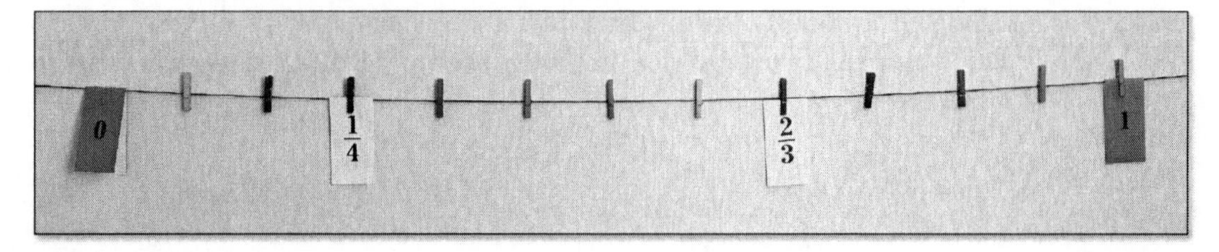

Teacher: Thank you, Adam. May I now have a volunteer group to pin the equivalent fractions with the common denominator?

Leighanne raises her hand.

Leighanne's group, please head to the clothesline.

Leighanne's group writes two equivalent fractions and pins them to the line (see Figure 5.43).

Figure 5.43 Leighanne's Placement

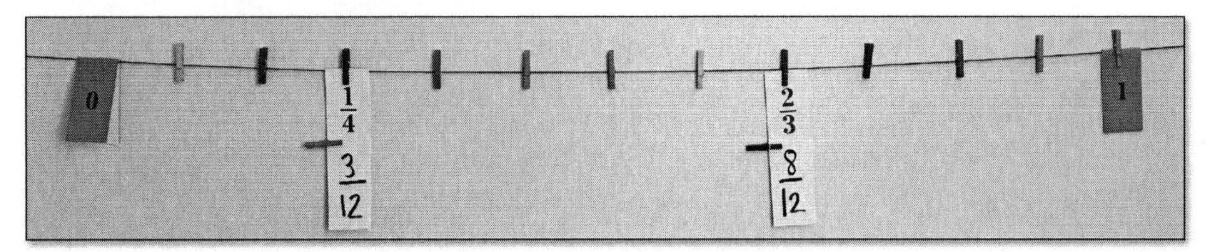

Thank you, Leighanne. How about a group to now place our sum, $\frac{1}{4} + \frac{2}{3}$?

Wilson raises his hand.

Wilson.

Wilson's group pins the expression to the left of 1, as shown in Figure 5.44.

Figure 5.44 Wilson's Placement

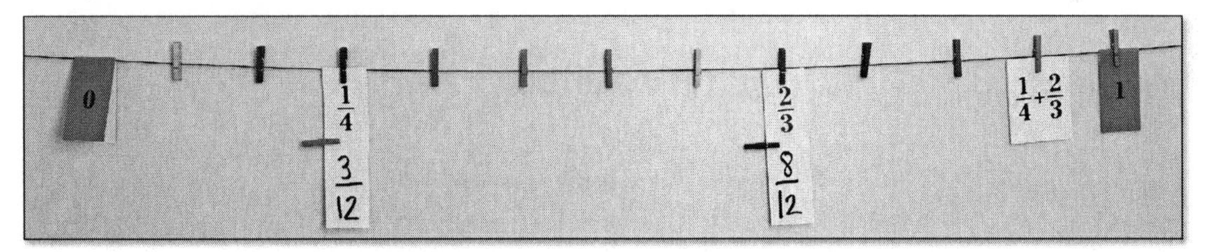

Class, write the answer on your lapboards. What is $\frac{1}{4} + \frac{2}{3}$? While you are writing, I need one last volunteer group.

Mariah raises her hand.

Mariah's group, thank you.

Mariah's group writes $\frac{11}{12}$ and pins it to the expression card (see Figure 5.45).

Figure 5.45 Mariah's Placement

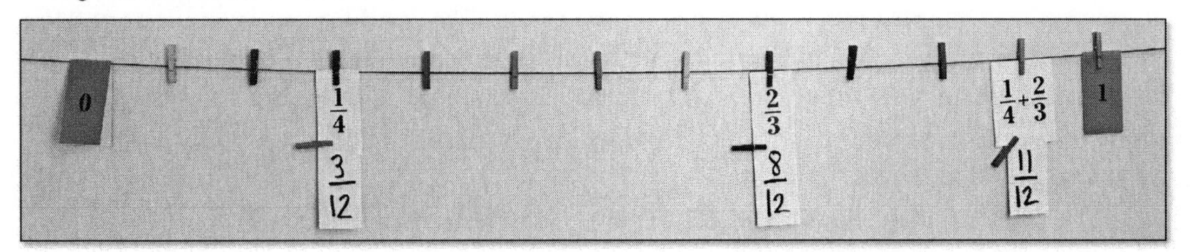

Teacher: Declare your boards.

Students show lapboards that match the same placement for $\frac{11}{12}$.

Mariah, the whole class agrees with your group. Let's summarize—the least common multiple of the denominators will determine the unit fraction and give us our common denominator for the equivalent fractions. That will be the same denominator for our answer. We then add the numerators of our equivalent fractions. Show me you understand this by applying the algorithm we discussed in the last problem to this one. Show the steps on your lapboard for how $\frac{1}{4} + \frac{2}{3} = \frac{11}{12}$.

Students work on their lapboards.

Declare your boards.

Students hold up their boards, with most of them showing the correct answer.

Overall, pretty good. Let me show you what it should look like while you write the example underneath your number line (see Figure 5.46).

Figure 5.46 Teacher's Equation

$$\frac{1}{4}+\frac{2}{3}=\frac{3}{12}+\frac{8}{12}=\frac{11}{12}$$

123 Why These Numbers?

To begin the progression of adding fractions with different denominators, the lesson begins with only having to change one of the denominators. The simplest example of this is $\frac{1}{2}$ and $\frac{1}{4}$. To keep mental computations easy, the next example in the progression would involve thirds. To include a numerator that demands more than to be simply changed to the other denominator, $\frac{2}{3}$ was chosen over $\frac{1}{3}$. The warm-up was utilized to connect this lesson to the prior experience of students by using the next simplest unit fraction of a fifth. Fourths and thirds were considered, but there would not have been enough uniqueness among the values to distinguish the answers in the warm-up from those in the lesson. The fifths example also allowed for the introduction of using clothespins to designate the unmarked fractions (e.g., "the other fifths").

- Why did you choose to place that card there on the number line?
- Does our procedure for adding fractions with common denominators apply to fractions with unlike denominators?
- Does our algorithm for adding fractions when one denominator is a multiple of the other still apply when that is not the case?
- How many sixths will be in a half?
- Can you designate the twelfths on the number line?
- Is there a multiple of 3 that is also a multiple of 2?

This lesson on unlike denominators is intended to follow some time after a lesson on adding fractions with common denominators, so the lesson began with the specific warm-up. Typically we use warm-ups to activate students' prior knowledge, but as the warm-up showed, not every student possessed that prior knowledge. However, everyone had the prior experience of adding fractions with common denominators on the clothesline. Revisiting that example first allowed the groundwork for the current lesson.

The common error of adding the denominators was anticipated. I wanted to get that misconception out of the way before we ran into any others. The strength of the clothesline in teaching is very apparent with the first example of $\frac{1}{4} + \frac{1}{2}$. The open number line allows students to add distances visually and *see* the sum must be $\frac{3}{4}$. It is not until after this revelation that students are presented with a pencil-and-paper algorithm. This process serves several solid pedagogical purposes:

1. **Access**—This is a terrific example of the phrase "low floor, high ceiling." Not all students can initially follow the symbolic manipulation of finding equivalent fractions with common denominators and then adding fractions. That is a lot of procedural knowledge to be crammed into a young student's head. However, all students can see where $\frac{1}{4}$ and $\frac{1}{2}$ are located on the number line. They can also visually add these distances with finger reasoning.
2. **Conceptual versus Procedural**—Conceptual understanding precedes procedural fluency. Students are taught the *why* before the *how.*
3. **See and Use Structure**—Other mathematical habits of mind from the new twenty-first century standards are evident in this lesson, such as using tools (i.e., the clothesline), reasoning quantitatively, and critiquing the reasoning of others. However, drawing the connection between the addition of fractions on the number line and the algorithm is the strength of the lesson. It is typically understood that any time students develop a new mathematical algorithm, they demonstrate use of structure and repeated reasoning.

4. **Greater Depth of Knowledge (DOK) Levels**—Students are being asked to think and make sense of the mathematics rather than simply memorize rules for the purpose of a correct answer.

There is another technique that should be highlighted in this lesson: the use of the clothespins to designate the unmarked fractions. Rather than writing on blank cards and cluttering the number line, the clothespins clearly show the unit fractions (e.g., $\frac{1}{6}$, $\frac{2}{6}$, or $\frac{3}{6}$). The absence of the numbers forces students to think about the distances these fractions represent.

Finally, a key component of this lesson is writing the algorithms to generalize student thinking. Students do not automatically make the leap from visual manipulations to symbolic representations, so teachers must make that connection for them. This is the reason I have students write the procedural example next to visual examples.

The cards for this lesson are available in the Digital Resources (lesson9.pdf).

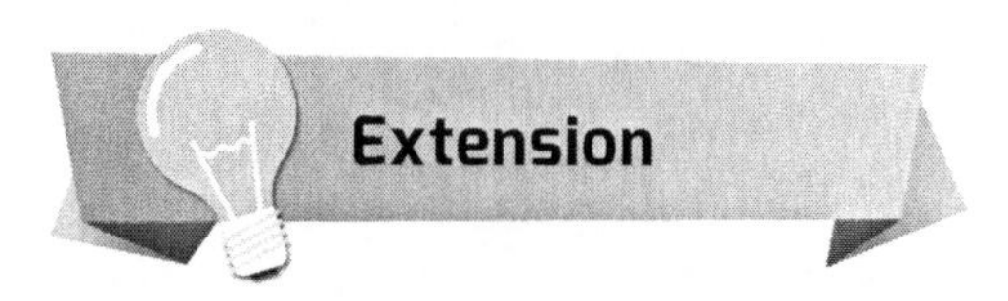

Extension

Add mixed numbers $1\frac{1}{2}$, $2\frac{2}{3}$, $3\frac{1}{4}$.

Algebraic Reasoning for Elementary School

Lesson 10: Order of Operations

14 | 20 | $2+3$ | 3×4

12 | $2+3 \times 4$ | $(2+3) \times 4$

Objective: Properly apply multiplication before addition in numeric expressions, unless parentheses dictate otherwise.

Teacher: **Good day, class. We have been studying order of operations and learning how certain operations need to be done before others. This is not because of random rules, but because of mathematical properties and because it makes sense when we think about the expressions in real life. I noticed on our last assessment that you are still struggling with what to do when addition is written before multiplication in an expression or when parentheses are part of an expression. Look at these two examples I have written on the board** (see Figure 6.1)**:**

Figure 6.1 Teacher Examples

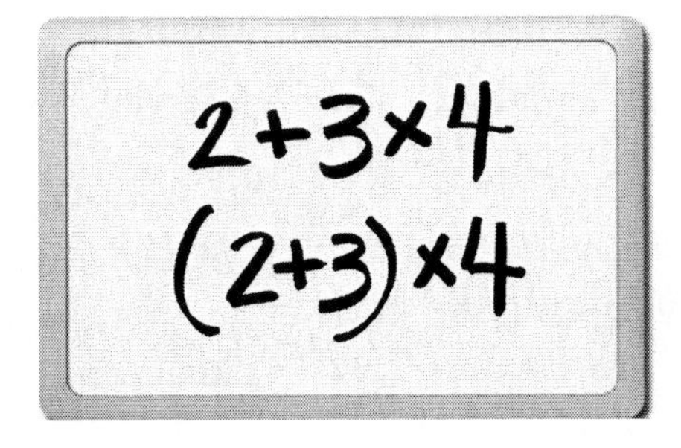

Teacher: **Today, we are going to fix that issue with a little help from the clothesline and some added context. Before we do, let's think about what the simpler parts of these expressions look like on the number line. I am going to place two benchmark cards on the clothesline** (see Figure 6.2).

Figure 6.2 Benchmark Cards

Kylia, your group is to place 2 + 3 on the number line. Everyone else, work with your partners on your lapboards.

Kylia's group grabs a 5 benchmark card and pins 2 + 3 to it, as shown in Figure 6.3.

Figure 6.3 Kylia's Placement

Kylia, I know your group understands simple addition, but how did you think about where to place the 2 + 3 when there were no other benchmarks on the number line?

Kylia: We just cut 2 in half to find 1, and then we counted 1, 2, 3 of those spaces from 2.

Teacher: **All in favor, say "aye."**

Class: Aye!

Teacher: **Given this new number line showing 4, Kylia, which group is going to share where to place 3 × 4?**

Kylia: Hmm...Jaylene!

Teacher: **Jaylene, it's your group's turn. Everyone else, erase your boards and draw this new number line.**

Jaylene's group chooses the 12 benchmark card off the table at the front of the class and pins 3 × 4 to it, as shown in Figure 6.4.

Figure 6.4 Jaylene's Placement

Teacher: **Jaylene, I know your group knows your multiplication tables, but how did you decide where to put 3 × 4 with no other benchmarks present?**

Jaylene: We counted the 4 three times.

Teacher: **Awesome. We can then think of 3 × 4 as 3 groups of 4. Thinking of multiplication as grouping will help a great deal today. I'm going to put back our previous problem so they are both on the same number line. Please do the same on your boards.**

Students fix their number lines on their lapboards while the teacher adds cards back to the clothesline (see Figure 6.5).

Figure 6.5 Teacher Placement

Now, let's decide what happens when we combine these expressions into one. Take a look at 2 + 3 × 4. We need to decide where this goes on the number line. Think for a moment while Jaylene chooses our next group.

Jaylene points to Jonsey.

Jonsey, your group is up. Everyone else, leave what is already on your lapboards, and place this expression.

Jonsey's group selects a 20 benchmark card to pin the expression card to and places them on the clothesline, as shown in Figure 6.6.

Figure 6.6 Jonsey's Placement

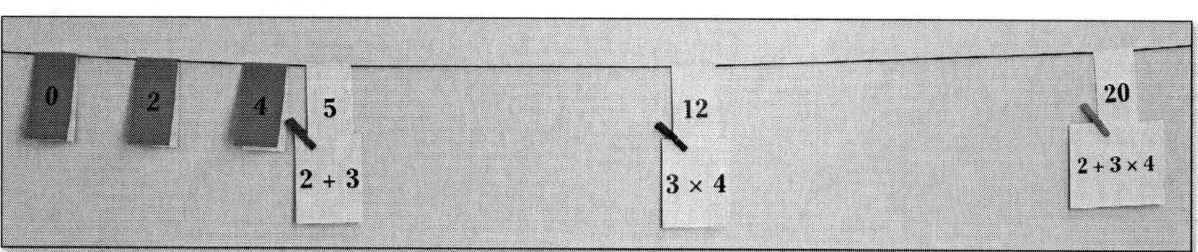

PART II – Chapter 6

Teacher: Class, declare your boards.

Students have varying answers on their boards. Only some boards match the clothesline at the front of the class.

Jonsey, most of the class agrees with you, but some disagree. Do you want to explain?

Jonesy: $2 + 3 = 5$, and $5 \times 4 = 20$.

Teacher: Thank you. Reza, you have something different on your board.

Reza: We did the multiplication first and then did $2 + 12 = 14$.

Teacher: Thank you. So, how are we going to settle this? How about context? Let's say I give you $2 for coming to school today. Then, I give you $3 for every math problem you do and you do 4 math problems. Think about that one and see whether you want to adjust your answer.

Amir raises his hand.

Yes, Amir.

Amir: Will you really give us money for our math homework?

Teacher: No, but your boss will someday. Alright class, declare your boards again.

Students present their boards to the teacher, showing more students have changed their answers.

Interesting, most of you changed your answers. Frankie, you did not. Are you saying that you make $5 for every math problem?

Frankie: No, I'm not.

Teacher: You added $2 + 3$ then multiplied 5×4. So, you are saying that I am giving you $5 per problem.

Frankie: Oh, I get it. Okay, wait a minute.

Frankie adjusts the math on his board to reflect the correct answer of 14.

Teacher: Everyone now seems to agree with you, Jonsey. Please come change your response on the clothesline (see Figure 6.7).

Figure 6.7 Jonsey's Second Placement

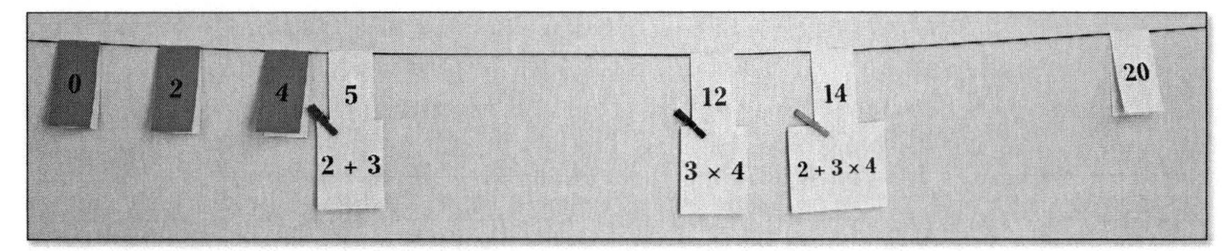

Teacher: **How did you decide where 14 was on the number line without the extra benchmarks?**

Jonesy: We have 3 × 4 already on the line. You just have to add 2.

Teacher: **Thank you, Jonsey. In summary of what we have done so far, we have these rules called the order of operations. This way, when we do not have the story to help us figure out the order, we can still get the correct answer. The rules tell us to multiply before we add because that is what we do in real life, in context. But what if we see parentheses? Does that change things? Let's see.**

Jonsey, which group is going to place (2 + 3) × 4 for us?

Jonsey selects Pedro's group.

Pedro, will your group please help us at the clothesline? Everyone else, place this on the same number line with the rest of your answers.

Pedro's group pins their card to the 20 benchmark, as shown in Figure 6.8.

Figure 6.8 Pedro's Placement

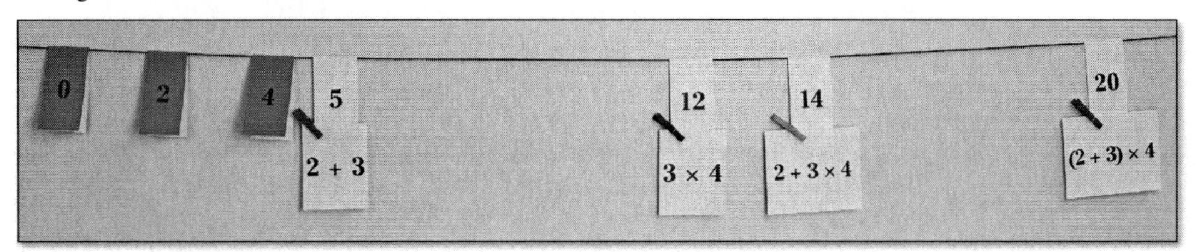

Teacher: **Pedro, why 20? I thought we already decided we don't add first.**

Pedro: Except when parentheses tell you to.

Teacher: **Oh, so sometimes we want to add first, but then we have to tell people to bend the order of operations a little bit. You are saying we communicate that with parentheses.**

Pedro: Yes.

Teacher: **Class, declare you boards.**

Students present their boards, showing they all agree with Pedro.

Teacher: **Interesting, everyone agrees with you. Let me give context and see whether they will still agree. I give $2 every morning and $3 every night for 4 days. Think about this for a moment. Discuss with your partners, and change your answer if you wish.**

There is a great deal of discussion, but no group changes answers.

I am glad to see I was able to give you something that made you think so hard and discuss so well. By the way, Pedro, 20 is correct.

Pedro: I knew it!

Teacher: **Funny how just because I gave a scenario to check your work, you thought I was questioning your answer. I just wanted to make sure your rule was correct. It was, so let's restate it—multiply before we add, unless the parentheses dictate otherwise. All in favor, say "aye."**

Class: Aye!

Teacher: **But, how did you know where the 20 was without the extra benchmarks?**

Serena: The 2 + 3 = 5 right there. We just counted four of those.

Teacher: **Oh, so our idea of grouping is helping again. Multiplying means grouping, and this problem means we want four groups of 2 + 3.**

Serena: Yes.

Teacher: **I would like someone to summarize two big ideas from our lesson. Who would like to explain why we have the order of operations?**

Makani raises her hand.

Makani.

Makani: So we can do it right without the story.

Teacher: **Thank you. Who would like to explain why we multiply before we add? Mollie.**

Mollie: Because that is the way we do it in real life.

Teacher: **Awesome. Record all of this on your activity sheet on one number line. That was some hard thinking today. Nice job.**

Why These Numbers?

This lesson focused on addition, multiplication, and the use of parentheses. Division and subtraction were saved for another day because it was important to connect how context dictates the rules we use and what to do when context is not available. Therefore, the lesson was kept simple. Parentheses were included to show when to add first, which so many students wanted to do all the time. The numbers 2, 3, and 4 gave us the easiest computational examples that could be created and contextualized.

The Key Questions

- How did you think about where to place $2 + 3$ and 3×4 when there were no other benchmarks on the number line?
- How are we going to decide which order is correct?
- Why do we have the order of operations?
- Why do we multiply before we add?
- How do parentheses change the order?

Analysis

The misconception that this lesson addresses is far too common during elementary and through high school. The results of this lesson showed that many students were making this error when initially introduced to the order of operations. Therefore, reengagement with the content was needed. My mantra is: when misconceptions are found, take them to Clothesline!

Rather than just restate the order of operations and have students practice it several times, this lesson drilled to the root cause: students don't understand why the order of operations is what it is. The lesson was designed to connect the rules to context.

The lesson also provided the opportunity to work on basic computation skills, such as addition and subtraction—skills many students still need to practice. It's important to have students first compute $2 + 3$ and 3×4. Then, have them employ some number sense to properly place these expressions on the number line.

For the sake of building number sense, most benchmarks were intentionally left off the number line. With only 0 and 2 offered, all other placements could be discerned. Students handled this very well.

Placing the expressions on the number line also reinforced the ideas of addition and multiplication as grouping, which was helpful when we inserted the parentheses at the end of the lesson.

The cards for this lesson are available in the Digital Resources (lesson10.pdf).

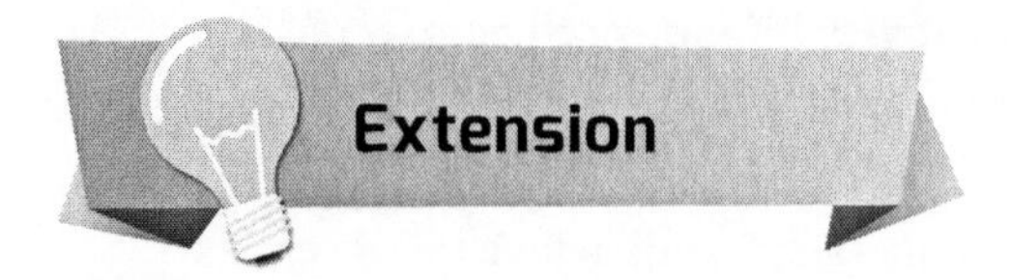

Extension

Would either of the expressions $2 + 4 \times 3$ or $3 + 2 \times 4$ be equivalent to the original expression of $2 + 3 \times 4$?

Lesson 11: Variables

b	*g*	**20**

Objective: Show that a variable may be used to represent multiple values.

Teacher: Good morning, everybody.

Class: Good morning!

Teacher: I am very excited to teach math today because we get to explore more about variables. This means you are awesome enough to use algebra. Remember, the root word of variable is "vary," which means the number can change. That's what our lesson today is going to be about—showing how a variable can be used to represent a set of numbers, not just one number. To do that, we need a scenario. I have written a scenario for today on the board (see Figure 6.9).

Figure 6.9 Scenario

The number of students in all classes at our school is 20. How many boys and how many girls are in each class?

Don't say anything about this just yet. Think for a minute. Some of you are wrinkling your foreheads—I am already making you think hard, which is very good. Talk to your partner; I want an answer from you both in 30 seconds. Raise your hands if your partner has figured it out.

Students discuss the problem on the board amongst themselves.

Kayla, what do you two think?

Kayla: We think there is not enough information to tell.

Teacher: Thank you. Larry?

Larry: We say it depends. There can be lots of answers.

Teacher: Thank you. Paulina?

Paulina: 12 boys, 8 girls.

Teacher: Why do you say that?

Paulina: I counted all the boys and girls in our class.

Teacher: There is logic behind all your answers. For example, is Paulina's answer true? Are there 20 students in here? Are there 12 boys and 8 girls?

Class: Yes!

Teacher: Do we have enough information to know how many are in the classroom next door?

Class: No!

Teacher: Could there be a different combination for each class at our school?

Class: Yes!

Teacher: Well, it sounds like everyone is correct in some way. Let's explore this idea further on the number line. We need to start by defining our variables. Today, we are going to use *b* to represent the number of boys, and *g* to represent the number of girls. Notice I said *b* will represent the number of boys, because *b* is not a label, it is a variable. Listen carefully as I say it again—*b* does not represent boys, it represents the number of boys. Finn, *b* does not represent boys, it represents...

Finn: I don't know.

Teacher: I will come back to you. Quinn, *b* does not represent boys, it represents...

Quinn: The number of boys.

Teacher: Finn, back to you—*b* does not represent boys, it represents...

Finn: The number of boys.

Teacher: Good. So class, if that is the case, then *g* does not represent girls, it represents...

Class: The number of girls!

Teacher: Terrific. Let's write these definitions at the top of our activity sheet. I will write it on the board (see Figure 6.10).

Figure 6.10 Definitions

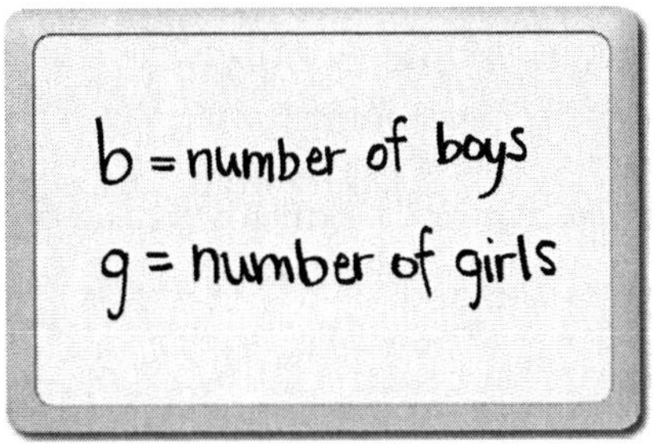

b = number of boys
g = number of girls

Teacher: **Time to put our variables on the clothesline. Paulina, your group offered this instance, so I am going to start us with this** (see Figure 6.11).

Figure 6.11 Teacher Placement

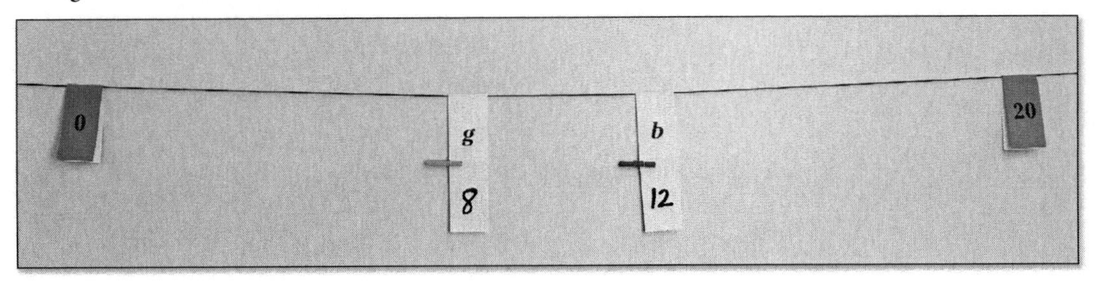

Paulina, is this the instance you and your partner were discussing?

Paulina: Yes.

Teacher: **Summer, what is this saying?**

Summer: There are 12 boys and 8 girls.

Teacher: **Everyone, let's record this on our first number line diagram on our activity sheets. Larry, you claimed there is more than one answer. Will you and your partner please show us another instance?**

Larry and his partner adjust the cards on the clothesline, as shown in Figure 6.12.

Figure 6.12 Larry's Placement

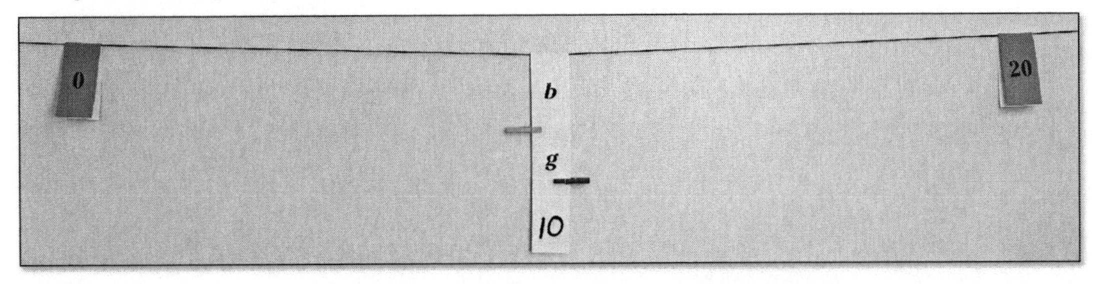

Carmen, What does Larry's instance represent?

Carmen: 10 boys and 10 girls.

Teacher: Thank you. Everyone, record this example while Carmen and her partner give us another instance.

Carmen and her partner move the variables to different benchmark cards, as shown in Figure 6.13.

Figure 6.13 Carmen's Placement

Lauren, why are you giggling at this one?

Lauren: I wish there were no boys.

Teacher: Hey, be nice. How many girls do you see for this one?

Lauren: Twenty.

Teacher: Let's record this one also. Now, I am going to really challenge you.

The teacher removes the g *variable and hangs the* b *variable on the clothesline (see Figure 6.14).*

Figure 6.14 Teacher's New Placement

Charlotte, where should this *g* go?

Charlotte and her partner head to the clothesline and place g, *as shown in Figure 6.15.*

Figure 6.15 Charlotte's Placement

Teacher: **Charlotte and Angel, please stay and explain.**

Angel: It looks like about 1, so we put the other one at 19.

Teacher: **The reason I have you staying at the clothesline is because I would like you two to stand to the left of the clothesline. One of you, reach and hold *b*. I am going to stand to the right and grab *g*. I am going to slowly slide *g* along the number line. As I do, I want you to slowly move *b* until *b* and *g* still have a sum of 20. Here we go.**

I slide g *to approximately 18. The students don't move* b.

Angel: We don't understand.

Teacher: **I will start again. I move *g* to about 18. Where should *b* go?**

The students move b *to 2. I continue to move* g *to the left as the girls move* b *to the right. We eventually cross paths as I continue* g *toward 0. The students move* b *toward 20. I stop* g *at 0. The students stop* b *at 20.*

Thank you, ladies. You may have a seat. The idea that Carmen, Angel, and I just demonstrated is the idea we started with at the beginning of the lesson. Variables represent quantities that may vary. Might the number of boys in each class vary?

Class: Yes!

Teacher: **Might the number of girls in each class vary?**

Class: Yes!

Teacher: **Then we can represent these numbers with variables. Please record two more instances of our scenario today on your activity sheet.**

Why These Numbers?

This lesson could have been conducted with one variable (e.g., number of boys only). The decision to go with two variables was made because simultaneously seeing different quantities reinforced the idea of assigning variables to those quantities. The idea of establishing a relationship between the quantities ($b + g = 20$) without formalizing it was done for two reasons. One, it allowed students to practice subtraction using mental math. Two, it helped lay the groundwork for writing equations later.

The Key Questions

- How many boys and how many girls are in each class?
- Do you have enough information to tell how many boy and girls are in the classroom next door?
- Could there be a different combination for each class at our school?
- What does this instance represent?

Analysis

This lesson started with a question that gets to the heart of the objective: How many boys and how many girls are in each class? The idea that a variable can be more than just one unknown number is a cornerstone of algebra. The question was posed to cause some deep thought, and students handled it well. The idea that there is not enough information to give just one instance is the flipside of the other response stating there can be several instances. The claim with the exact number of boys and girls in the class missed the mark on variability, but it arranged the conversation nicely by starting the discussion with values for b and g with which students could relate.

We sampled a few more concrete examples before getting very abstract by sliding the cards along the number line. This idea came to me when I was deciding how many examples to show before the idea that the variables represented "a lot" of different classes.

Time was also taken to define the variables. This is a very important step for the mathematical modeling students will be doing throughout the rest of their schooling. The idea that variables represent numbers and not just labels is key to understanding equations, so I chose to bring that point home with the statement "*b* does not represent 'boys,' it represents 'the number of boys.'"

The cards for this lesson are available in the Digital Resources (lesson11.pdf).

Extension

Rather than a discrete scenario like the boys and girls in class, offer a continuous scenario in which the number of instances is infinite. For example, 10 gallons of water are split between two buckets. Let a represent the number of gallons in the first bucket and b represent the number of gallons in the second bucket.

Lesson 12: Expressions

x | $2x$ | $x + 1$ | $x + 2$

Objective: Show understanding of algebraic expressions by properly placing them on the number line in relation to the variable.

Teacher: **Hello, class. We have learned about variables and how they may be used to represent a number or, even more powerfully, an infinite set of numbers. Today, we are going to build upon that idea with expressions, which means we are going to add and multiply our variable by a number. Look at the three examples I wrote on the board** (see Figure 6.16).

Figure 6.16 Teacher Examples

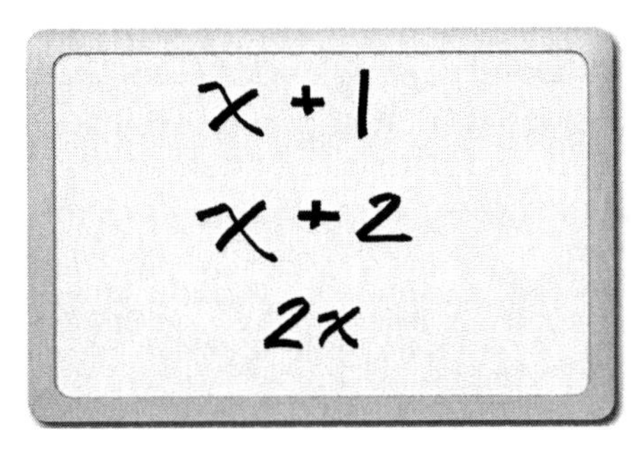

Say them after me: *x* plus one.

Class: *x* plus one!

Teacher: *x* plus two.

Class: *x* plus two!

Teacher: Two times *x*.

Class: Two times *x*!

Teacher: When you see 2 and *x* written with no operation in between, we often call it "two *x*," but that means to multiply. When we see 2*x*, what are we going to do?

Partial Class: Multiply.

Teacher: That was only some of you. When we see 2*x*, what are we going to do?

Class: Multiply!

Teacher: Good. I have set up the clothesline with 0, 1, and x (see Figure 6.17).

Figure 6.17 Teacher Placement

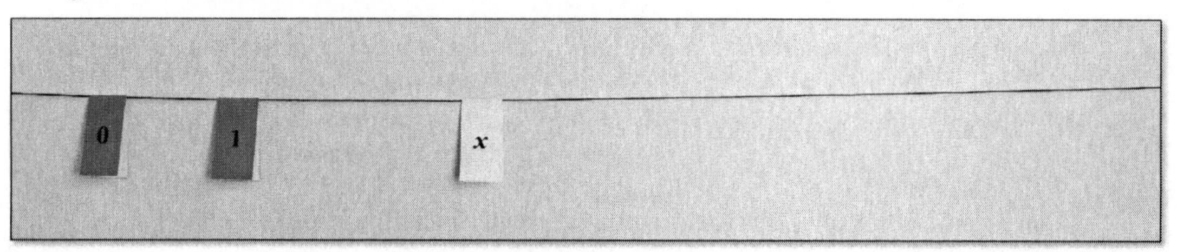

Notice that we have established how long a unit of 1 is. Also notice that I do not have a number pinned to x. That is because x gets to represent any value we want. The beauty of algebra is that it gives us the superpower to think about all numbers at the same time. To understand how the expressions work, though, we will first think about x as one number at a time and then apply that to all numbers. Let's get started. With 0, 1, and x shown here, Connor's group is going to show us where $x + 1$ and $x + 2$ go while the rest of you work with your partners on your lapboards. Go.

Connor's group places the two expression cards on the clothesline, as shown in Figure 6.18.

Figure 6.18 Connor's Placement

Connor, please have your group explain to us why you placed $x + 1$ and $x + 2$ in those locations.

Connor: We added 1 and 2 to where the x was.

Teacher: Can you show me with your hands how long 1 is?

Connor shows the distance between x *and* x + 1*.*

Huh. All in favor, thumbs-up. If you disagree, thumbs-down.

Students unanimously show thumbs-up.

The class agrees with your group. Let me show you what I was thinking, if I may. This distance is 1, so I am going to add 1 to x. So, the $x + 1$ would go here (see Figure 6.19).

Figure 6.19 Teacher's Hand Reasoning

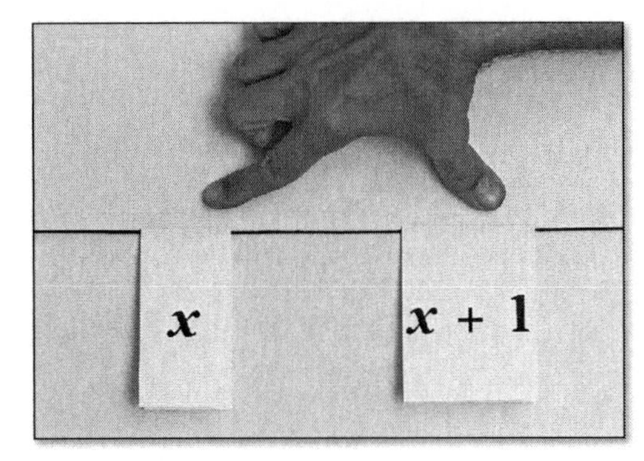

Class: No way! Move it back!

Teacher: Wow, that is a very strong reaction. Why do you disagree so passionately?

Kwan: Because you have to add 1.

Teacher: But I did.

Kwan: No you didn't, this is 1!

Teacher: No, that it x. This is 1. Watch. This is x, plus 1…

Kwan: Nuh, uh!

Teacher: Hold on, plus 2…

The teacher holds up his hands to show the distance between x + 1 *and* x + 2 *(see Figure 6.20).*

Figure 6.20 Hand Reasoning between Expression Cards

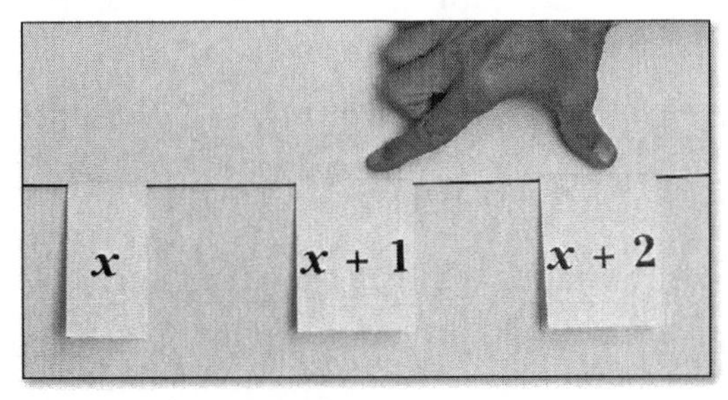

Kwan: Ohhh!

Teacher: You were thinking about $x + x + x$, or x, $2x$, and $3x$. So, let's discuss that now. Kwan, who is going to place $2x$ for us?

Kwan chooses Janessa.

Janessa, your group is heading to the clothesline. Everyone else, fix $x + 1$ and $x + 2$ on your number lines and place $2x$.

Janessa's group places 2x*, as shown in Figure 6.21.*

Figure 6.21 Janessa's Placement

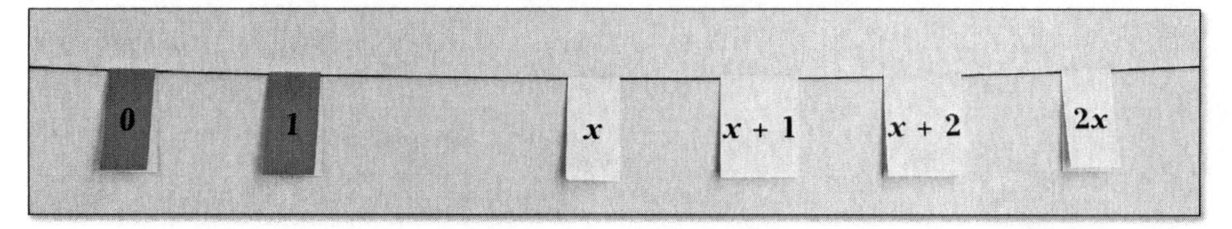

Teacher: Janessa, can you explain this one?

Janessa: It's what we just talked about—you need 2 times the x, so we just doubled the x.

Teacher: Can you show us with your hands how long x is and how you doubled it?

Janessa holds her hands up to show her reasoning, as shown in Figure 6.22.

Figure 6.22 Janessa's Hand Reasoning

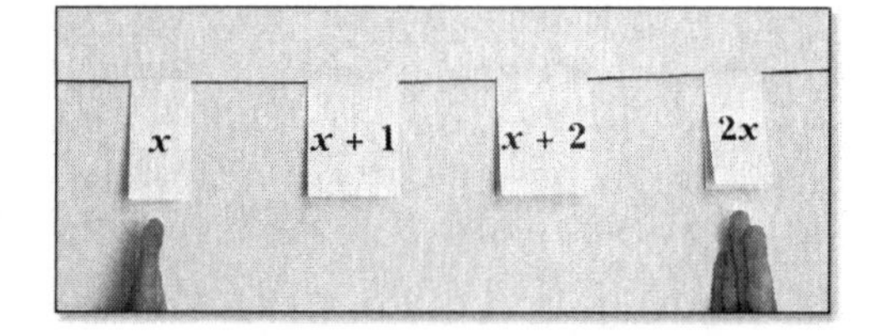

Teacher: All in favor, clap twice.

Students clap twice. Jonah keeps clapping after everyone stops.

Jonah, I said two claps. Please stop applauding and record this on your activity sheet. But, since you are so enthusiastic about our use of algebraic expressions, I am going to move *x* and ask your group to adjust our expressions. Everyone else, erase your boards, and draw this next one.

Jonah's group moves the 2x *card, as shown in Figure 6.23.*

Figure 6.23 Jonah's Placement

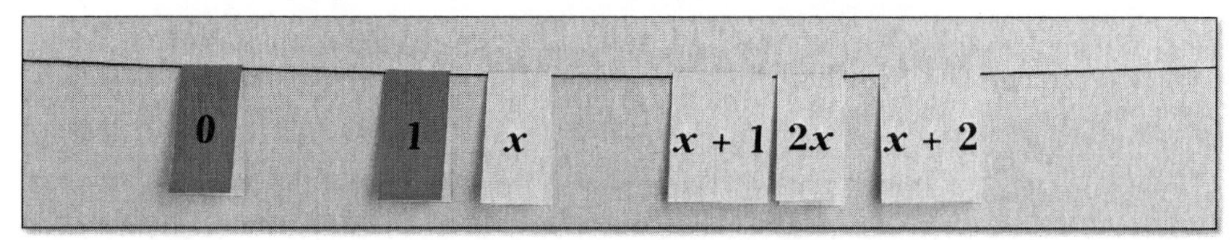

Jonah, please explain you group's thinking.

Jonah: We added 1 with our fingers like we did last time and then added it again.

Teacher: To find what?

Jonah: To find $x + 1$ and $x + 2$.

Teacher: So, you added 1 and 2 to our new value of x?

Jonah: Yes.

Teacher: Thank you. Keep going.

Jonah: Then, we placed $2x$ at what would be two x's.

Teacher: Class, this is because $2x$ means what?

Class: Multiply!

Teacher: We call the number we are multiplying the coefficient. The prefix co- means "together," like cooperate means to operate together. As Jonah said, it tells us how many x's we have, and it means to multiply. In the other expressions of $x + 1$ and $x + 2$, we call these numbers constants. That is because this distance of 1 stays constant. Did you notice the distances of $x + 1$ and $x + 2$ from x did not change after I moved the x? If something does not change, we call it constant. The x is a variable because it varies, which means we can change it.

There are nods around the classroom showing that students feel comfortable with what the teacher is saying.

Fantastic work, class. You did some very good algebraic thinking today, so you are now officially my Algebra Crew. Give yourselves a big hand, and then record your second example.

123 Why These Numbers?

This is an early introduction to algebraic expressions, so the simplest expressions were chosen. The expressions $x + 2$ and $2x$ were chosen to help students discern the difference of sums and products in algebra. The placement of the x here is carefully chosen. To avoid having $x + 2$ and $2x$ equivalent, x must be placed such that it is obviously *not* 2. While the placement of 2 only requires the location of 0 and x, the expressions $x + 1$ and $x + 2$ require the definition of a unit measure, so the location of 1 is necessary.

The Key Questions

- When we see $2x$, what does that mean?
- So, you added 1 and 2 to our new value of x?
- After I moved x, what did you notice about the distances of $x + 1$ and $x + 2$ from x?

Student understanding of the symbolic representations of mathematical relationships is often thwarted by reading issues. In other words, they just don't understand what an expression like $2x$ actually means. Therefore, I started the lesson with reading the expressions properly and having students repeat them aloud, making a point that $2x$ means "two times x."

The vocabulary terms *coefficient* and *constant* are not vital at this time, so their meanings were addressed only briefly. The lesson would have been effective without mentioning them at all, but the idea of constant was so apparent after the replacement of x that I couldn't pass up the opportunity.

The real fun was with students' defense of the erroneous placement of $x + 1$ and $x + 2$. They were so sold on their thinking that they were willing to argue with me. This was a good sign of independent thinking and academic confidence, but a bad sign for algebraic reasoning. The good news in this is their introduction to expressions, so the error was expected, even if their passion for the wrong answer was not. The debate heightened the engagement and made for more lasting understanding of what it means to add a constant to a variable. That discussion also made the placement of $2x$ easy, because students already had the idea of $2x$ in their heads. They were simply assigning the wrong expression to it.

On the second placement of x, I moved the variable closer to 1. I did this because logistically moving to the left made sliding the other expressions less awkward than if I had gone to the right. More interesting, though, was the fact that $2x$ would be less than $x + 2$. This made the conversation about multiples and constants more engaging and tied in well to the potential extension prompt.

The cards for this lesson are available in the Digital Resources (lesson12.pdf).

Prompt students to place x such that $x + 2 = 2x$.

Lesson 13: One-Step Equations

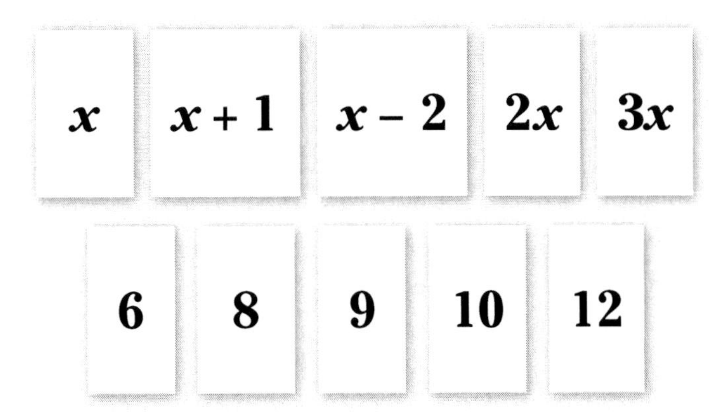

Objective: Solve one-step equations of the forms $ax = b$ and $x + a = b$ by applying the inverse operation to both sides.

Teacher: **Good day, my young mathematicians. Are you ready to do some serious algebra today?**

Class: Yes!

Teacher: **We started off working with variables. Then, we used those variables to build expressions. Now, we get to use the understanding of those expressions to solve equations. Are you ready to impress the world with how well you think algebraically?**

Class: Yes!

Teacher: **Excellent.**

Dwayne raises his hand.

Yes, Dwayne?

Dwayne: What does alge…bray…

Teacher: **Algebraically?**

Dwayne: Yes. What does that mean?

Teacher: **Terrific question. It means "having to do with algebra." It's a word we usually use when we have a variable. So algebraic thinking means we will be thinking about variables, expressions, and equations.**

Dwayne: Thank you.

Teacher: **You are welcome. Thank you for asking such an important question. Let me show you how algebraic thinking applies to what you will be doing today. We have studied variables like x and how they may represent either one unknown number or many numbers. We also have expressions, which are variables combined with numbers, like $x + 1$. Today, we will be thinking about equations. The root of the word *equations* is *equate*, or *equals*. So, if we have an expression equal to something, then we have an equation. Solving an equation means figuring out what value for the variable will make the equation true. Look at this first example of an equation that I have placed on the clothesline** (see Figure 6.24):

Figure 6.24 Teacher Placement

What equation does this represent? Diego.

Diego: $x + 1 = 9$.

Teacher: **Thank you. Everyone, on a lapboard with your partner, draw this diagram, and show where x would go if $x + 1 = 9$ while Diego's group goes to the clothesline.**

Diego: Why us?

Teacher: **Because I know you can do it. Come on up.**

Diego's group places x *to the right of the 9 benchmark, as shown in Figure 6.25.*

Figure 6.25 Diego's Placement

Diego, please explain.

Diego: The expression says $x + 1$, so we added 1 and put x there.

Teacher: **Thank you. Shani, your group has *x* on the other side. Do you want to explain your thinking?**

Shani: No, we think Diego is right.

Mateo's teammates raise their hands.

Teacher: **Okay, but it looks like Mateo's group wants to challenge. Mateo.**

Mateo: The $x + 1$ is after you add the 1, so before that you need to be at 1 less.

Teacher: **Pablo, you wanted to say something?**

Pablo: One way gives you 10, but 10 plus 1 is not 9. The other way gives you 8, and 8 plus 1 is 9, so that one is correct.

Teacher: **Let's put it to a vote. Class, you have heard several very intelligent explanations. If you agree with placing *x* to the right of $x + 1$, then point to the right. If you agree with placing *x* to the left of $x + 1$, then point to the left.**

All students point to the left.

Wow, everyone says to the left, which means *x* is less than $x + 1$. I agree with that also. Mateo and Pablo each spoke of an important point when it comes to solving equations. Let's look at them one at a time. Mateo claims "this is where it goes after you add," so we have to think about where *x* was before we added 1. Mateo is talking about doing the inverse operation. Everyone say, "inverse operation."

Class: Inverse operation.

Teacher: ***Inverse operation*** **means "the opposite operation of the one you are looking at." So, if we add 1 in the expression, then we must subtract 1 to solve. Now that we know where *x* goes, let's declare its value. Notice what I'm doing, though—we subtracted 1 from $x + 1$ to equal just *x*, and we also subtracted 1 from the 9. Take a look at the cards I have pinned to the clothesline that show this (see Figure 6.26).**

Figure 6.26 Teacher's New Card Placement

PART II – Chapter 6

Teacher: Let's draw that on our activity sheet. Then, I will show you how the algebra looks on paper. Do the inverse operation to both the expression on one side and the number on the other side, like we did on the number line.

Students write the equations on their lapboards.

Great job, everyone. I will write the equations on the board as well (see Figure 6.27).

Figure 6.27 Equations

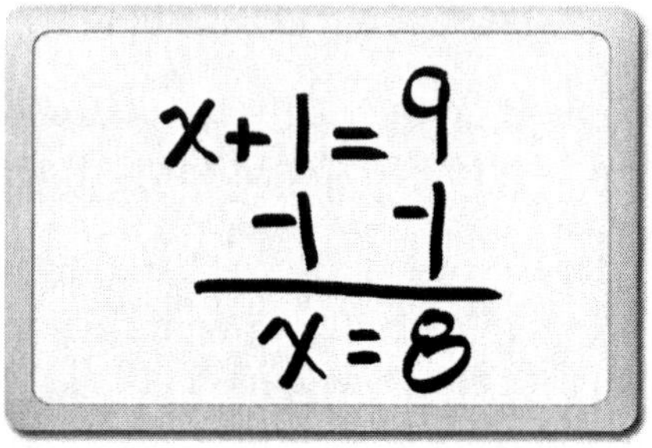

$$x + 1 = 9$$
$$-1 \quad -1$$
$$x = 8$$

We are doing algebra on the number line first so we understand what we are doing and why we are doing it. But, notice how much faster and easier it is to solve for x when we write our algebraic thinking like I did on the board. Now, let's go back to what Pablo said. He was talking about checking our answers. If we substitute 8 for x, then the check looks like this (see Figure 6.28):

Figure 6.28 Teacher Check

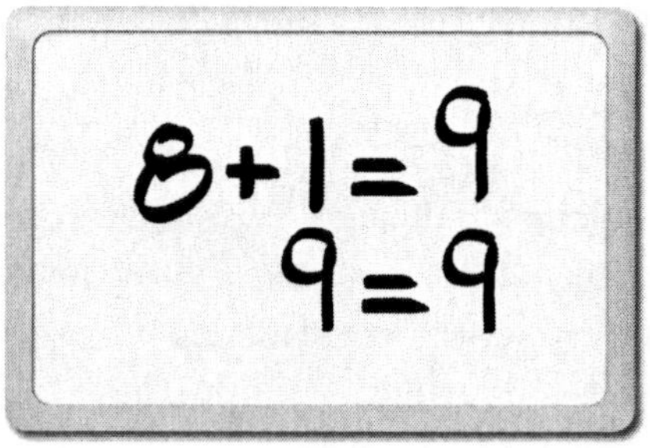

$$8 + 1 = 9$$
$$9 = 9$$

Teacher: If we did it wrong, the check would look like this (see Figure 6.29):

Figure 6.29 Incorrect Check

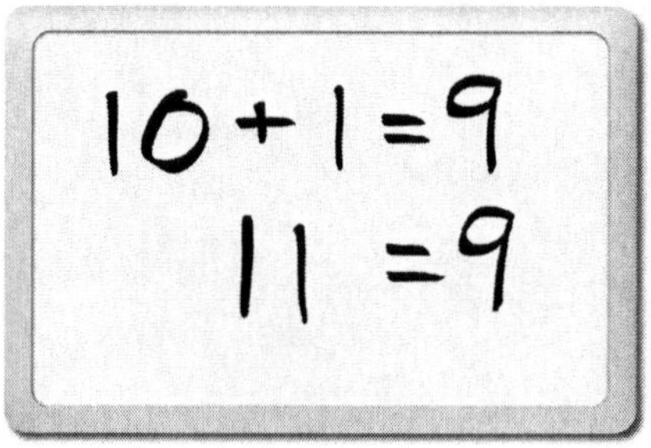

$$10 + 1 = 9$$
$$11 = 9$$

Teacher: **We are looking for a value of x for which the statement is true. So, which statement is true: $9 = 9$ or $11 = 9$?**

Shani: Nine equals 9!

Teacher: **Correct. Substitution can easily tell us whether our answer is correct. Record this example of the check while I talk to you. Substitution means to replace. Just like when a teacher is absent and a substitute teacher arrives that day, when we substitute for a variable, we put a number in place of the variable. So far, we have discussed inverse operations and checking by substitution. Those are so important that we will continue to talk about them throughout the lesson. Let's thank Mateo and Pablo for sharing such important ideas.**

Students applaud.

Teacher: **Are you ready for our next equation?**

Class: Yes.

The teacher changes the cards on the clothesline, as shown in Figure 6.30.

Figure 6.30 New Cards

Teacher: **Diego, you get to pick who solves our next equation on the clothesline.**

Diego: I pick Hanna.

Teacher: **Hanna, time to head to the clothesline with your group.**

Hanna's group adds two cards to the clothesline (see Figure 6.31).

Figure 6.31 Hanna's Placement

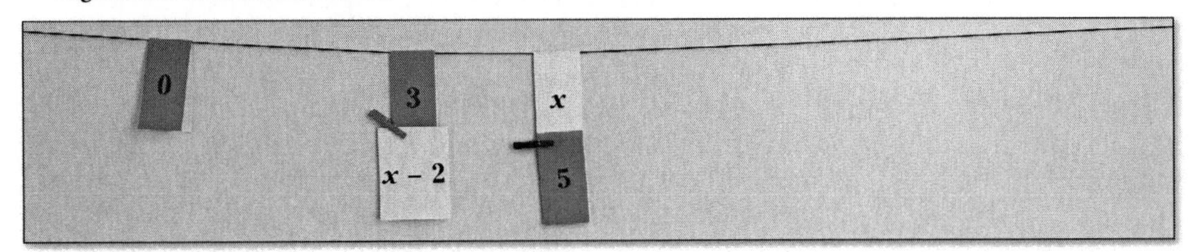

Teacher: Hanna, thank you for also pinning the value to *x*. Will you explain your group's thinking?

Hanna: We did the opposite of subtraction. We added 2 to 3 and got 5.

Teacher: The opposite of subtraction. Again, we call that the inverse operation of subtraction. What do we call it?

Class: Inverse operation.

Teacher: That was not everyone. What do we call it?

Class: Inverse operation!

Teacher: Good. And, the inverse operation according to your group is addition. Dalia, how should we check our answer?

Dalia: Substitution.

Teacher: Good word. What does that mean to do here?

Dalia: Substitute 5 for *x*.

Teacher: Please walk us through that.

Dalia: $5 - 2 = 3$.

Teacher: Thank you. Let's copy this diagram onto our activity sheet and also show the solve and check with our mad algebra skills. I will also write it on the board (see Figure 6.32).

Figure 6.32 Teacher's Board Work

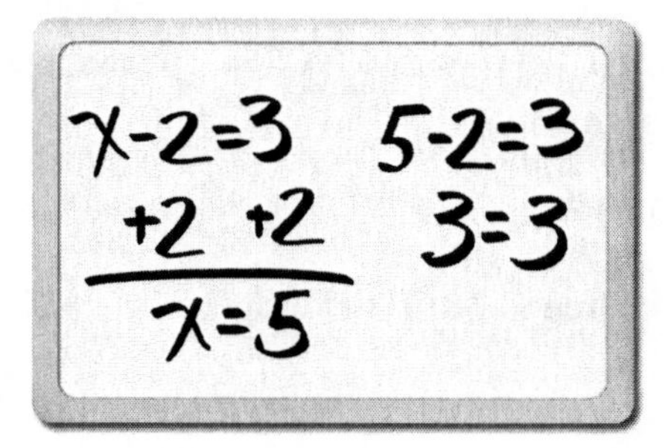

Teacher: Well done. Since you are doing so well, I'm going to change gears on you. I want to hear what you think about this one.

The teacher places all new cards on the clothesline (see Figure 6.33).

Figure 6.33 New Placement of Cards

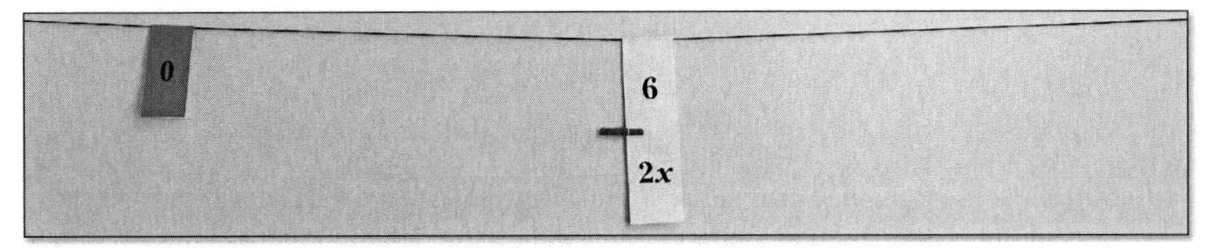

Teacher: Hanna, who gets this one?

Hanna points to Pham.

Pham, time for your group to go to the clothesline.

Pham's group adds x and 3, as shown in Figure 6.34.

Figure 6.34 Pham's Placement

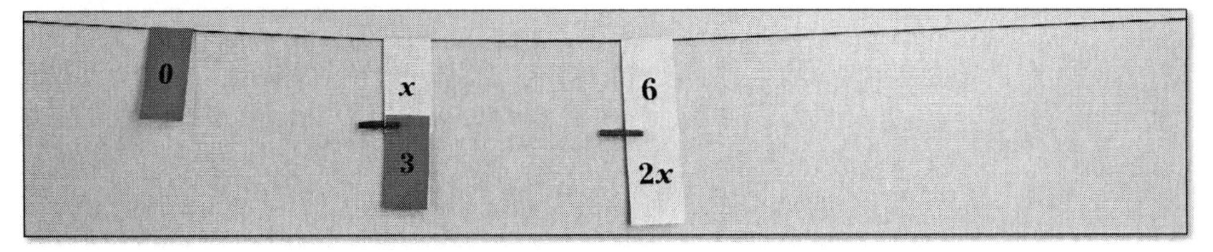

What do you think about this one, Pham?

Pham: Six was $2x$, so we just went halfway on both.

Teacher: Interesting. Why did you not add or subtract on this one?

Pham: Because we are not adding.

Teacher: What are we doing instead?

Pham: Multiplying?

Teacher: Are you sure?

The group confers with one another.

Pham: Yes.

Teacher: So, what is the inverse operation of multiplication?

Pham: Division!

Teacher: Did you check it?

Pham: Yes, $2 \times 3 = 6$.

Teacher: Sweet. Your algebra and check would look like this then (see Figure 6.35).

Figure 6.35 Algebra and Check

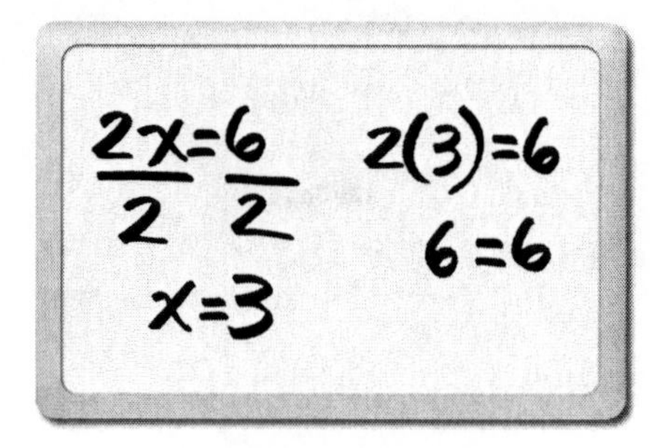

Teacher: Copy our diagram, the algebra, and the check on your paper. Notice the use of the parentheses on the check. Now, for our finale.

The teacher places the final set of cards on the clothesline (see Figure 6.36).

Figure 6.36 Final Teacher Card Placement

On your lapboards, I want you and your partner to show where x and $2x$ go. Also, make sure to show the algebra that is represented by your number line diagram and your check. I will call someone to the clothesline in just a moment.

A few moments pass, and the students work on their lapboards.

Billy, will your group please show us your thinking on the clothesline? Angie, you and your partner write the algebraic solution on the board. Payne, write the check on the board.

Billy's group adds three cards to the clothesline, as shown in Figure 6.37.

Figure 6.37 Billy's Placement

Angie and her partner write the following algebraic solution on the board (see Figure 6.38).

Figure 6.38 Angie's Solution

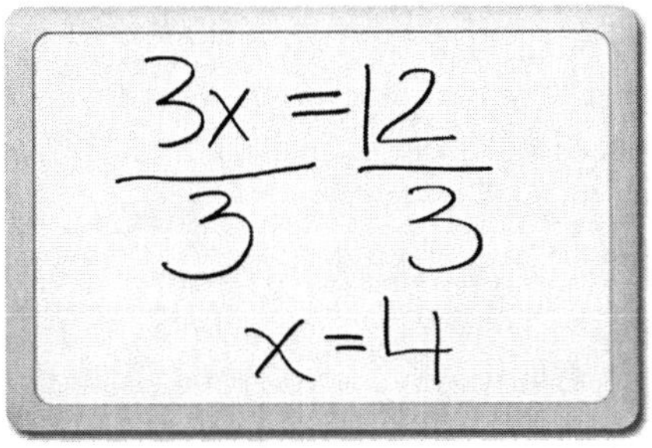

$$\frac{3x}{3} = \frac{12}{3}$$

$$x = 4$$

Payne writes a partial check on the board (see Figure 6.39).

Figure 6.39 Payne's Partial Check

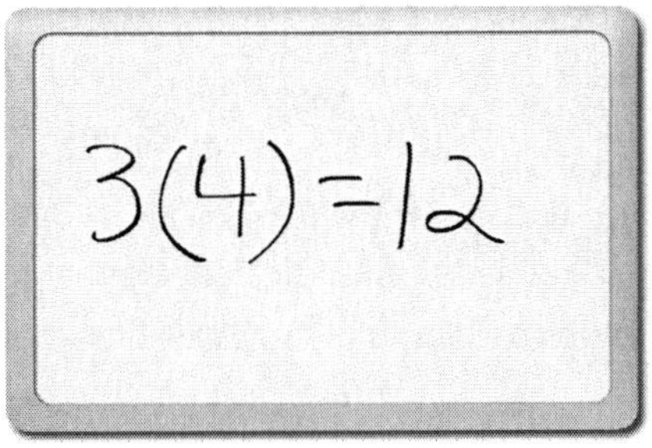

$$3(4) = 12$$

Teacher: Payne, please finish your check.

Payne: What? $3 \times 4 = 12$.

Teacher: Notice what we have been doing, working it out until we have the same number on both sides. You are correct, and I agree, it might seem like extra work now, but that habit will be important for us later. Thank you.

Payne: But...

Teacher: Thank you for doing that.

Payne returns to his work on the board and adds a second line (see Figure 6.40).

Figure 6.40 Payne's Full Check

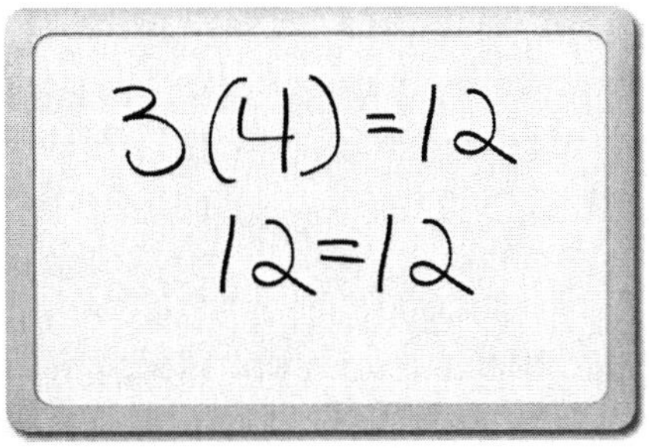

$$3(4) = 12$$

$$12 = 12$$

What were the two big ideas we discussed today about solving equations? Nelson.

Nelson: Do the inverse operation.

Teacher: Thank you. Russ?

Russ: Check your answers.

Teacher: By doing what? Diego.

Diego: Substitute.

Teacher: By substituting, very good. Thank you for applying your algebraic thinking today.

Since this is the first time students have been asked to analyze equations on a number line, the one-step equations were kept simple. The equations $x + 1 = 9$, $x - 2 = 3$, $2x = 6$, and $3x = 12$ require simple computations, as well. The real estate required on the number line is compacted with these numbers, so there is not a great deal of benchmark adjusting as the lesson moves from benchmark to benchmark.

- What equation is represented by this number line diagram?
- What do we call it when we do the opposite operation?
- Did you check your answer?
- How do we check our solution?
- What does substitution look like in this case?
- What were the two big ideas that we discussed today about solving equations?
- Why did you not add or subtract?
- What is the inverse operation of multiplication?

This lesson started with connections to prior knowledge by explaining to students that equations are built upon expressions, which are built upon variables.

The first equation was one of the simplest. Posting it to the clothesline and asking students to make sense of it exposed valuable insights into students' thinking regarding the symbols. Fortunately, Diego's group made the common mistake of adding 1 instead of subtracting 1, which was the catalyst for the desired conversation. Had the group correctly solved on the number line, I would have called on a group that made the error to incite the debate.

I found it very interesting that Shani's group had the correct answer on their lapboards but were swayed by the argument for the incorrect method. Once both methods were shared, the class converged on the correct method, which was the right time to introduce the term *inverse operation* because the intellectual need generated from the discussion demanded it.

I had intended to bring up the idea of checking our work through substitution, but Pablo set me up for it anyway. It is important to note that students find the algebraic thinking on the number line far easier than symbolic thinking of algebra on paper. Yet, that symbolic manipulation contains the immense power of algebra. In the end, the symbolic methods are far more efficient than the visual methods on the clothesline.

This dynamic brings up three very important points for using the clothesline to initially teach the solving of equations, even though it will eventually be abandoned for the algebraic method. The first is that we want to teach conceptual understanding before procedural fluency. This lesson demonstrated the importance of that. Students need to know why we do what we do when solving equations. The second point is that algebraic reasoning is not used when implementing algebraic procedures in the same way our rules for operations with fractions eliminate all thinking about fractions and reduce the math to whole number arithmetic. That is the beauty of algorithms: they eliminate the need to think, so we can use our mental energies to think about other things. The third is a very important concept—students do not automatically make the cognitive leap from the number line representation to the symbolic representation of equations. This is why teachers get paid the big bucks; we need to make that connection for students. That is the reason I had students write the number line solution *and* the algebraic solution for each equation. That connection will need to be made over a period of weeks before the visual will be left behind and students will naturally use standard algorithms for solving equations.

The cards for this lesson are available in the Digital Resources (lesson13.pdf).

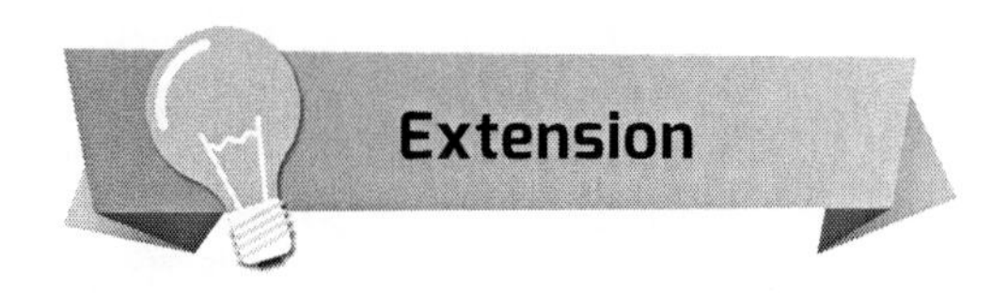

Extension

Solve $\frac{1}{2}x = 4$ or $2x + 1 = 7$. Both of these equations are slightly beyond expectations for the grade level but still within reach of students that need the extension.

Clothesline Lessons for Middle and High School

Arithmetic for Middle School

Lesson 14: Multiplying Fractions

$\frac{1}{4}$ $\frac{2}{4}$ $\frac{3}{4}$ $\frac{1}{3}$ $\frac{2}{3}$ $\frac{1}{2}$ $\frac{2}{6}$ $\frac{3}{8}$

$\frac{1}{2} \bullet \frac{3}{4}$ $\frac{1}{2} \bullet \frac{2}{3}$ $\frac{3}{4} \bullet \frac{1}{2}$ $\frac{2}{3} \bullet \frac{3}{4}$ $\frac{3}{4} \bullet \frac{2}{3}$ $\frac{2}{3} \bullet \frac{1}{2}$

Objective: Multiply two fractions.

Teacher: Hello, all. We have been learning how to multiply fractions. Today, we are going to look at it in a different way, so we can see why our rules for multiplying fractions work. We are also learning a new method to make it even easier to multiply fractions in certain situations. You may have noticed that I have a double clothesline on the board today. Let me show you how the clotheslines will help us understand multiplication of fractions. Let's start with $\frac{1}{2} \bullet \frac{2}{3}$. I will start with $\frac{1}{2}$ on the top line, and place $\frac{2}{3}$ on the bottom (see Figure 7.1).

Figure 7.1 Double Clothesline

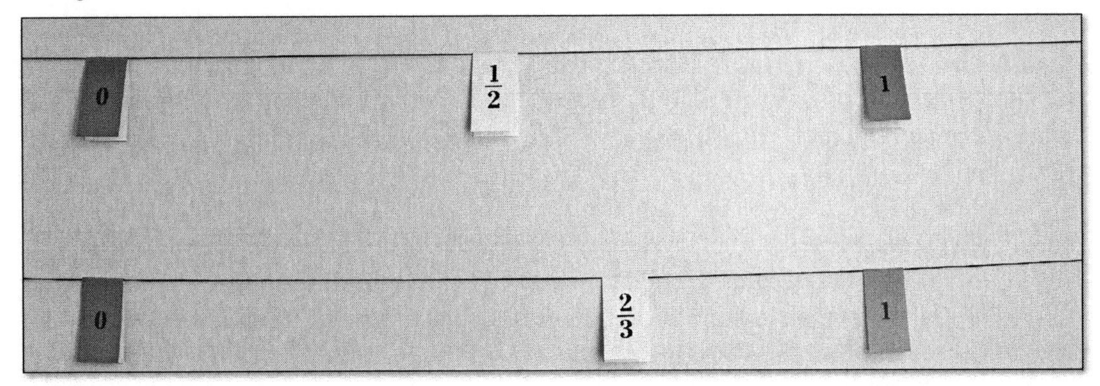

Teacher: Trystan, your group is to place $\frac{1}{2} \cdot \frac{2}{3}$ on the bottom clothesline while everyone else does this with their partners on their lapboards.

Trystan's group places the number cards, as shown in Figure 7.2.

Figure 7.2 Trystan's Placement

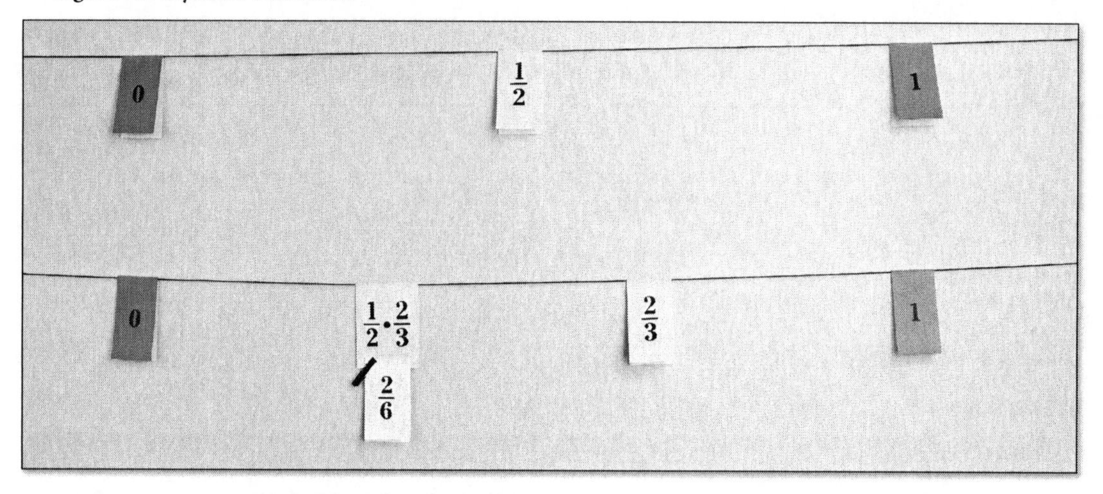

Trystan, why there?

Trystan: Well, we multiplied $\frac{1}{2}$ times $\frac{2}{3}$ by just multiplying straight across, and we got $\frac{2}{6}$. So, we visualized breaking up the 0 to 1 into sixths and put it at 2 of those sixths.

Teacher: Thumb vote.

Most students vote thumbs-up, except Mason.

Mason, you disagree?

Mason: We need to simplify. It should be $\frac{1}{3}$.

Several students nod in agreement.

Teacher: I see nods around the room so others agree. Time to thumb vote again.

Students vote thumbs-up.

Everyone seems to agree with you, Mason. I'm curious, though—did anyone think differently? Richley.

Richley: Since you want half of $\frac{2}{3}$, you can just pick a spot half way between 0 and $\frac{2}{3}$.

Teacher: Interesting. Trystan multiplied the numerators and denominators. Then, Mason helped simplify *after* the multiplication. Richley chose to simplify *before* the multiplication. Let me show you on the board (see Figure 7.3).

Figure 7.3 Simplifying Fractions

$$\frac{1}{2} \cdot \frac{2}{3} = \frac{2}{6} = \frac{1}{3}$$

$$\frac{1}{\cancel{2}_1} \cdot \frac{\cancel{2}^1}{3} = \frac{1}{3}$$

Teacher: Record this double-line diagram, and write both ways of multiplying the fractions with it. As we continue, keep these concepts in mind. We can simplify before we multiply, just like we can simplify after we multiply.

Let's do another problem to test the commutative property of multiplication. If I reverse the order, do I get the same answer? I'm going to switch these by placing $\frac{2}{3}$ on the top clothesline, and $\frac{1}{2}$ on the bottom. Richley, who is going to place $\frac{2}{3} \cdot \frac{1}{2}$?

Richley selects Ingrid's group.

Ingrid, your group needs to head to the clotheslines. Everyone else, erase your boards and work on this one.

Ingrid's group quickly places their two cards on the bottom clothesline (see Figure 7.4).

Figure 7.4 Ingrid's Placement

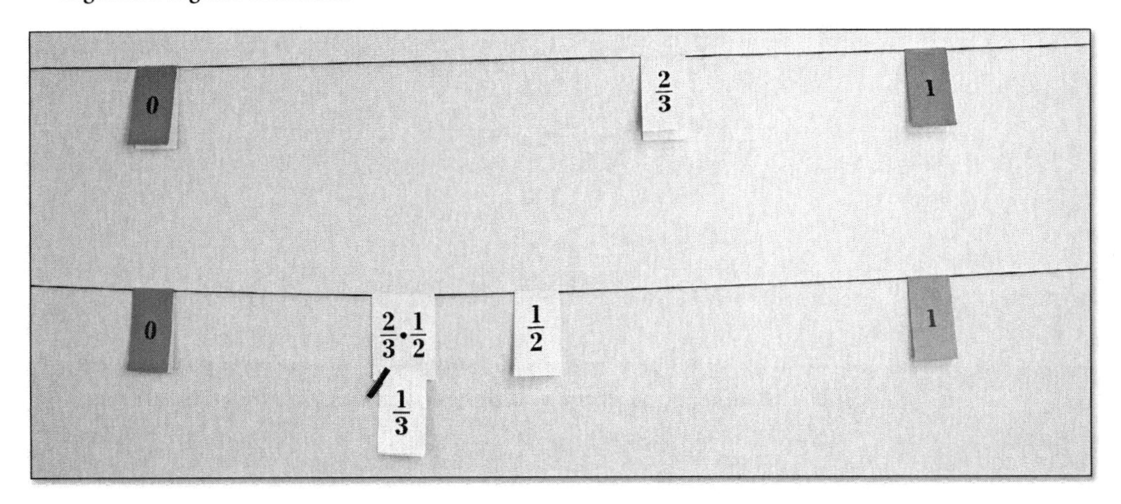

That was fast.

Ingrid: We knew it would be the same answer, so we just measured thirds.

Teacher: How did you know it would be the same?

Ingrid: Because of the communication property?

Teacher: Commutative property.

Ingrid: Yes.

Teacher: Say *commutative.*

Ingrid: Commutative.

Teacher: Well done. That would be one reason. I wonder whether anyone else in the room did it by one of the other two methods we just discussed.

Stevie raises his hand.

Stevie.

Stevie: If you multiply straight across, you get $\frac{2}{6}$ again. Then, you simplify to $\frac{1}{3}$.

Teacher: Thank you. Did anyone visualize $\frac{2}{3}$ of $\frac{1}{2}$? In other words, you broke the $\frac{1}{2}$ into three parts and placed this product at two of the three?

Students nod and present thumbs-up in agreement.

From all the hands, I see most of you did it that way. Here is how we would show both of those methods (see Figure 7.5).

Figure 7.5 Teacher Explanation

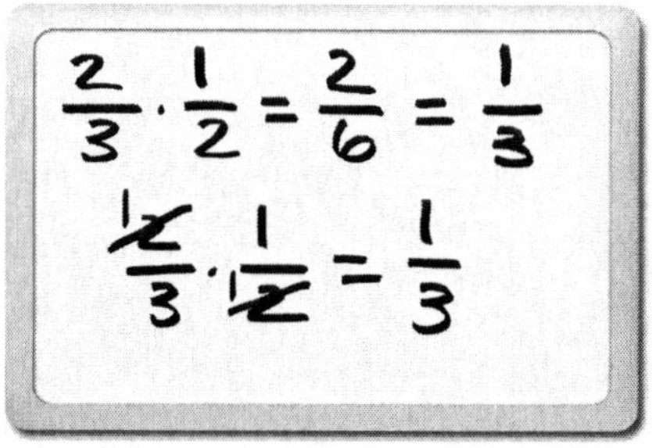

$$\frac{2}{3} \cdot \frac{1}{2} = \frac{2}{6} = \frac{1}{3}$$

$$\frac{\not{2}^{1}}{3} \cdot \frac{1}{\not{2}_{1}} = \frac{1}{3}$$

Record both of these as well as the diagram.

Our next problem is $\frac{1}{2} \cdot \frac{3}{4}$. I'm going to have someone else besides me start this problem for you on the number line. Ingrid, who is next?

Ingrid chooses Kim's group.

Kim, have your group head to the clotheslines. Everyone, wipe your boards. Determine which card goes on the top line according to the previous diagrams.

Kim's group places the card, as shown in Figure 7.6.

Figure 7.6 Kim's Placement

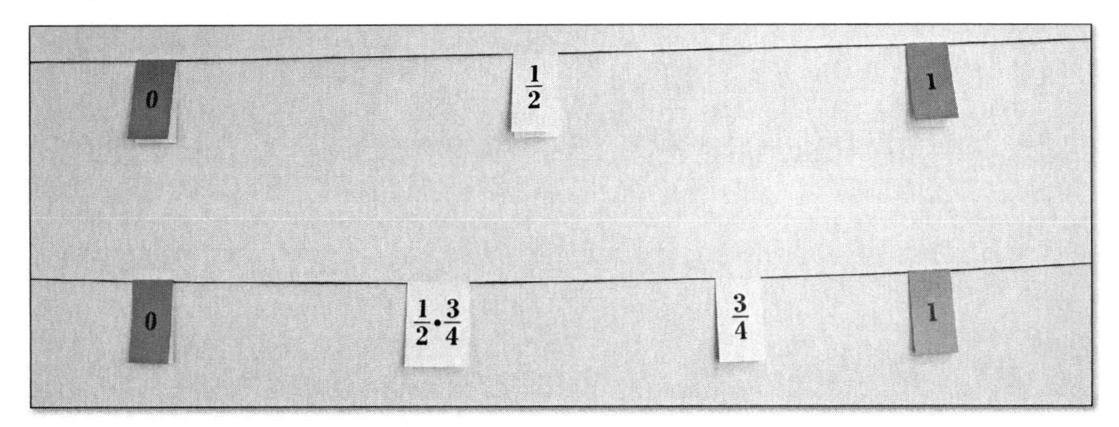

Teacher: Kim, please explain.

Kim: We noticed you always set the first fraction on the top so you can take that much of the second fraction on the bottom.

Teacher: Nice job recognizing the pattern.

Kim: So, we just put it halfway between for $\frac{3}{4}$.

Teacher: Thank you. I noticed Kim's group did not pin a value to this. Did anyone here determine the value would be at $\frac{1}{2}$ of $\frac{3}{4}$? Yun.

Yun: If you multiply the tops and the bottoms you get $\frac{3}{8}$.

Teacher: What do we call the top and the bottom of fractions?

Yun: You multiply the numerators and the denominators.

Teacher: Thank you. Willy, we haven't simplified this one yet. Will you do that for us?

Willy: You can't.

Teacher: Correct. Remember, sometimes we can simplify, sometimes we can't. Since 3 and 8 do not have common factors, the fraction cannot be simplified. Is there another strategy other than multiplying across? Yolanda.

Yolanda: We marked fourths on our line, but then we noticed the answer was between $\frac{1}{4}$ and $\frac{2}{4}$, so we cut all the fourths in half and got eighths. Then, we noticed our answer was on the third eighth.

Teacher: Nice explanation, thank you. So, as we can see, multiplying the denominators tells us how many parts we will have total, and multiplying the numerators tells us how many of those parts we want. Record this, please. Show the multiplication with it. Kim, which group is putting our next one up?

Kim chooses Janks's group.

Teacher: Janks, your group has $\frac{3}{4} \cdot \frac{1}{2}$ while everyone else draws the new diagram at their desks.

Janks and his group place the card (see Figure 7.7).

Figure 7.7 Janks's Placement

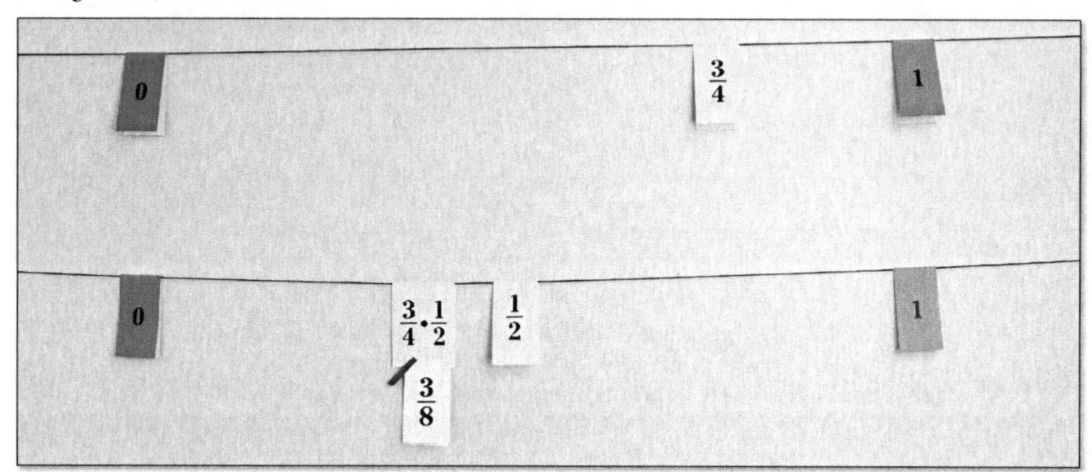

Janks. How did you figure this one out?

Janks: We started with $\frac{3}{4}$ on the top line and we visualized that much the half on the bottom clothesline. If you chop a half into fourths, you get eighths, so we get the same $\frac{3}{8}$ as last time.

Teacher: Thank you. Our commutative property holds true once again and so does our method of multiplying fractions. Janks just did a nice job of explaining how multiplying the denominators gives the denominator of our answer when he said "we chop the half into fourths to give us eighths." Let's record that before we do our last two. Janks, who gets $\frac{2}{3} \cdot \frac{3}{4}$?

Janks selects Penny's group.

Penny, have your group come to the number line. Everyone else, work on your lapboards.

Penny's group places their card on the number line, as shown in Figure 7.8.

Figure 7.8 Penny's Placement

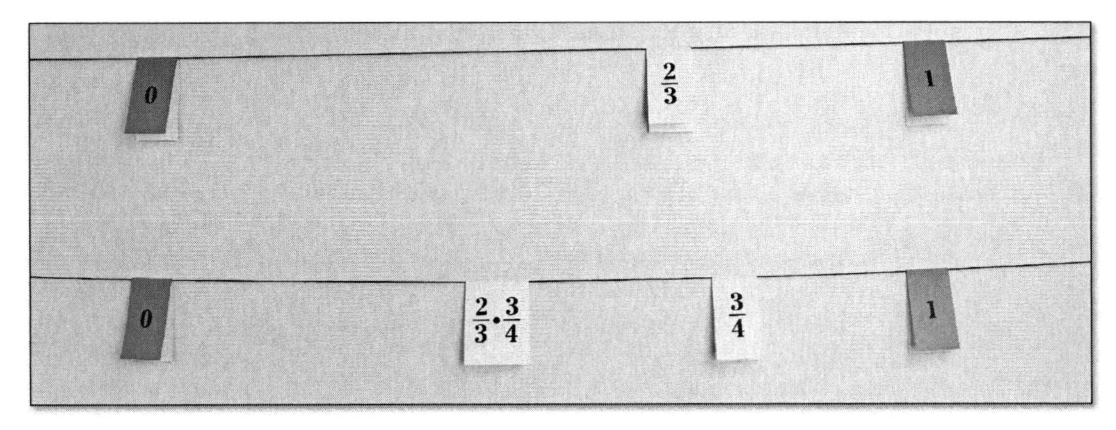

Teacher: Penny, tell us what you were thinking.

Penny: Sure. We started with $\frac{2}{3}$ and realized we needed to chop the $\frac{3}{4}$ into three parts. The fourths was already kinda chopped into three parts because we have three out of the four. Do you know what I mean?

Teacher: Class, if you understand Penny so far, clap twice.

Most, but not all students, clap twice.

Who needs Penny to explain further?

Several hands go up.

Penny, can you go back up to the clothesline and show us? Place some benchmarks on the parts you are talking about.

Penny: Oh yeah, that would help.

Penny returns to the clothesline and adds benchmark cards (see Figure 7.9).

Figure 7.9 Penny's Second Placement

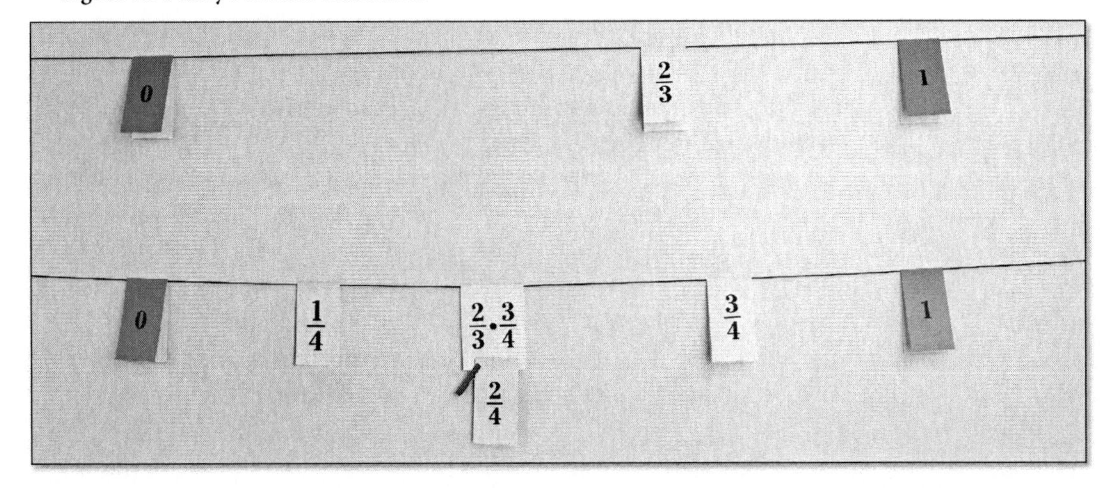

Teacher: **You see, this is $\frac{1}{4}$, $\frac{2}{4}$, and $\frac{3}{4}$. The fourths are already broken into thirds, so we just needed to put our answer at the two of those thirds.**

Several nods follow Penny's explanation.

Teacher: **Does $\frac{2}{4}$ simplify?**

Penny: Yes, you can see that it is at $\frac{1}{2}$.

Teacher: **Thanks. That was a lot of extra work to explain. Let's give Penny some applause.**

Students applaud enthusiastically.

Paul, you can stop applauding now. But, since you are so happy about this, can you tell us if we multiplied the numerators and the denominators, would we get the same thing?

Paul: Yes, because that's the way I did it.

Teacher: **Really?**

Paul: Yes, you get $\frac{6}{12}$, which is also the same as $\frac{1}{2}$.

Teacher: **Where would we see the twelfths up here?**

Paul: I'm not sure.

Teacher: **Can anyone help out Paul on this one? Sandy.**

Sandy: You cut all the fourths into thirds. That is why you multiply the 3 and the 4 in the denominator.

Teacher: **Paul, does that make sense?**

Paul applauds.

You like applauding today. Good job, though. Let me show how all this talk ties into both methods of simplifying *after* we multiply and *before* we multiply. Take a look at what I've written on the board (see Figure 7.10).

Figure 7.10 Teacher's Notes

$$\frac{2}{3} \cdot \frac{3}{4} = \frac{6}{12} = \frac{1}{2}$$

$$\frac{\overset{1}{\cancel{2}}}{\underset{1}{\cancel{3}}} \cdot \frac{\overset{1}{\cancel{3}}}{\underset{2}{\cancel{4}}} = \frac{1}{2}$$

Teacher: Our last one is $\frac{3}{4} \bullet \frac{2}{3}$. We are going to handle it differently, so I will take a volunteer group for this.

Jesus raises his hand to volunteer.

Jesus, here is want I want your group to do—I'm going to pin this one to our answer from the previous problem, $\frac{2}{3} \bullet \frac{3}{4}$, since we have proven several times already that the commutative property will hold true. I would like to arrange $\frac{3}{4}$ and $\frac{2}{3}$ in such a way that we actually *see* $\frac{3}{4} \bullet \frac{2}{3}$ will still be at $\frac{1}{2}$.

Jesus and his group nod. The group heads to the clothesline and places the final two cards, as shown in Figure 7.11.

Figure 7.11 Jesus's Placement

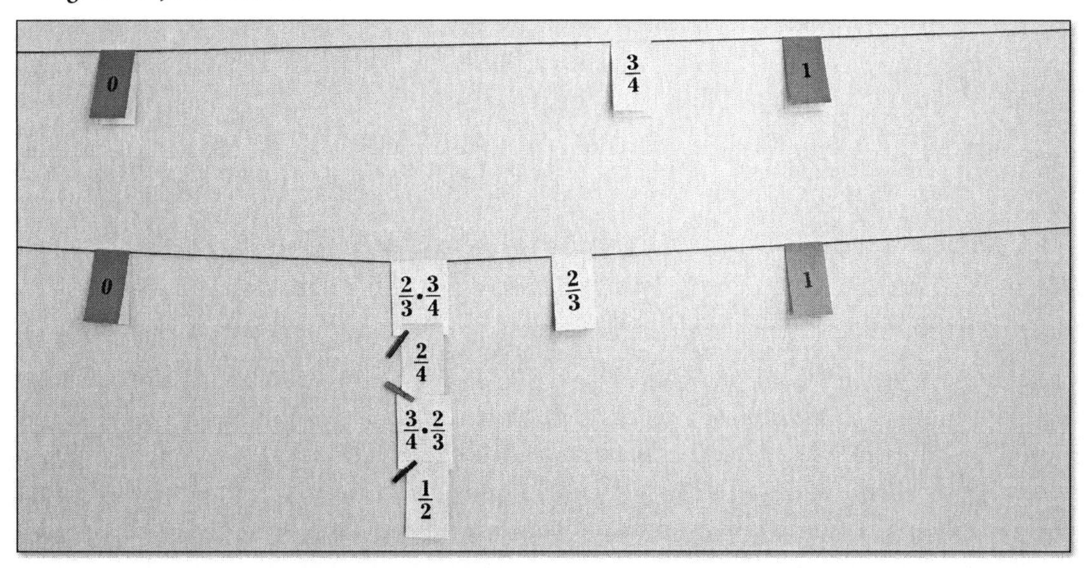

All in favor, say "aye."

Class: Aye!

Teacher: Wow, that was some seriously tough thinking today. Record this last placement, and let's spend the rest of our time practicing multiplying fractions.

123 Why These Numbers?

The values of $\frac{1}{2}$, $\frac{2}{3}$, and $\frac{3}{4}$ were used because they are the three simplest denominators to see on the number line. The numerators also allow us to simplify in two of the three combinations. This allows for the discussion of simplifying, as well as showing that products do not always simplify.

The Key Questions

- Did anyone think differently?
- Is there another strategy other than multiplying straight across?
- If I reverse the order, will I get the same answer?
- Where would you see twelfths?
- Does that $\frac{2}{4}$ simplify?
- Who needs this explained further?
- How did you know they would be the same?

Analysis

The double clothesline was utilized for easier visualization. It wasn't important if students knew where $\frac{1}{2}$ is in relation to $\frac{2}{3}$ as much as it was critical for them to be able to visualize $\frac{1}{2}$ *of* $\frac{2}{3}$.

There were very few computational mistakes made during this lesson. This is once again another validation to the visual power of the clothesline. Visual diagrams of operations on fractions are easier to understand than symbolic representations. Since computational accuracy was not the issue, the lesson focused on tying the strong conceptual understanding of students to procedural algorithms, both for multiplying fractions as well as for simplifying during the process. This is why I wrote the algorithm next to students' number line responses.

I also wanted students to understand that while there are multiple strategies for thinking about multiplication of fractions on the number line, there are also multiple strategies for symbolic manipulation. Specifically, simplifying after we multiply versus simplifying before we multiply. Most importantly, students needed to grasp that sometimes fractions do not simplify at all.

This lesson relied on the teacher making the connection for students between the visual representation and the symbolic representation and between the strategies on the number line versus the procedures on paper.

The cards for this lesson are available in the Digital Resources (lesson14.pdf).

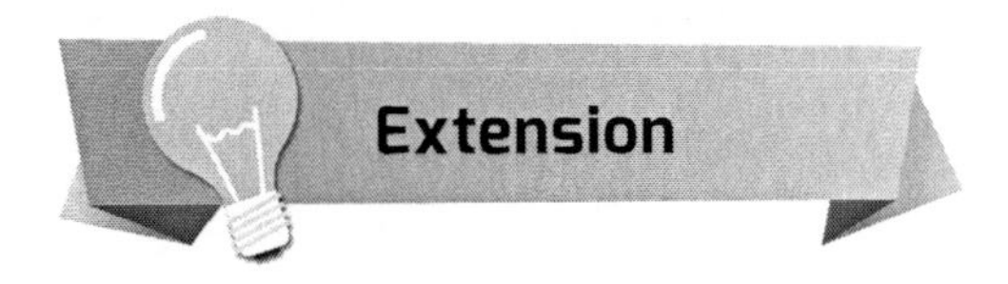

Multiply whole numbers by fractions and improper fractions by mixed numbers.

Lesson 15: Converting Fractions, Decimals, and Percentages

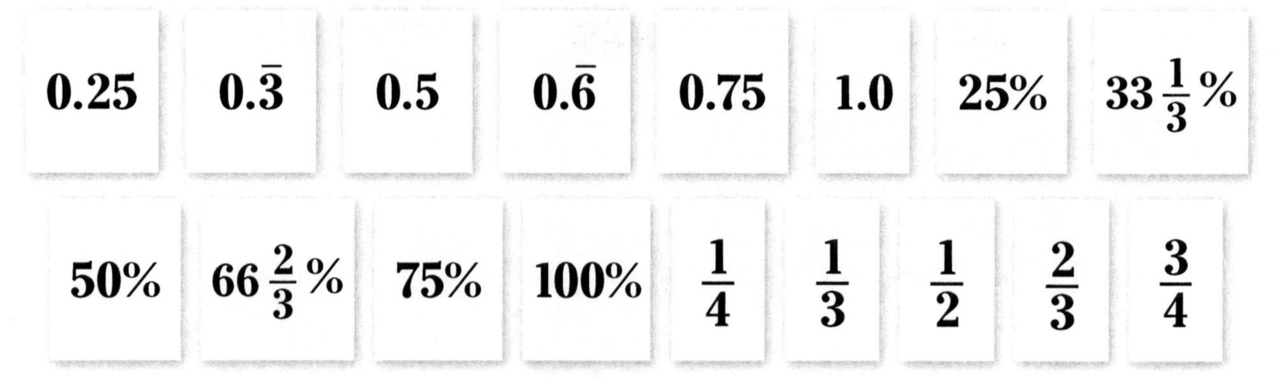

Objective: Fluently convert the most common fractions to decimals and percentages.

Teacher: Hi, class. We have been studying a great deal about fractions, decimals, and percentages and how to convert between the various forms. There are some values, however, that are so commonly used we don't want to do any converting. Our goal is to know them automatically in their different forms. As you can see, the most important one to know is already on the clotheslines (see Figure 7.12).

Figure 7.12 Triple Clothesline

Teacher: Today, our clothesline activity is going to run a little differently. I will put one of these values from the table on the clothesline. Then, I will call on any two students to choose the corresponding forms of that value. For example, if I had chosen 1, then the other two people would place 1.0 and 100%. After those two cards are placed, the two students will place another card on the correct line in the correct place and call on two other people to continue the same process. Therefore, I am going to place $\frac{1}{2}$ on the number line and ask Anthony and Dora to place their equivalent values.

Anthony places 0.5 on the middle clothesline, while Dora place 50% on the bottom clothesline (see Figure 7.13).

Figure 7.13 Anthony and Dora's Placements

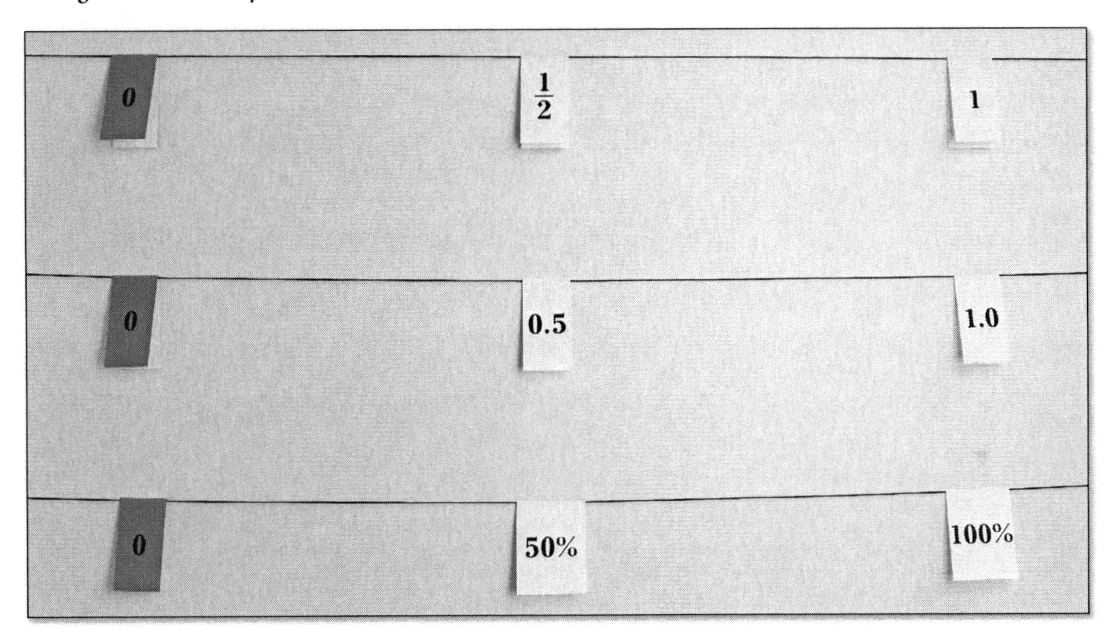

Thank you. Now, choose another value and two more people.

Anthony: We pick $\frac{1}{4}$ and Sven and Darren.

Dora places $\frac{1}{4}$ and then returns to her seat. Sven and Darren place the next two cards, as shown in Figure 7.14.

Figure 7.14 Sven and Darren's Placements

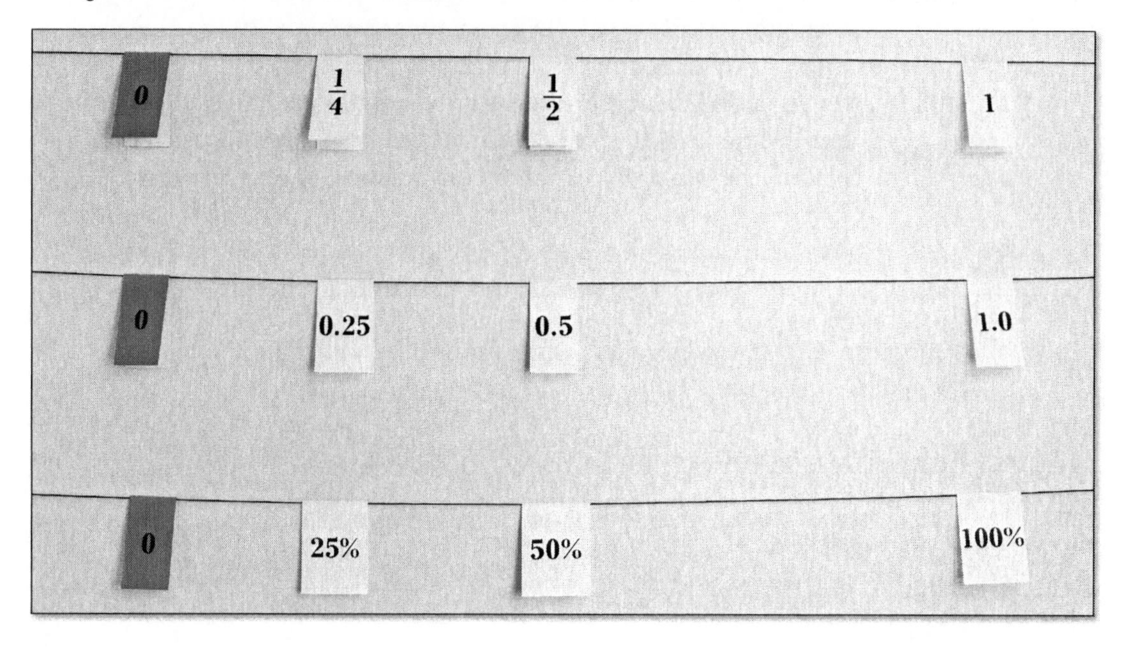

Teacher: Thank you. Now, choose another value. Pick something this time that is not a fraction and tell us which two students will place the equivalents.

Darren: $33\frac{1}{3}$%, and Bert and Carl are next.

Bert and Carl place one card on the middle line, but pause before placing the final card (see Figure 7.15).

Figure 7.15 Bert and Carl's First Placement

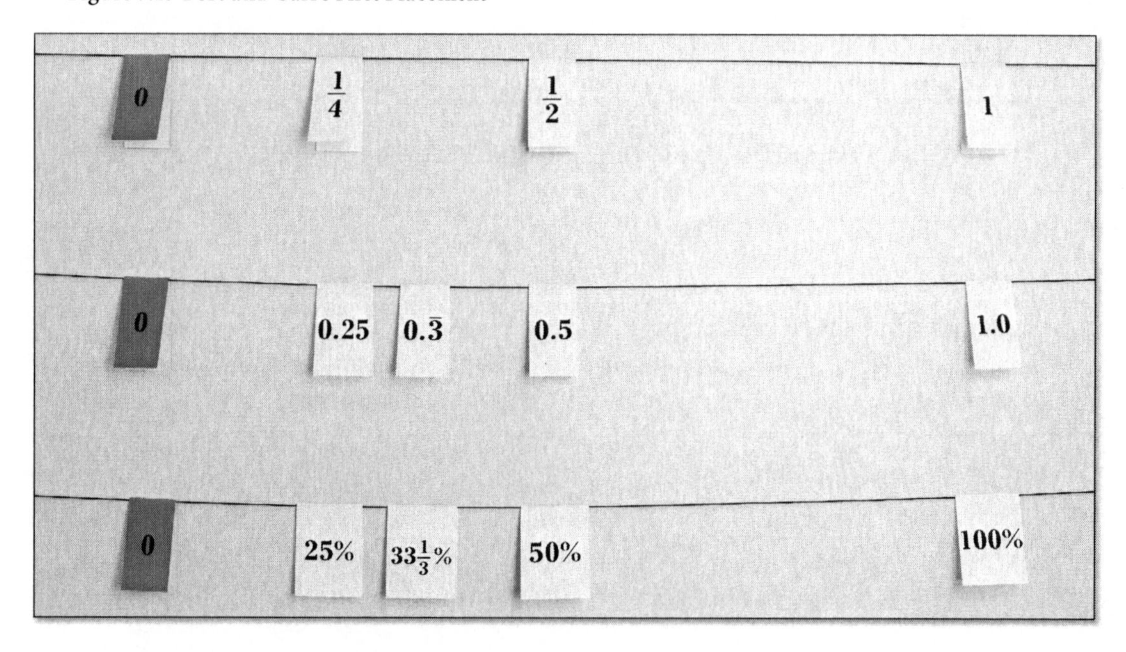

Carl: We are stuck on the fraction.

Teacher: Good, then we are all ready to learn something. Rather than doing any kind of conversion technique, just look at where you placed your decimal and percent cards on the number lines. Can you estimate what fraction of the entire unit from 0 to 1 that is?

Bert and Carl pick up $\frac{1}{3}$ and hang it on the top line, as shown in Figure 7.16.

Figure 7.16 Bert and Carl's Second Placement

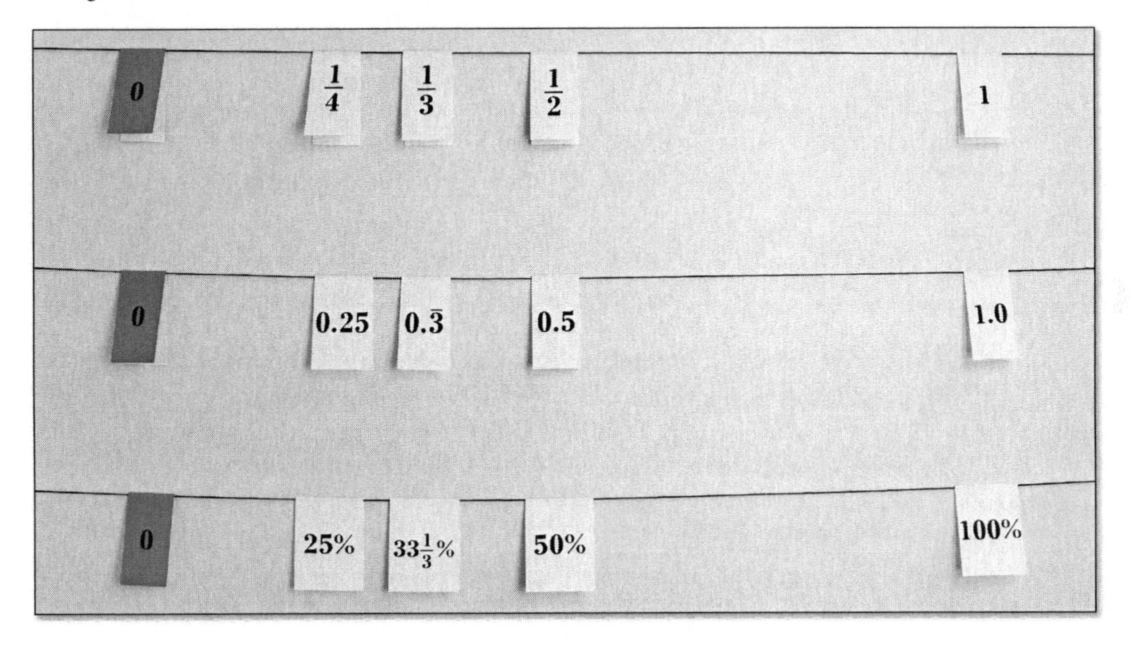

Teacher: Good thinking. What and who's next?

Bert: Quinn and Joseph for $66\frac{2}{3}\%$.

Quinn and Joseph place their two cards correctly on the clotheslines (see Figure 7.17).

Figure 7.17 Quinn and Joseph's Placements

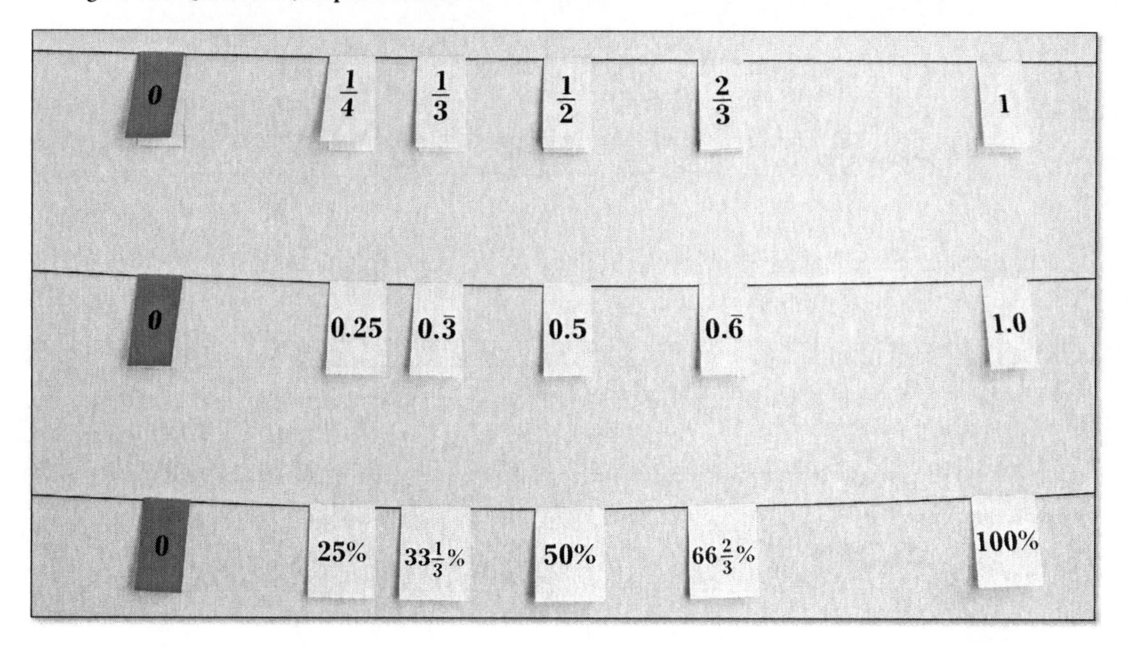

Teacher: We are down to the last set of three. Mona and Timothy, the clothesline is yours.

Mona and Timothy place the final two cards, as shown in Figure 7.18.

Figure 7.18 Mona and Timothy's Placements

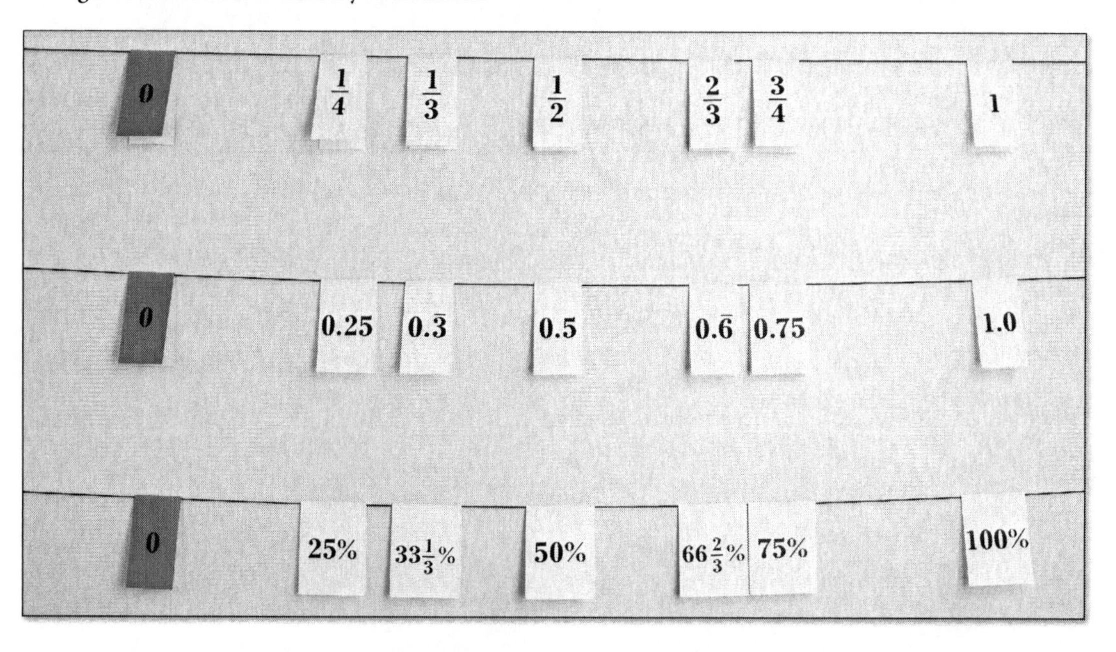

Good teamwork here. These are the most common fractions we need to know and their percentages, so let's record them on the activity sheet. Rather than three different lines, simply stack each set of three numbers on one. This will help us reinforce the idea that they are equal.

Why These Numbers?

The most common fractions and percentages were chosen here. These are everyday values that students should be able to fluently move between forms without needing to follow an algorithm.

The Key Questions

- Can you estimate what fraction of our unit from 0 to 1 is represented by that percentage?
- Given the location of the fraction, can you estimate decimal and percentage equivalent?

Analysis

This particular lesson has the least amount of discussion among students. In fact, the only real conversation we had surrounded the fraction equivalent of $33\frac{1}{3}\%$, which is a typical sticky spot. The reason for the lack of discussion was that this activity serves more as a practice task than a learning task. The students have done their share of converting fractions to decimals by division, as well as converting decimals to percentages. This task was to give an overall visual of where these values reside in relation to each other, which helped the struggling students settled the issue of $\frac{1}{3}$.

The cards for this lesson are available in the Digital Resources (lesson15.pdf).

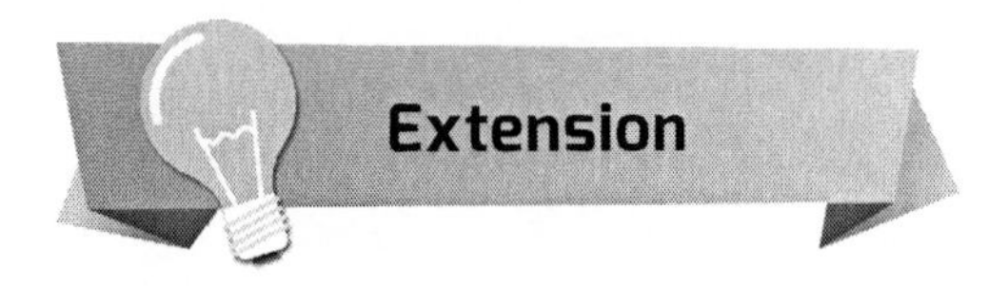

Extension

Other common fractions that are often converted to percentages are $\frac{1}{5}$, $\frac{1}{8}$, $\frac{1}{10}$, or values greater than 1.

Lesson 16: Operations with Integers

$3 + {}^{-}5$	${}^{-}5 + 3$	$3 - {}^{-}5$	${}^{-}5 - 3$	${}^{-}5$	8	${}^{-}8$

Objective: Evaluate expressions involving addition and subtraction of signed numbers.

Teacher: **Hello, all. We have been studying signed numbers, also known as integers. Let's warm-up today by reviewing what integers are. On your individual lapboards, use set notation to show the set of integers. Your hint is that your answer should look something like this** (see Figure 7.19):

Figure 7.19 Brackets

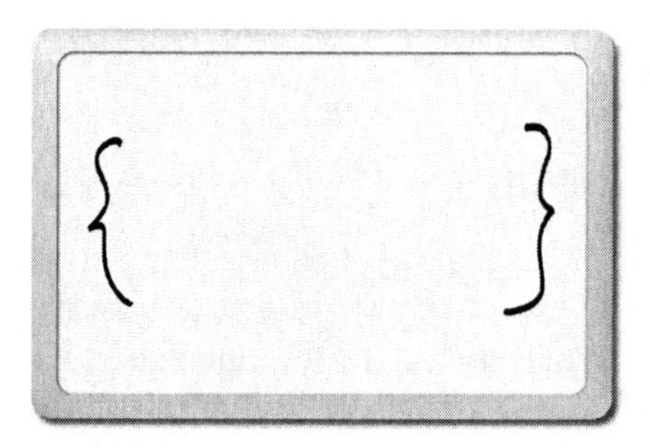

Write something inside the brackets so we understand what integers are.

Students write integers on their boards.

Now, rotate boards clockwise in your groups. If you agree with what's on the board, give it a check. If you want to make any modifications, write them.

Students review the boards, occasionally marking suggestions for classmates.

Rotate again.

Students rotate boards and repeat the process.

Rotate one last time.

Students rotate boards and repeat the process.

Now, take back your original board, and see what your group thinks of your response. Declare your boards.

Students hold up their boards to show the teacher their results.

Teacher: Nice work, class. So, we have a set that looks like this: {...-3, -2, -1, 0, 1, 2, 3...}.

There is a pattern shown here. From that pattern, I want you to write any two examples of a number that is NOT an integer.

Students write their examples.

Declare.

Students show boards. Nearly all have fractions or decimals.

Well done, crew. We see in our set that integers are all whole numbers and their opposites. Any decimal or fraction that does not simplify to a whole number or its opposite would be a good example of a non-integer. Today, we are looking at adding and subtracting integers. I am giving only two integers, 3 and -5. I have placed them on the number line as you can see up here (see Figure 7.20).

Figure 7.20 Teacher Placement

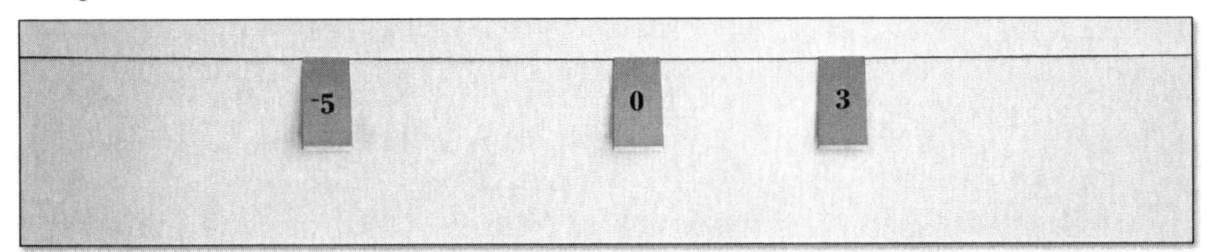

I'm going to hand out four expressions today. When I do, your group will place the expressions on the clothesline and then choose a group that will give us a money scenario for it. Bella, your group is first. Where will 3 + -5 go? Everyone else, discuss potential scenarios in your groups. Be ready to share.

Bella's group places two cards on the clothesline, as shown in Figure 7.21.

Figure 7.21 Bella's Placement

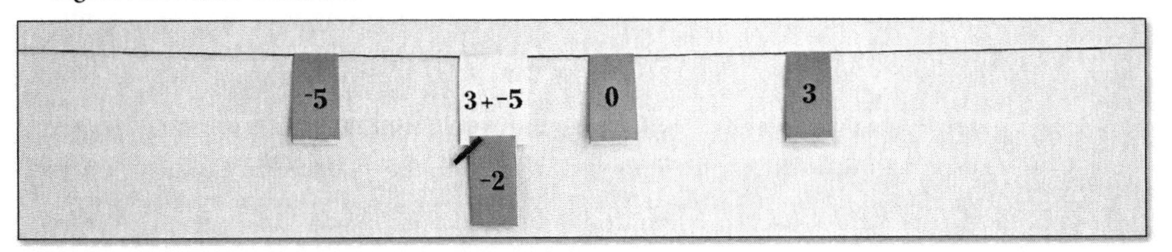

Bella, can you choose a group to offer a scenario for us?

Bella selects Landon's group.

Landon, what is your group's scenario?

PART III – Chapter 7

Landon: You have $3, you spend $5, so you owe $2.

Teacher: Does anyone want to challenge?

No hands raise.

Let me confirm with some distance reasoning. Watch my hands—from 0 to -5 is this far. If I start at 3 and move this negative direction, do I end at -2?

Class: Yes!

Teacher: Then who's next, Landon?

Landon selects Patti's group.

Patti, your group gets -5 + 3. Everyone else, discuss a potential scenario.

Patti's group places the card to the left of -5, as shown in Figure 7.22.

Figure 7.22 Patti's Placement

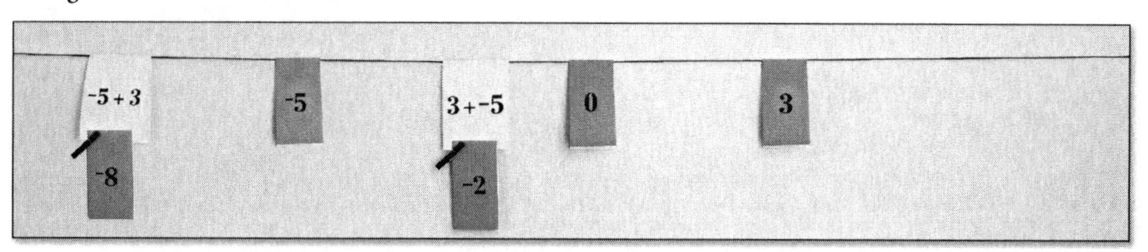

Patti, your group says -8. Whose scenario do you want to hear?

Patti: Shandra's group.

Shandra: If you already owe $5 and somebody gives you $3, you now owe $2.

Teacher: Back to Patti's group then because your group claims you would owe $8.

Patti and her group discuss the scenario.

Patti: We think they are right. It's not -8, it's -2.

Teacher: Let's do some distance reasoning to check. My hands are showing how far it is from 0 to +3. If I add that to -5, do I finish at -2?

Class: Yes!

Teacher: All in favor that I should move this to -2, say "aye."

Class: Aye!

The teacher pins the card below -2, as shown in Figure 7.23.

Figure 7.23 Teacher's Adjustment

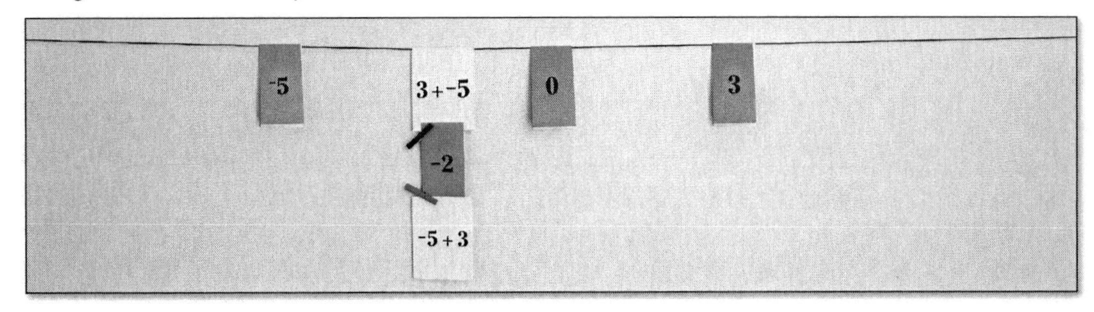

Teacher: Shandra, who is next?

Shandra points to Billy's group.

Billy, your group has 3 – -5. Everyone else, brainstorm your money scenario.

Billy's group places their card to the far right of the clothesline (see Figure 7.24).

Figure 7.24 Billy's Placement

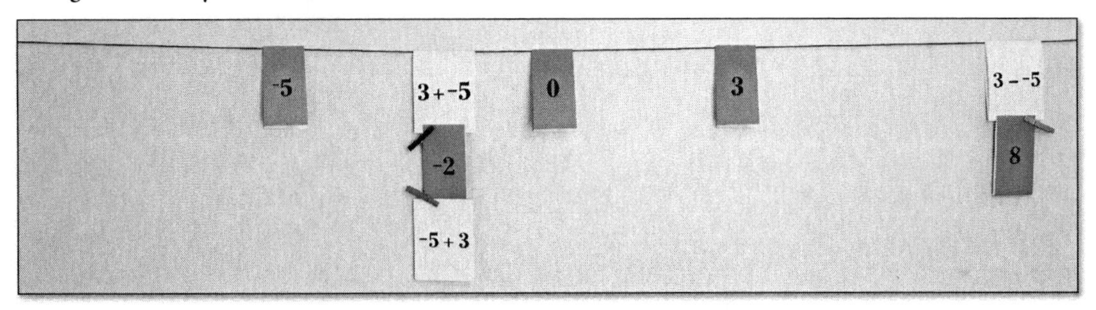

Billy, whose scenario do you want to hear?

Billy: Blake.

Blake: You have $3 and…umm…take away a bill of $5, which is like getting money, so you will have $8.

Teacher: Blake, your group's explanation agrees with the response of Billy's group. Let's do some distance reasoning. My hands represent the distance from 0 to -5. Apparently, this was added to something to equal 3, so I'm moving it over to start at 3. And now I'm taking it away, so I will drop my left hand and see where my right hand remains. Does this equal 8?

Class: Yes!

Teacher: Last one then, Blake.

Blake selects Chance's group.

Chance, your group gets -5 – 3. Everyone else, brainstorm scenarios, please.

Chance's group places the final card (see Figure 7.25).

Figure 7.25 Chance's Placement

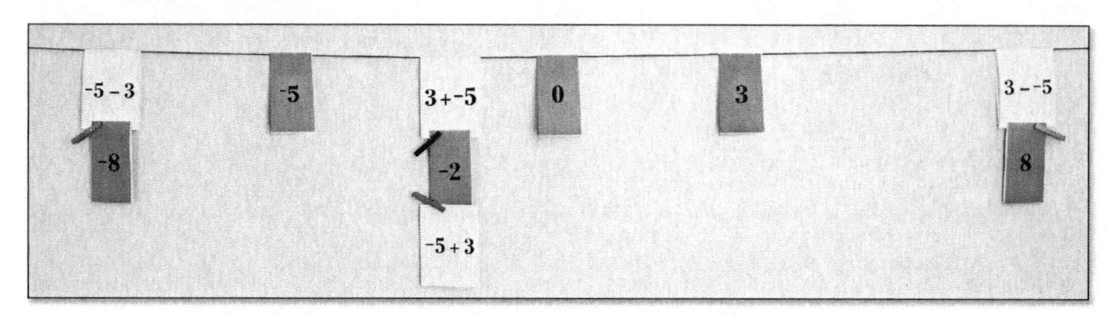

Teacher: Chance, who gets to share their scenario?

Chance: Eve.

Teacher: Eve, what do you have to share with us?

Eve: We started with -5. Then, we subtracted 3 to get positive 8 because two negatives make a positive.

Teacher: Did your scenario involve money?

Eve: Oh…no.

Teacher: Let's begin again. How much money do you start with?

Eve: We owe $5.

Teacher: Are we adding or subtracting?

Eve: Subtracting.

Teacher: How much?

Eve: Three dollars.

Teacher: So how do I owe $5, spend $3 more, and end up with $8 dollars in my pocket?

Eve: You don't. You owe $8.

Teacher: Let's use distance reasoning to prove it. This is the distance from 0 to +3. Starting at -5, we subtract this +3 distance, and finish at what?

Eve: Negative 8!

Teacher: Even though we had to adjust a few scenarios, you did terrific with the computations. Let's record each of these four on the same number line. Be sure to include a scenario for each.

Why These Numbers?

By choosing one negative and one positive integer, there is opportunity to arrange the integers in four different ways with adding and subtracting. These four different arrangements allow for various discussions, such as subtracting a negative.

The Key Questions

- What is your group's scenario?
- Does anyone want to challenge?
- Are we adding or subtracting?
- If I add this distance to this value, where do I finish on the number line?

Analysis

Even though this is a number line activity for operations on integers, I started with the definition of integers to solidify the vocabulary term in students' minds and familiarize them with set notation.

Since I didn't want to simply recite tricks or rules ("when subtracting a negative, just change the signs"), I chose not to have students explain their reasoning but to create scenarios instead. This drove the class to think on a conceptual level. The choice to have other groups create the scenario is important for two reasons. The first is engagement. The second is time—it would be too much to ask the group at the clothesline to calculate and create. Distance reasoning, or hand reasoning, added one more layer of understanding to the verbal and symbolic representations. Figure 7.26 is the clothesline version of an arrow diagram. Distance reasoning replaces the need for an arrow diagram with the clothesline.

Figure 7.26 Arrow Diagram

6 – 9 = -3

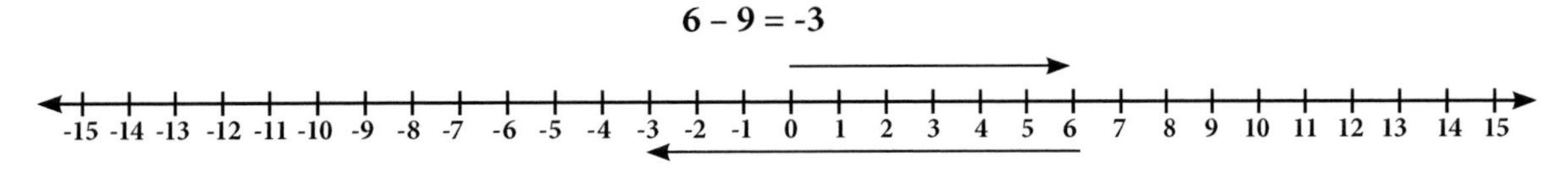

The cards for this lesson are available in the Digital Resources (lesson16.pdf).

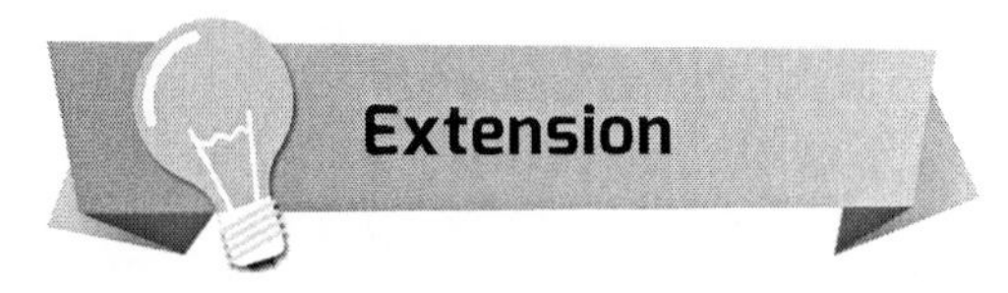

Extension

Place the same four expressions using -3 and +5 instead.

Algebra

Lesson 17: Evaluating Expressions

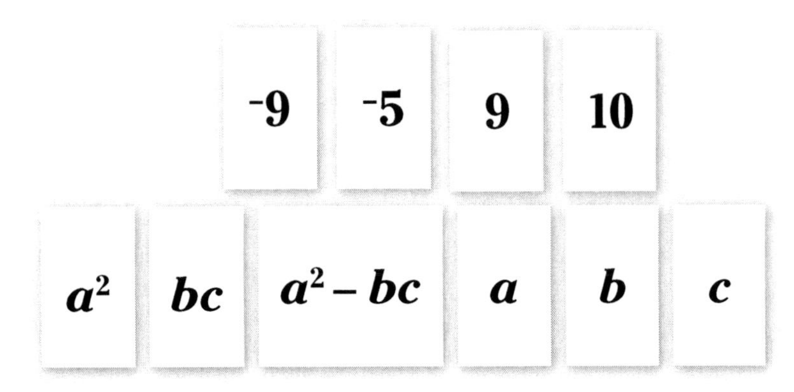

Objective: Evaluate expressions having both linear and quadratic terms with signed numbers.

> Teacher: **Good morning, class. I noticed you are very good at evaluating expressions when we substitute positive values. You still have a challenge when we substitute negative numbers. So, let's take your challenge to our open number line. I have placed *a, b,* and *c* at -3, -5, and -2, respectively** (see Figure 8.1).

Figure 8.1 Teacher Placement

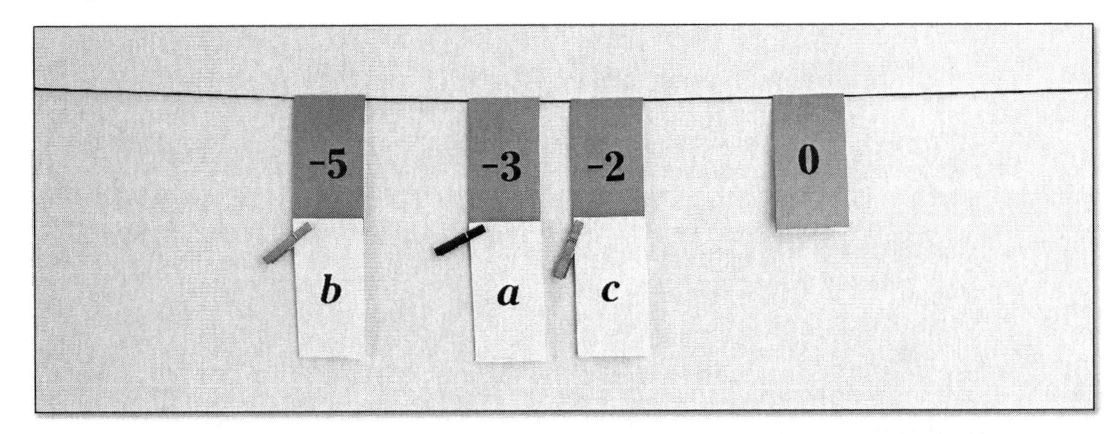

> Teacher: **Our first group will place a^2 for us. Marvin, it's your group's turn for the clothesline. The rest of you, work on your lapboards.**

PART III – Chapter 8

Marvin's group chooses -9 to pin the expression to, as shown in Figure 8.2.

Figure 8.2 Marvin's Placement

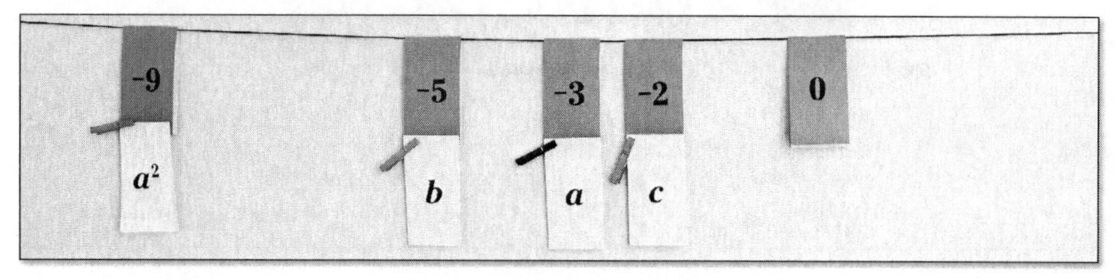

Teacher: Marvin, why did your group choose -9?

Marvin: Because -3 squared is -9.

Teacher: Jordan, your group disagrees.

Jordan: Yes, 3 squared is 9. So, it's not negative.

Teacher: I noticed a difference on your lapboards in the way you wrote your answers. Some of you wrote $(-3)^2 = 9$, while others wrote it as $-3^2 = -9$. You both have correctly evaluated how you wrote it. Marvin, yours says square the -3. Jordan, yours says square the 3 and then take the opposite. The question is: which one is written correctly? Which way does the person who wrote this expression want us to think about it? Yes, PJ.

PJ: They want you to put -3 in for x, so you should put -3 in the parentheses.

Teacher: Yes, since the expression asks us to square a, we must square whatever we substitute. Everyone correct your boards, if necessary, and now place bc on the number line. Marvin, who is next?

Marvin selects Deni.

Deni, your group is up.

Deni's group adds positive benchmark cards to the clothesline and pins the two expressions to them (see Figure 8.3).

Figure 8.3 Deni's Placement

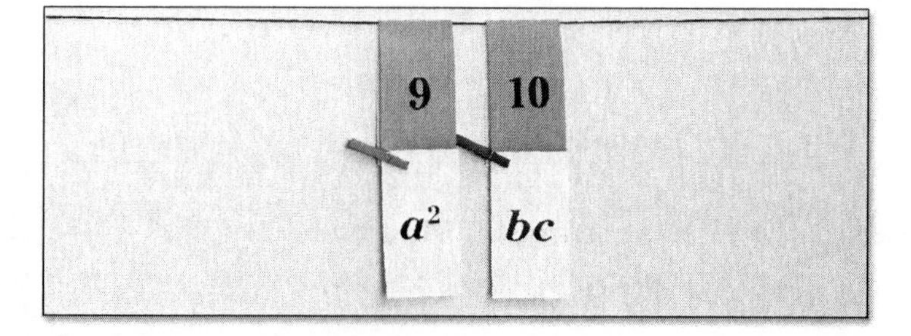

Teacher: So Deni, I gave your group two negative numbers, and you placed their product on the positive side of 0.

Deni: Yes, negative times negative is positive. So, -5 times -2 equals positive 10.

Teacher: Let's see who agrees. Class, declare your boards. Deni, everyone seems to agree. Now, let's place the entire expression, $a^2 - bc$. Deni, who will do that?

Deni selects Charlie.

Charlie, your group is up.

Charlie's group pins its expression to 1 (see Figure 8.4).

Figure 8.4 Charlie's Placement

Teacher: Charlie, what does your group say?

Charlie: We say 9 minus 10 is 1.

Teacher: Let's see who agrees. Everyone, declare.

Students are divided on their answers.

Charlie, the class is split. I see two different challengers. Chantal, can you explain your response?

Chantal: Like what was said before, negative times negative is positive. So, we added positive 10 to the 9 and got 19.

Teacher: Thank you. For those of you who disagree with both of these answers still, who would like to challenge? David.

David: You still have to subtract the positive 10, and 10 – 9 is positive 1, but 9 – 10 is -1.

Teacher: Interesting. Watch my hand reasoning. How far is this?

Salma: Ten.

PART III – Chapter 8

Teacher: **I'm going to subtract that much starting at 9. Where do I end up?**

Salma: Negative one.

Teacher: **All in favor of moving this to -1, clap twice.**

Students clap twice unanimously. The teacher moves the expression to -1 (see Figure 8.5).

Figure 8.5 Teacher's Placement

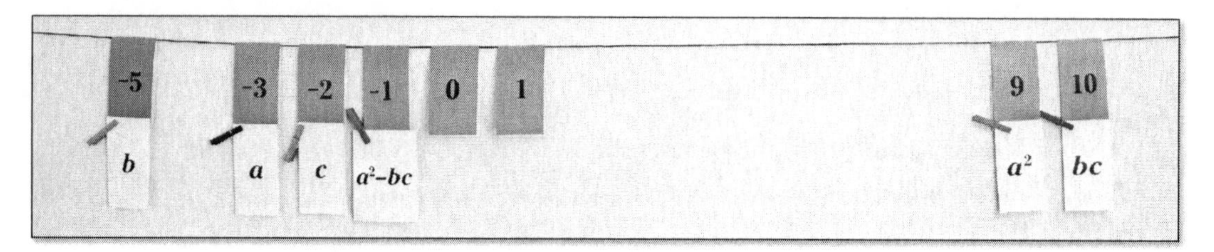

Teacher: **Okay, let's record our thinking from today. Draw this number line on the activity sheet. Be sure to include any notes regarding squaring the negative and subtracting a larger positive number from a smaller one.**

Why These Numbers?

On a recent assessment, this particular class had trouble evaluating $x^2 - 3x + 5$, for $x = -2$. Therefore, I chose the expression $a^2 - bc$ for the Clothesline activity because it mimicked the assessment question while having flexibility for students to change all expression values. To confront errors when calculating negative values, the terms *a*, *b*, and *c* were plotted on negative values on the clothesline. The common errors of squaring a negative number and multiplying two negative numbers were addressed by having students place the results of evaluating a^2 and *bc* before evaluating the entire expression. The -5 and -2 were chosen to force the issue of subtracting a greater positive number (10) from a lower positive number (9).

The Key Questions

- Why did you choose 9?
- Why did you choose -9?
- How does the person who wrote this expression want us to think about it?
- Why do you challenge this?
- What is the distance between these numbers on the number line?
- If I subtract this distance from 9, where do I end up?

This activity was generated from a true formative assessment. Even though students should have mastered this skill before they reached an algebra course, half of them answered the following question incorrectly on the assessment: Evaluate $x^2 - 3x + 5$, for $x = -2$. The number of students lacking such a critical skill demanded that I have them reengage.

The two biggest mistakes made on the assessment were squaring a negative number and dealing with the -3(-2). Students multiplied the two negatives to equal a positive but then still subtracted, rather than subtracting a -6 or adding a positive 6, depending on how they saw the problem. Although, I wondered after the assessment whether I should have chosen a positive value for *b* for this day's expression to more closely resemble the assessment, I chose to push the issue of correctly evaluating signed numbers.

As it turned out, some students still struggled with adding 10 to 9 or subtracting 10 from 9, and we had additional conversations regarding 9 – 10 (versus 10 – 9). This was an extra treat I had not expected, but it offered an opportunity to show how to subtract distance on the number line, which is why I used the hand reasoning to show that subtracting a greater positive number from a lower yields a negative value. That one was fun.

Breaking the expression into various terms and dealing with them individually was fruitful. The anticipated mistake of squaring -3 and getting -9 showed up, and it allowed for an opportunity to discuss the helpfulness of using the parentheses appropriately.

The cards for this lesson are available in the Digital Resources (lesson17.pdf).

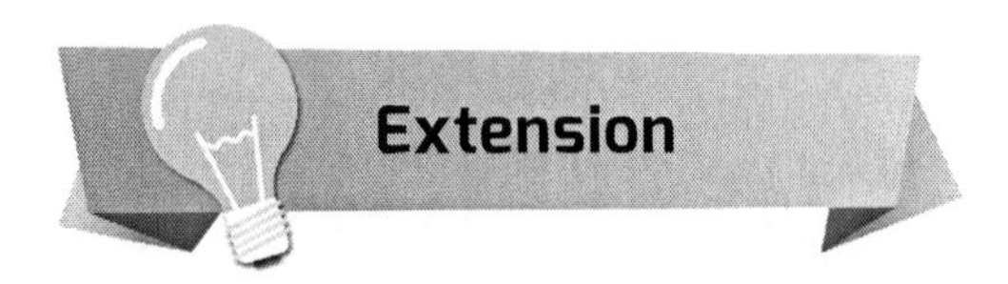

Mix the values of *a*, *b*, and *c* so some are negative and some are positive.

Lesson 18: Variable on Both Sides

$3x - 6$ $2x + 9$ x $2x$ $3x$

Objective: Solve an equation with the variable on both sides.

Teacher: **Good day, class. Take a look at what I have waiting for you on the clothesline today** (see Figure 8.6).

Figure 8.6 Teacher's Placement

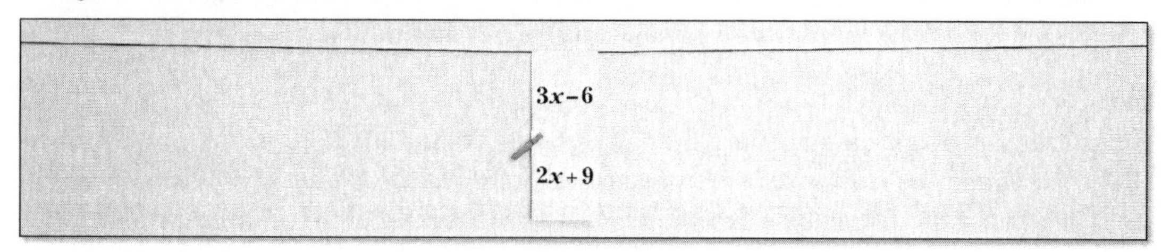

On your own, think about what you notice and wonder. Once everyone in your group has at least one notice and one wonder, share with each other. Everyone in the group should speak. Once you have all shared, send one person from your group to the board to write one notice and one wonder.

Students write their responses on the board (see Figure 8.7).

Figure 8.7 Student Responses

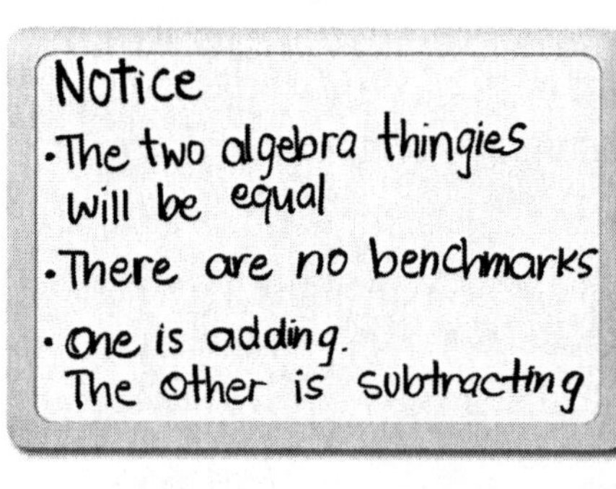

Wonder
• What is x?
• What is the whole thing equal to?
• Do we need more numbers?
• Is it positive or negative?
• Where is 0?

Okay, it looks like everyone agrees that we are dealing with expressions that are equal to each other. That's what those "thingies" are called: *algebraic expressions*. One of these expressions involves addition and the other subtraction. We see that there are no benchmarks offered including 0, and we wonder whether it is a big deal. We also agree that since these expressions are equal, we can solve for *x*.

Teacher: Before we start solving for x, I want to challenge you all. Tanya, please choose the group that will place $2x$ and $3x$ on the clothesline.

Tanya selects Holly.

Holly, your group is up. The rest of you, work in pairs to write these expressions on a number line on your lapboards. Don't forget to include $2x$ and $3x$.

Holly's group places two cards (see Figure 8.8).

Figure 8.8 Holly's Placement

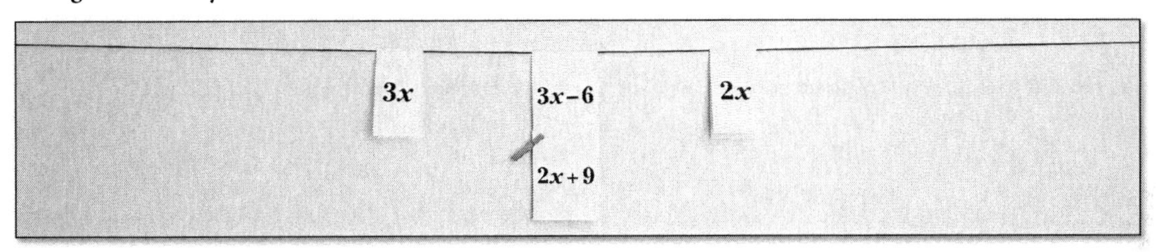

Teacher: Declare your boards please. Holly, some of your classmates agree with you, some do not. Please argue for your group's answer here.

Holly: Since it says $3x$ minus 6 we placed $3x$ in the negative direction, and since it says $2x + 9$ we put $2x$ in the positive direction.

Teacher: Any counter-arguments? Sammy.

Sammy: $3x - 6$ is where you are after you subtract 6, so the $3x$ has to be greater. And the reverse is true of the other one. $2x + 9$ is where you are after adding 9, so the $2x$ starts out less than.

Teacher: Let me switch them and see what the class thinks about your reasoning (see Figure 8.9).

Figure 8.9 Teacher's Placement

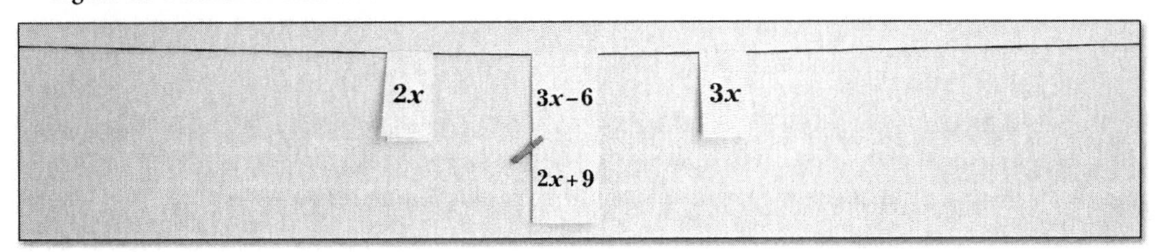

Teacher: Thumb vote.

Some students vote thumbs-down.

Some of you are still voting thumbs-down. Tanya, back to you. You disagree?

Tanya: Not with the order, just the spacing.

Teacher: Show us, please.

Tanya adjusts the spacing of the cards, as shown in Figure 8.10.

Figure 8.10 Tanya's Placement

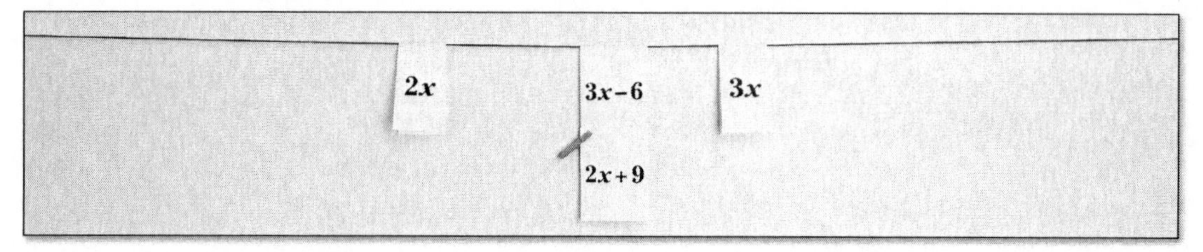

Teacher: What was your group being so particular about?

Tanya: Well, 6 is two 3s, and 9 is three 3s, so we're making sure that 6 is only $\frac{2}{3}$ of the 9.

Teacher: Everybody, finger reasoning. Double-check your work. Is the 6 equal to $\frac{2}{3}$ of 9?

Students use finger reasoning.

Class: Yes.

Teacher: Now that we have the spacing between $2x$ and $3x$ finished, please watch my hands because I am going to ask you a question about the distance from $2x$ to $3x$.

The teacher places his left hand at 2x and his right hand at 3x. He turns to the class with hands up, representing the distance from 2x to 3x.

Teacher: What is the distance? Everyone.

Class: Fifteen!

Teacher: Okay, I am going to measure it again, and this time I want a different answer. What is this distance?

Thoughtful silence.

Teacher: Let me measure again. What is this distance? Anyone.

Jameson: x?

Teacher: Jameson, what makes you say that?

Jameson: Because it is $3x$ minus $2x$.

Teacher: All those who agree, say "aye."

Class: Aye!

Teacher: Well now, you all are confusing me. Is the distance 15, or is it x?

Class: Both!

The teacher speaks as he writes the solution of the equation on the board.

Teacher: You say both. So, let me see whether I have this correct. You started by saying that $2x + 9 = 3x - 6$. You then added 6 plus 9 to equal 15 but subtracted $2x$ from $3x$ to equal x (see Figure 8.11).

Figure 8.11 Board Work

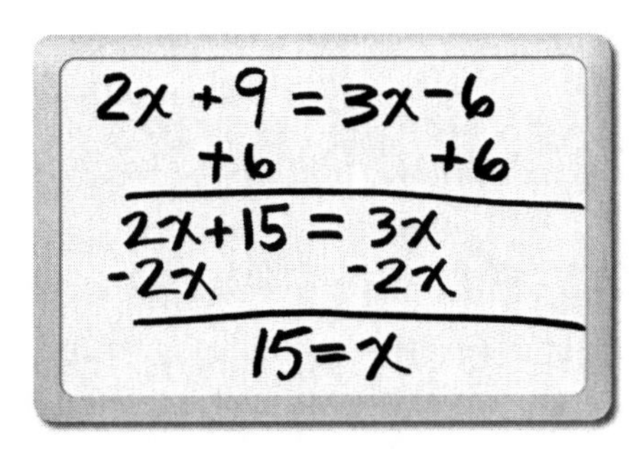

That is amazing everyone. You just made sense for me why we solve equations the way we do with a variable on both sides of the equation. When we solve, we are actually finding the distance between the variable terms. In this case, between $2x$ and $3x$. It is both 15 and x; therefore, x actually is equal to 15.

Zara: That's cool!

Teacher: I agree, so I'm going to pin 15 to x and place them on the clothesline halfway between $2x$ and…oh, I don't have 0 yet. I wonder what I should do.

Ren: Count x's to the left.

Teacher: Ren, what do you mean?

Ren: Start with $3x$ and count backward to $2x$. That same distance will be x. Then, count again for 0. They are all x apart.

Teacher: Like this?

The teacher moves the cards, as seen in Figure 8.12.

Figure 8.12 Teacher's Adjusted Placement

Ren: Yes!

Teacher: Sounds good. Everyone now, determine the values of $2x$ and $3x$. Go.

Students work in pairs on their lapboards.

Declare your boards.

Students hold up their boards.

Everyone is in agreement that $2x = 30$ and $3x = 45$. Now, find the values of our two expressions: $2x + 9$ and $3x - 6$. Ren, while the class is working on their boards, which group is working on the clothesline?

Ren selects Oliver's group.

Oliver, your group is up.

Oliver's group pins 39 to the given expressions (see Figure 8.13).

Figure 8.13 Oliver's Placement

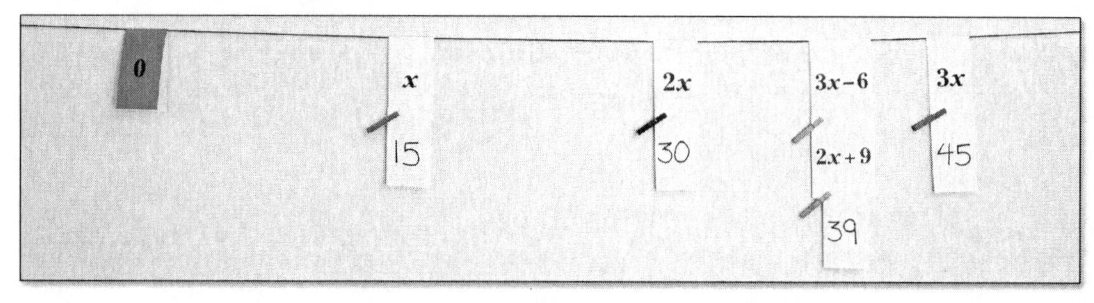

Teacher: Class, check the work of Oliver's group. If you agree, clap twice.

Students clap twice.

Teacher: That was some really heavy lifting that our brains just did. Let's record all that amazing thinking and learning on our activity sheet. I suggest that you also record this example of our algebraic solution to the equation, so later you can remember this important connection that we made today.

The big reveal in this lesson is that $x = 15$. The values were chosen so the difference between the two variable terms $2x$ and $3x$ was x, rather than $5x$ and $2x$, for example, which would have required another step to solve for x. The terms $3x$ and $2x$ were also chosen to place on the number line to reveal more understanding of the algebraic properties, such as the addition property of equality.

- Why should $2x$ be to the left of the expressions and $3x$ to the right?
- Why were you so particular about your spacing of $2x$ and $3x$?
- What is the distance between $2x$ and $3x$?
- Can you give me a different answer for that distance?
- Why do you say that the distance is equal to x?
- What is the value of these two equivalent expressions?

If I had simply asked students to solve for x, the lesson would have gone much quicker. Just work out the algorithm. Done. However, the process I led students through, particularly the placement of $2x$ and $3x$, and their correspondence to x and 15, was well worth the time invested. The misconception of $2x$ being to the right of the expression because there was an addition sign in the expression tells a great deal about why students have problems adding or subtracting correctly on both sides. The conversation about the proper spacing of $2x$ and $3x$ with $2x$ being "three 3s" away while $3x$ was "two 3s" away was good for students' proportional reasoning, but it also set the hook for realizing that x was going to be equal to 9 plus 6. The placement of x at 15 and the equivalent expressions at 39 allowed discussions about solving for x and finding the numerical values to check for accuracy.

The cards for this lesson are available in the Digital Resources (lesson18.pdf).

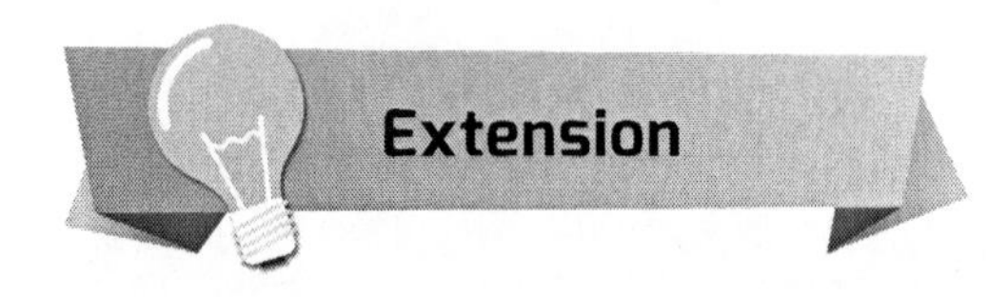

Pin the expressions $3x + 4$ and $2x - 1$ to each other on the negative side of 0, and have a terrific conversation about why $2x$ is now to the right of $3x$.

Lesson 19: Two-Step Equations

x	$2x$	$3x$	$3x + 5$	8	12	17

Objective: Solve a two-step equation.

Teacher: Hello, class. You already know how to solve one-step equations. This year, you are ready to learn about two-step equations. A couple of examples of one-step equations are on the board (see Figure 8.14).

Figure 8.14 One-Step Equation Examples

$3x = 15$

$x + 5 = 15$

We call these one-step because we can solve for x in one step. Today, I will be challenging you by combining these types of equations into one. So, you will have to do two steps in order to solve for x. Today's equation is $3x + 5 = 17$. We will solve this on the number line first. Then, I will show the algebra on paper. Here is what our equation looks like on the number line (see Figure 8.15).

Figure 8.15 Teacher Placement

Teacher: Notice the way I have this pinned. The clothespins represent equivalency. So, this is saying that $3x + 5$ and 17 are equal. The left side and the right side of the equation are equal, so we show they are equal on the number line. The next thing to place on the number line is $3x$. Rowena, have your group place this card. Everyone else, work on lapboards with your partners.

Rowena's group places 3x, as shown in Figure 8.16.

Figure 8.16 Rowena's Placement

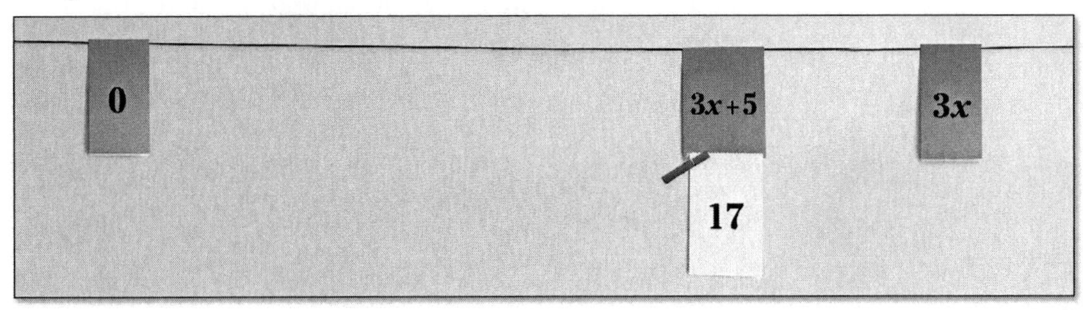

Rowena, why there?

Rowena: The equation says $3x + 5$, so we added 5 and put $3x$ there.

Teacher: Thank you. Trini, your group has something different.

Trini: That says 3 plus x plus 5, so we have to subtract to go backward to 5.

Teacher: Here?

The teacher moves 3x to the left of the equation, as shown in Figure 8.17.

Figure 8.17 Teacher's Adjustment

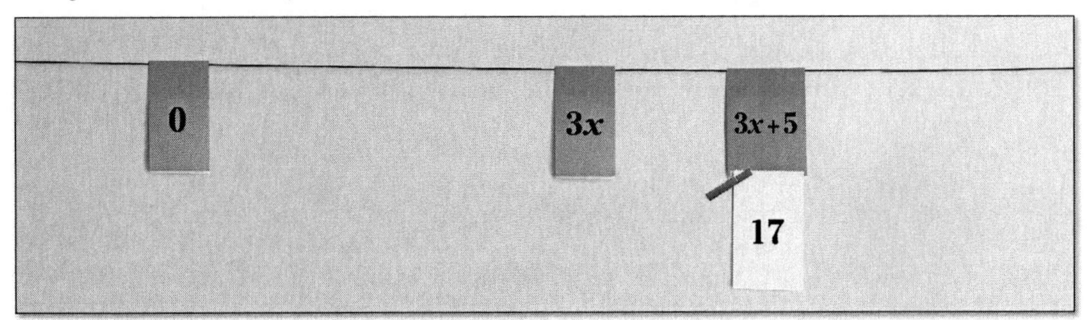

Trini: Yes.

Teacher: Well, now we have a fun debate! All those who think $3x$ should be to the left of the expression, point left. All those who think $3x$ should be to the right of the expression, point right.

All the students point to the left, including Rowena and her group.

Teacher: **Rowena, you are voting against your group's own logic. You changed you mind?**

Rowena: Yes, the other explanation makes more sense.

Teacher: **Also, remember how we spoke of inverse operations when solving? Now you can see why we must do that. We are looking at an equation that says to add 5. To know what $3x$ equals, we must complete the inverse operation of addition, which is subtraction. Are we good with my spacing? Does that look like about 5?**

Lin: It needs to move over more.

Teacher: **Which way, class?**

Students point to the left, and the teacher moves $3x$ farther to the left.

Good?

Class: Good!

Teacher: **What value should we pin to $3x$? Write it on your board just under $3x$.**

Students write on their lapboards.

Declare.

Students hold up lapboards with "12" written on them.

Everyone claims $3x = 12$. I will pin that for you (see Figure 8.18).

Figure 8.18 Teacher Including 12

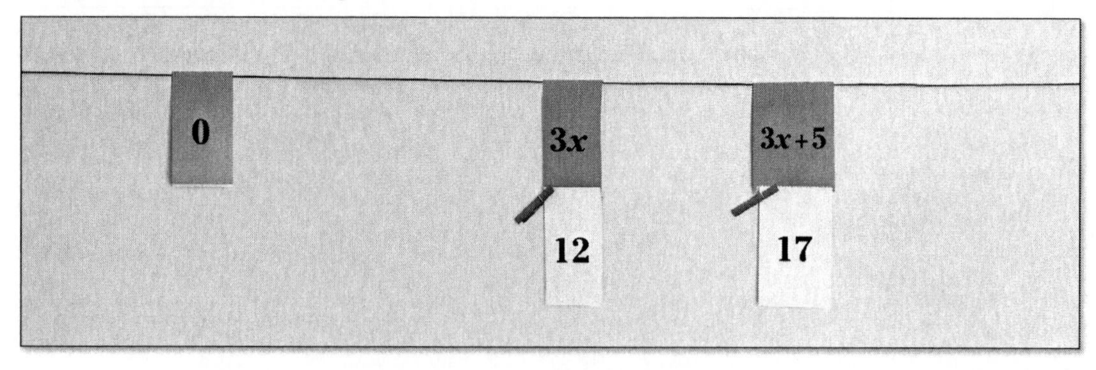

Teacher: **Now that we know where $3x$ is located, where should we place $2x$ and x? Rowena, whose group is up next?**

Rowena: I liked Trini's explanation; I think she should go next.

Teacher: **Okay, Trini have your group head to the clothesline. Class, approximate $2x$ and x on your lapboards.**

Trini's group places x and 2x, as shown in Figure 8.19.

Figure 8.19 Trini's Placement

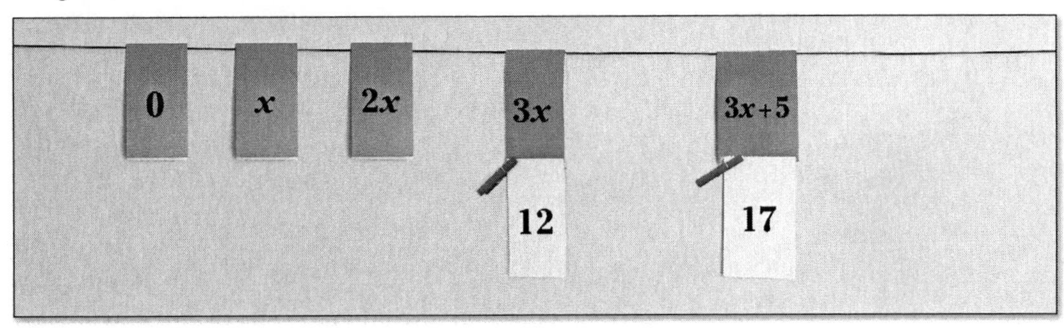

Everyone, declare you boards.

Students hold up lapboards that match the work Trini's group did on the number line.

Trini, the class seems to agree with you. Please explain.

Trini: We just took $3x$ and chopped it into three parts to get x and $2x$.

Teacher: **Class, write the value of x on your number line. Ready? Declare.**

Students show boards with different answers.

Some of you say 4. A couple of you say 36. Johnny, why 36?

Johnny: Three times 12 is 36.

Teacher: **Thank you. Mary, why 4?**

Mary: Because $3x = 12$ means 3 times x equals 12, and 3 times 4 equals 12.

Teacher: **Class, you have heard both arguments. I will give you a moment to change the answer on your board if you want.**

The teacher pauses while students work.

Ready? Declare.

Students hold up their boards, and everyone has 4 as the answer.

Teacher: Everyone agrees that x is 4. Mary's argument deals with the idea of inverse operations again. The inverse operation of multiplication is division...12 divided by 3 is 4. Let's check. Listen as I point to and read what's written on the card.

The teacher points to 4.

Four.

The teacher points to 3x.

Times 3.

The teacher points to + 5.

Plus 5.

The teacher points to 17.

Equals 17! Let's commit it to our number line now that we have checked it and are confident in our answer. On our activity sheet, take a minute to record this number line diagram. Next, I am going to take what you said about inverse operations of subtracting 5 and dividing by 3, and I am going to translate that into algebra on the board (see Figure 8.20).

Figure 8.20 Algebraic Form of the Clothesline

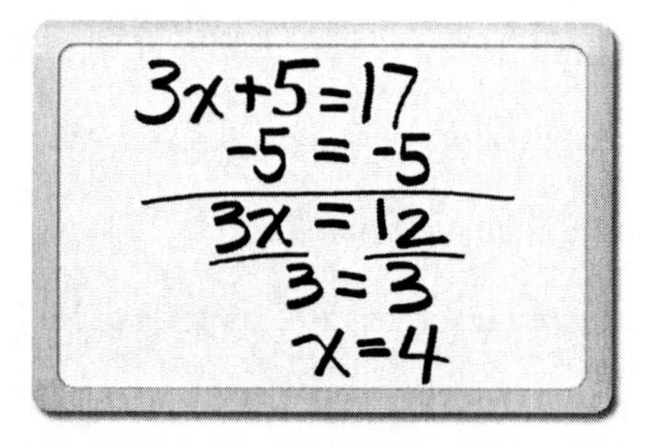

Write what's on the board next to your diagram on your activity sheet. See, the algebra was already inside your heads. I just had to teach you how to write it when it comes out of your heads. Strong work today.

Why These Numbers?

A simple two-step equation was chosen for this lesson for which each step was equal to a whole number that could easily be shown on the clothesline: $3x + 5 = 17$, $3x = 12$, and $x = 4$.

The Key Questions

- Why did you place $3x$ there?
- On which side of $3x + 5$ should we place $3x$?
- Is the spacing for $3x$ correct?
- How did you get 36 for x? How did you get 4?
- Knowing $3x$, how did you determine $2x$ and x?

Analysis

Rather than starting with teaching kids the algebraic procedure, this lesson ends with that, but only after starting with the conceptual underpinnings of inverse operations. The focus of the lesson was getting students to understand why we make the moves that we do. This particular lesson began by tying in the objective to the previous learning, letting students know that we will be combining two prior concepts in one. The pinning of $3x + 5$ to a value like 17 is a powerful visual of the clothesline that is often underappreciated. Instead of reading the equation like students so often do as "$3x + 5$ makes 17" or "gives us 17," pinning a value to it reinforces the idea that $3x + 5$ is *equal* to 17.

The next big concept is why it's important to subtract 5 from both sides, even though students were not asked to add or subtract. This is a critical first move—executing the inverse operation is the generalized procedure that produces the result we want, but lets us see what result we want first.

Then comes another important concept: why do we divide by 3? On the number line, no student subtracted 3 as so many do when first learning this concept on paper. Again, we let students partition $3x$ into 3 equal parts, each equivalent to x. Then, we generalize by mentioning the inverse operation technique they learned with one-step equations.

Finally, what every math teacher wishes students did more consistently—check the work. We started with $3x + 5 = 17$ and worked our way backward, solving for x. Then, we started with x and worked our way forward, evaluating for $x = 4$.

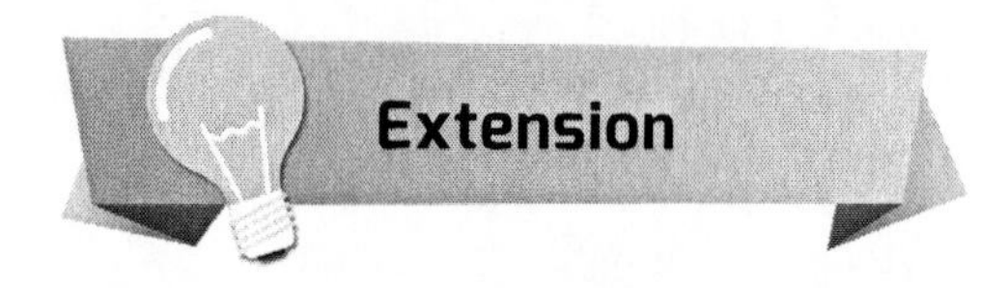

Use the equation $3x + 6 = 21$ and attempt to show the following common errors on the clothesline (see Figure 8.21):

Figure 8.21 Common Clothesline Errors

$$\frac{3x+6}{3}=\frac{21}{3}$$
$$x+6=7$$
$$-6\quad -6$$
$$x=1$$

$$\frac{3x}{3}+\frac{6}{3}=21$$
$$x+2=21$$
$$-2\quad -2$$
$$x=19$$

$$3x+6=21$$
$$+6\quad +6$$
$$\frac{3x}{3}=\frac{27}{3}$$
$$x=9$$

Lesson 20: Rules of Exponents

6	7	8	9	$\sqrt{42}$	$\sqrt[3]{8}$	$\frac{3}{2}$	$\frac{1}{25}$	$16^{\frac{1}{4}} + 32^{\frac{1}{5}}$

2^{-2}	2^{-1}	2^0	$2^{\frac{1}{2}}$	2^1	2^2	2^3	$4^{\frac{3}{2}}$	5^{-2}	$8^{-\frac{1}{3}}$

Objective: Evaluate expressions involving exponents, including negative and rational exponents.

Teacher: **Good morning, all. We spent this unit studying various rules of exponents. I thought it would be good to take time to review them together. Let's warm-up by having you recall the following rules at the top of your individual lapboards that I have written on the board. Copy this down and fill in the blanks** (see Figure 8.22). **Leave room at the bottom for a number line.**

Figure 8.22 Rules of Exponents

$$x^a = x \cdot x \cdot x \cdots x$$

$\underline{\quad}$

$x^0 = \underline{\quad}$ $x^{-a} = \underline{\quad}$

$x^{\frac{1}{a}} = \underline{\quad}$ $x^{\frac{a}{b}} = \underline{\quad}$

Teacher: **Declare your boards.**

Many of the blanks are filled in correctly, but several errors exist.

We are looking pretty good here, but we do have a few things to clear up. Leave your responses on your boards. You may confirm or change them as we apply each one on the clothesline. Today, we will do a student chain instead of a group chain. Let's start with you, Chelsea. Where would you place 2^3?

Chelsea places the cards, as shown in Figure 8.23.

Figure 8.23 Chelsea's Placement

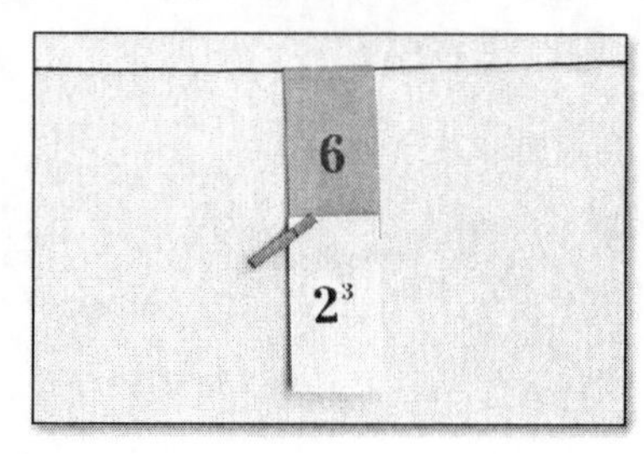

Teacher: Why 6, Chelsea?

Chelsea: Because 2 times 3 is 6?

Teacher: Thank you. Dev, you get the next card, which is 2^2. If you agree with what is already up on the number line, then place this card. If you disagree, change it, and hand this card to someone else in the class. Go ahead.

Dev moves Chelsea's card onto 8, as shown in Figure 8.24.

Figure 8.24 Dev's Placement

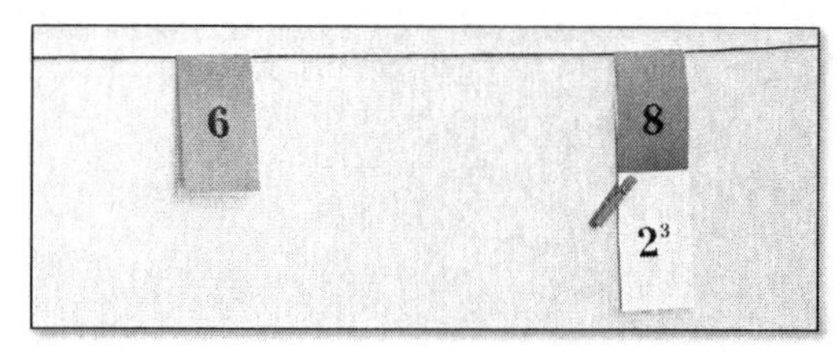

Teacher: Dev, why did you move 2^3 to 8?

Dev: Because it is not 2 times 3, but 2 times itself three times. Two times 2 times 2 equals 8.

Teacher: Okay, so hand 2^2 to another classmate.

Dev hands it to Molly.

Molly, you're up.

Molly pins 2^2 to 4 (see Figure 8.25).

Figure 8.25 Molly's Placement

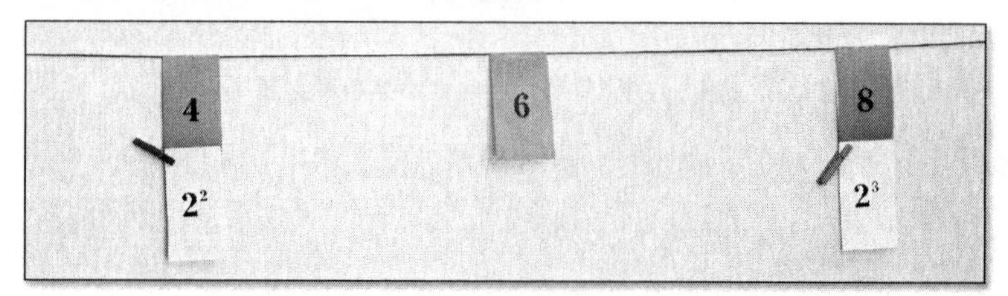

Teacher: Molly, tell us why.

Molly: Because 2 times 2 is 4.

Teacher: Do you mean 2 times itself, or 2 times the exponent?

Molly: Two times itself.

Teacher: Thank you. Gerry, 2^1 is yours.

Gerry pins his card to 2 (see Figure 8.26).

Figure 8.26 Gerry's Placement

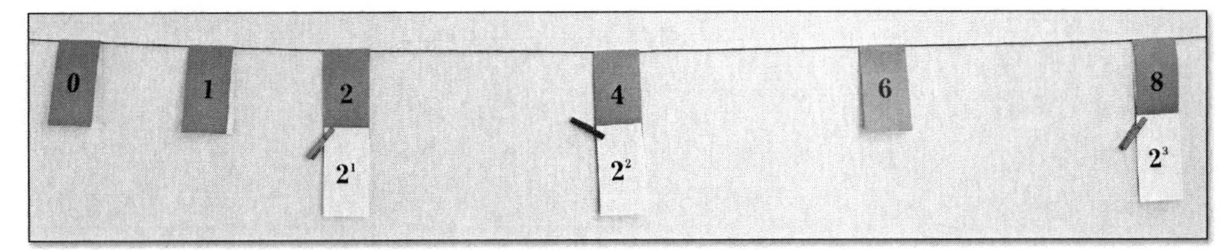

Teacher: **Gerry, explain.**

Gerry: Easy, 2 to the first power is 2.

Teacher: **So let's summarize our first rule from the warm-up. The expression x to the a means multiply x to itself a number of times.**

The teacher writes the following information on the board (see Figure 8.27):

Figure 8.27 Teacher's Equation

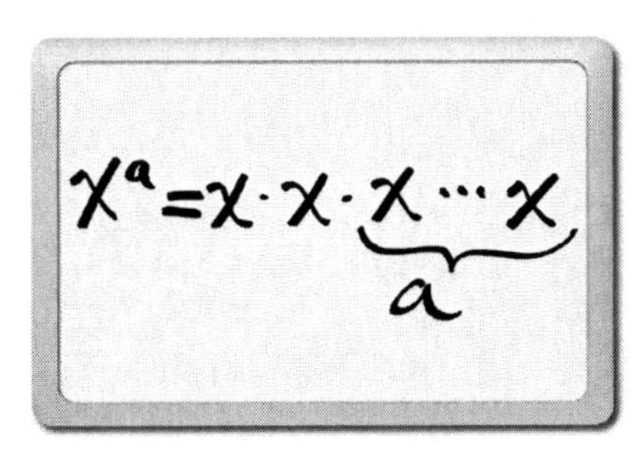

Our next one, 2^0, goes to…Owen.

Owen pins his card to 0 (see Figure 8.28).

Figure 8.28 Owen's Placement

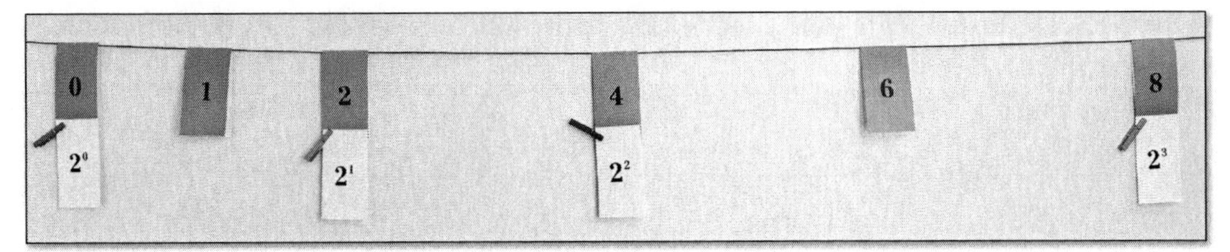

Owen, you are claiming that 2 to the 0 power is 0. Why?

Owen: Anything to the 0 power is 0.

Teacher: **Thank you. Hunter, you get the next one, 2^{-1}. Place it on the number line, unless you disagree with anything on the number line; in which case, you may change it and pass this one onto another student.**

Hunter moves cards on the clothesline, as shown in Figure 8.29.

Figure 8.29 Hunter's Placement

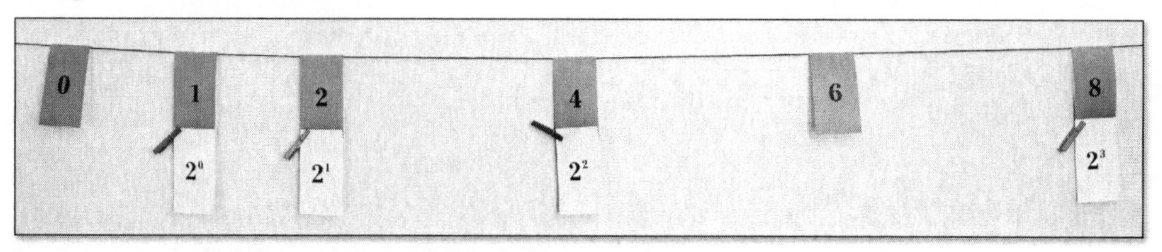

Teacher: **Hunter, you changed 2^0 to equal 1 instead of 0. Why?**

Hunter: We saw this pattern before with tables. 8 divided by 2 is 4, divided by 2 is 2, divided by 2 again is 1, not 0.

Teacher: **Thank you, Hunter, for tying this into something that we have studied previously. At that time, we also discussed how that pattern you just discussed applies to all numbers. So, let's summarize our second rule from the warm-up. Anything to the 0 power is 1. Hunter, who gets your card, then?**

Hunter gives his card to Benny.

Benny, you get to place 2^{-1}.

Benny hangs his card on the clothesline instead of pinning it to a benchmark (see Figure 8.30).

Figure 8.30 Benny's Placement

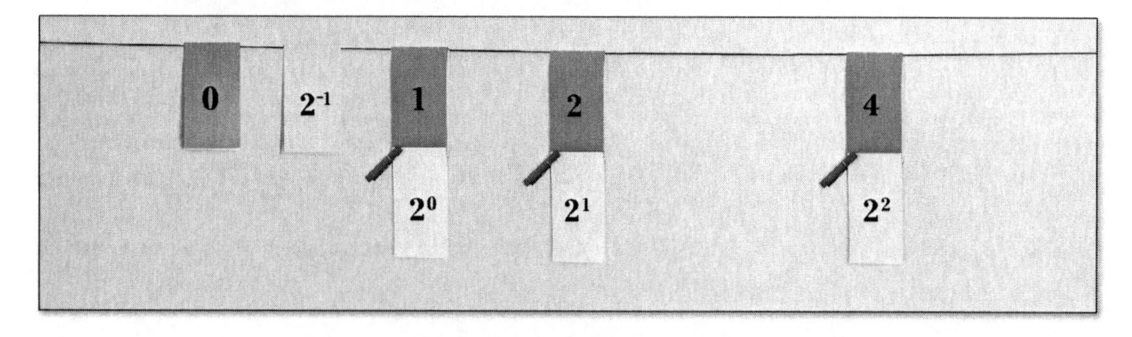

Teacher: **Benny, why did you place it at $\frac{1}{2}$?**

Benny: Well, at first I thought -2, but then I thought about what we just talked about. If you continue the pattern, then we should take half of 1.

Teacher: **Thank you. Let's continue this pattern one more time. Chelsea, you are up again. This time, you get 2^{-2}.**

Chelsea places her card to the left of 2^{-1} (see Figure 8.31).

Figure 8.31 Chelsea's Placement

Teacher: **Chelsea, where did you place it and why?**

Chelsea: According to the pattern, we should just keep cutting in half. Half of half is $\frac{1}{4}$.

Teacher: **Thank you, Chelsea. Let's summarize what we know then about negative exponents. Two to the -1 power is $\frac{1}{2^1}$. Two to the -2 power is equal to $\frac{1}{2^2}$. Therefore, the expression x to the negative a equals 1 divided by x to the positive a.**

Figure 8.32 Teacher's Equation

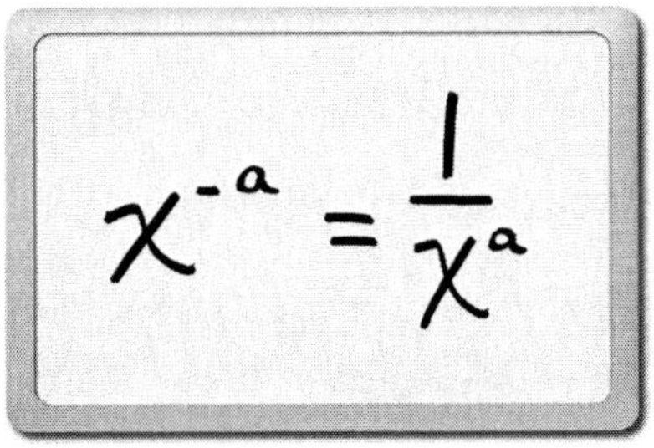

$$x^{-a} = \frac{1}{x^a}$$

Let's turn our attention now to rational exponents. We have learned that rational exponents have something to do with fractions and roots. So, let's do a quick review of those. Jantz, you get the next card, $\frac{3}{2}$.

Jantz places the card, as shown in Figure 8.33.

Figure 8.33 Jantz's Placement

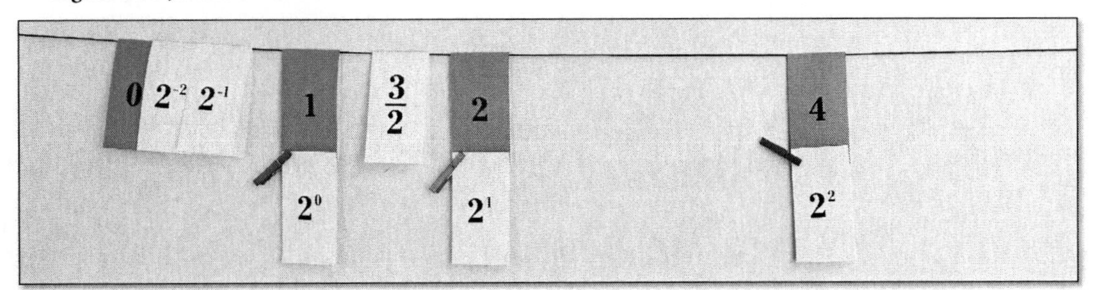

Teacher: Jantz, Can you tell us why you chose one and a half?

Jantz: Three divided by 2 is 1.5.

Teacher: Thank you. Elyssa, you're next. You get $\sqrt{42}$.

Elyssa: Can I squish everything?

Teacher: Why?

Elyssa: I need to place this at 21.

Teacher: Why?

Elyssa: Because 21 times....oh no...wait.

Elyssa places the card in a different location, as shown in Figure 8.34.

Figure 8.34 Elyssa's Placement

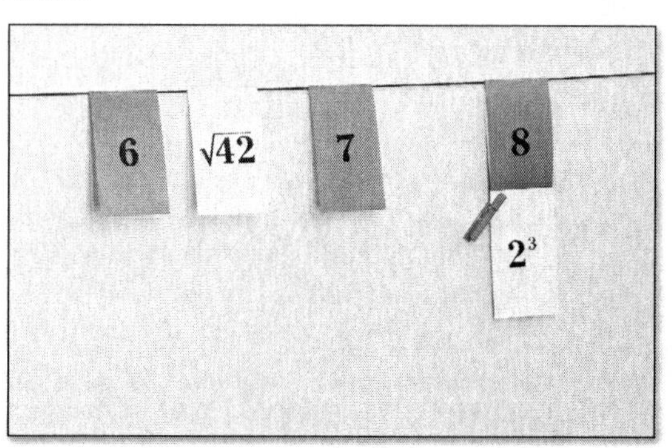

Teacher: You changed your mind mid-thought. Why?

Elyssa: Square root means what times itself. Seven times 7 is too much, 6 times 6 is not enough, so I went in between 6 and 7.

Teacher: Thank you, Elyssa. Let's stick with the roots to see whether the next person agrees with you. Avery, you get $\sqrt[3]{8}$ (see Figure 8.35).

Figure 8.35 Avery's Placement

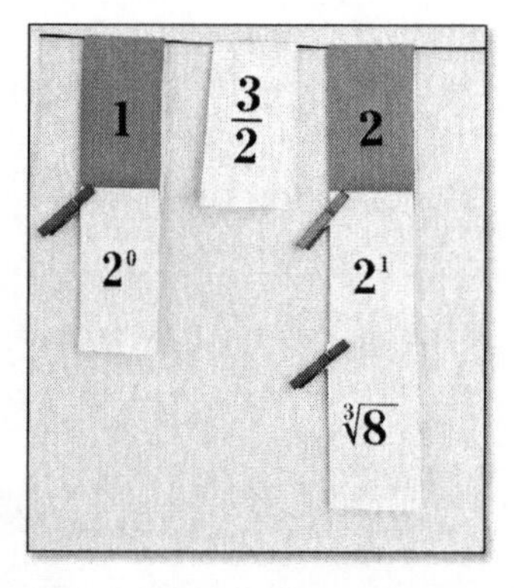

You are claiming that $\sqrt[3]{8} = 2$. Why?

Avery: We need to figure out what times itself three times equals 8. Since 2 times 2 times 2 equals 8, the cube root of 8 equals 2.

Teacher: Thank you, Avery. Now that we understand rational numbers, otherwise known as fractions, and also understand square roots and cube roots. Let's tackle rational exponents. Jemison, you get $2^{\frac{1}{2}}$.

Jemison pins the card to $\frac{3}{2}$ (see Figure 8.36).

Figure 8.36 Jemison's Placement

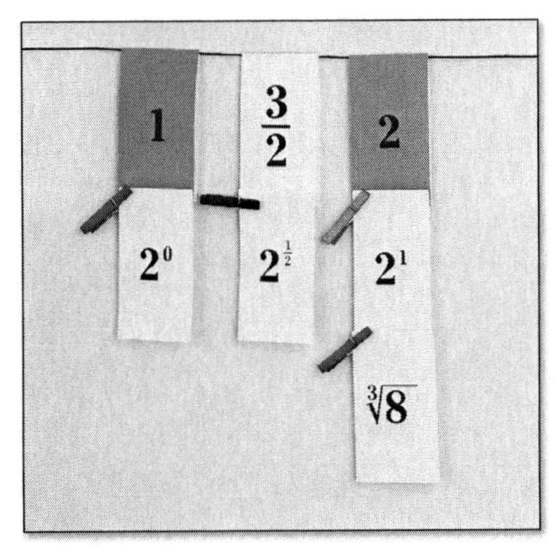

Jemison. Why 1.5?

Jemison: Because $2^1 = 2$ and $2^0 = 1$, so $2^{\frac{1}{2}}$ is halfway between those two.

Teacher: Interesting logic, thank you.

Jemison: Is it correct logic?

Teacher: Let's have the next person decide. Avery, back to you—you get $4^{\frac{3}{2}}$.

Avery moves $2^{\frac{1}{2}}$, as shown in Figure 8.37.

Figure 8.37 Avery's New Placement

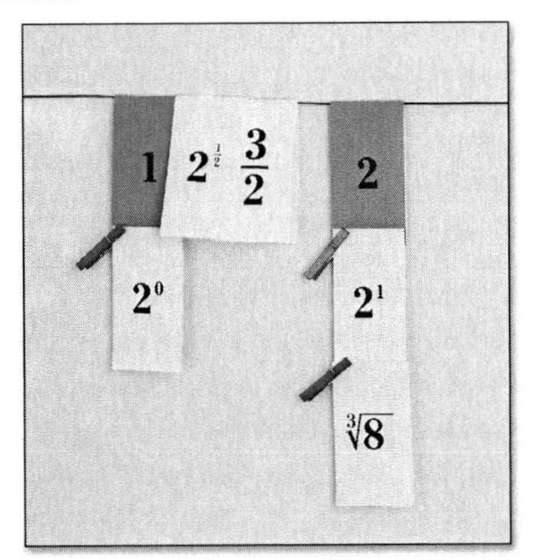

Teacher: Avery, you adjusted the $2^{\frac{1}{2}}$ ever so slightly.

Avery: I remember that the square root of 2 is about 1.4, and I think 1.5 times 1.5 is greater than 2, so I only adjusted it a little bit down.

Teacher: Thank you. Who gets $4^{\frac{3}{2}}$ then?

Avery selects Lucas.

Lucas, it's your turn. I think your classmates will need extra time on their lapboards for this one, so you will have plenty of time to think this one through.

Lucas pins his card to 9 (see Figure 8.38).

Figure 8.38 Lucas's Placement

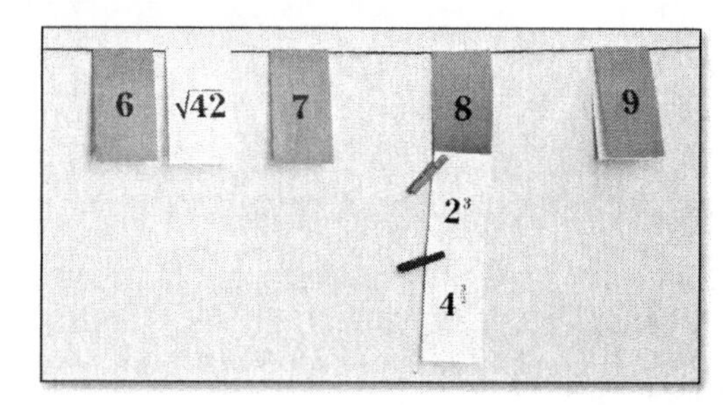

Why 9, Lucas?

Lucas: I used the same logic as before. Three halves is 1.5, so $4^{1.5}$ should be between 4^1 which is 4 and 4^2 which is 16. In between is 10, but we went a little less on the last one, so I fudged it down.

Teacher: I like that verb, *fudge*. Thank you. I'm going to pause things right here and say that Lucas has some strong reasoning and his estimation is very close. Does anyone remember a way that we can be even more accurate? Kat.

Kat: We can think of $4^{\frac{3}{2}}$ as $4^{\frac{1}{2}}$ first which is the same as the square root of 4, which is 2. Then, we raise 2 to the third power.

Teacher: What would that equal?

Kat: Eight.

Teacher: So, like I said, Lucas was very close with his estimation. I'm curious, though, whether Kat's reasoning would hold true if I reversed the order of the exponents. In other words, if I raised 4 to the third power, then took the square root, would I get the same number? Take a minute to calculate that mentally.

Now, all who think that we get the same value, thumbs-up. If not, thumbs-down.

Students vote thumbs-up unanimously.

Teacher: Leon, how did you calculate this?

Leon: Four cubed is 64. The square root of 64 is 8. It's the same both ways. Let me show you.

Leon pins the card to 8 instead (see Figure 8.39).

Figure 8.39 Leon's Placement

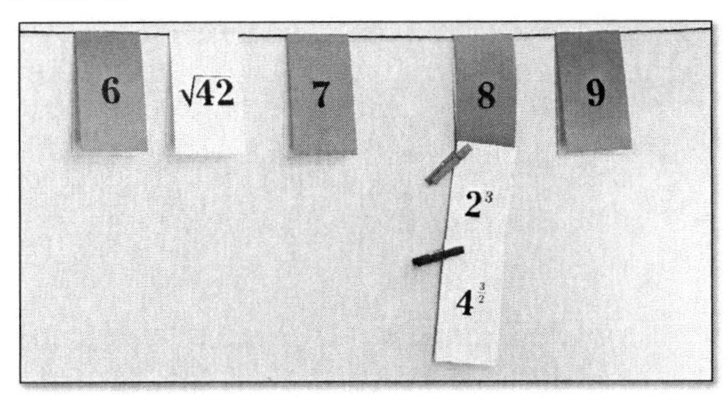

Teacher: Before we conclude, we need to summarize what we know about rational exponents, which are exponents that are fractions. You are all claiming that an exponent of $\frac{1}{2}$ is a square root, which is correct. In fact, an exponent of $\frac{1}{3}$ similarly means cube root. In other words:

The teacher writes the following information on the board (see Figure 8.40):

Figure 8.40 Rational Exponent with a Numerator of 1

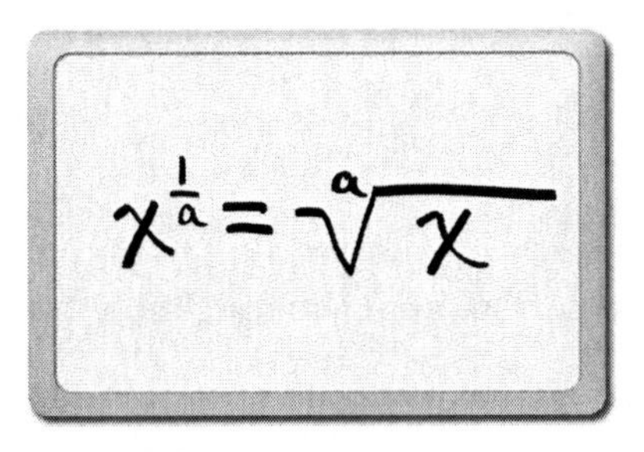

You are also saying that:

The teacher writes the following information on the board (see Figure 8.41):

Figure 8.41 Rational Exponent, More General

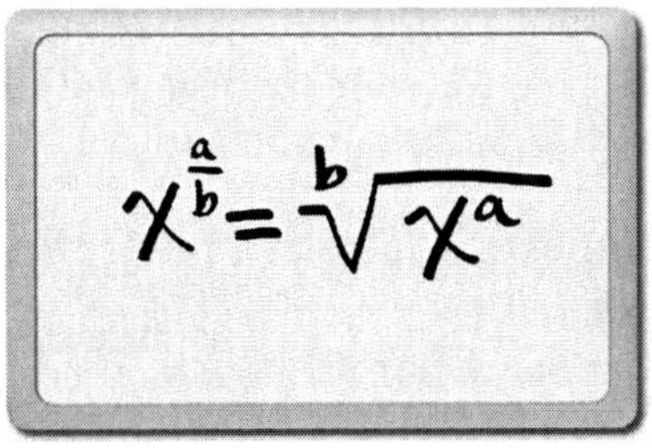

$$x^{\frac{a}{b}} = \sqrt[b]{x^a}$$

Teacher: **All in favor, say "aye."**

Class: Aye!

Teacher: **Then, let's write these rules and take the first three volunteers to put up our last three values, which are $8^{-\frac{1}{3}}$, 5^{-2}, and $16^{\frac{1}{4}} + 32^{\frac{1}{5}}$.**

Three students volunteer and place the final cards, as shown in Figure 8.42.

Figure 8.42 Final Cards Placed

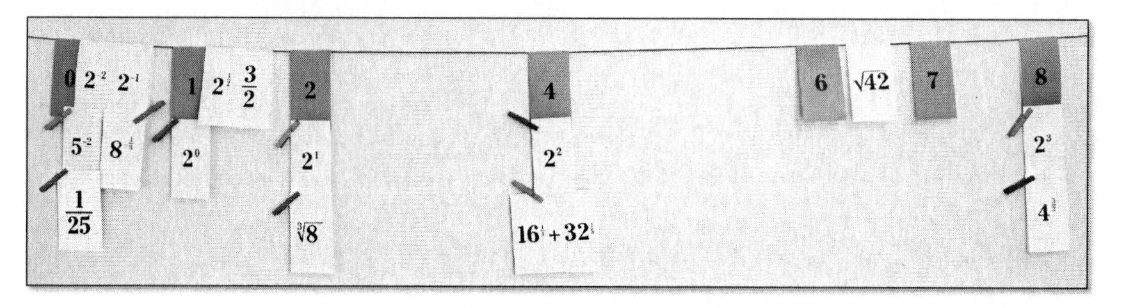

Our volunteers are claiming that $8^{-\frac{1}{3}} = \frac{1}{2}$, and that $5^{-2} = \frac{1}{25}$, and this beast $16^{\frac{1}{4}} + 32^{\frac{1}{5}} = 4$. Everyone double check our rules we revisited today, and I will take any challengers.

Al, do you agree?

Al: Yes.

Teacher: **How did you get 4 for this last one $16^{\frac{1}{4}} + 32^{\frac{1}{5}}$.**

Al: I don't know.

Teacher: **Did you just say "yes" hoping I would leave you alone and move on?**

Al: Pretty much.

Teacher: **You get points for being honest, but let's think our way through it.**

Al: I can do it. I just didn't.

Teacher: **Please, then. Share.**

Al: The fourth root of 16 is 2, because 2 raised to the fourth power is 16, and times 2 again would give you 32, so the fifth root of 32 is also 2. Two plus 2 is 4. So yes, I agree.

Teacher: **Thank you for joining in. I think everyone benefitted from your explanation. Ursula, can you explain how this person got $5^{-2} = \frac{1}{25}$?**

Ursula: The negative exponent means 1 over that exponent, so it should be 1 over 5 squared.

Teacher: Okay, then 1 divided by 5 squared. I see how that equals $\frac{1}{25}$. Luis, how about our last one? Can you explain why this person is claiming $8^{-\frac{1}{3}} = \frac{1}{2}$?

Luis: 8 to the $\frac{1}{3}$ is the same as the cube root, which is 2. The negative exponent means you put it on the bottom.

Teacher: Bottom of what?

Luis: In the denominator.

Teacher: So are you saying I'm supposed to divide 1 by the cube root of 8?

Luis: Not really like that, but yes, that is what you do.

Teacher: Thank you. Class, let's record these. You may do so all on one number line, or you may break it up on different number lines. Make sure you also record the rules of exponents. Those were very serious mental muscles you flexed today, everyone. Nice job.

Why These Numbers?

These numbers were chosen to build a progression. Knowing this was a review and that most students knew them but some still did not, I started with the most basic definition of an exponent and built ideas from there. Since I was revisiting the ideas of why a 0 power yields 1 and a negative exponent yields a multiplicative inverse, I chose base 2 for the sake of ease of computation. The value of three-halves and the two examples of roots were used to ease the work with rational exponents. The last three examples were similar to those students would see on the upcoming assessment, so I used them for the final discussion of the day.

The Key Questions

- Why did you choose that value?
- Why do you think this person chose this value?
- Do you mean 2 times itself or 2 times the exponent?
- Would I get the same value if I computed the exponents in the reverse order?
- Does anyone remember a way that we can be even more accurate?

Students often remember the rules of exponents in isolation but get baffled when asked to recall them in a mixed exercise. Therefore, while I knew this was a review of units, I anticipated that some students would still struggle, which turned out to be true. I wanted some scaffolding on the abstract rules, so I started with the rules and allowed students to correct their mistakes as we progressed through the lesson. This also required me to pause to restate the abstract rules after we played with the numbers. The pattern of watching an exponential expression, with base-2 getting cut in half at each successive decrement of the exponent, was a solid visual for the students. It was also a nice review of a zero exponent and negative exponents.

The intent of placing $\frac{3}{2}$ on the number line was to help below-level students who still had trouble understanding that $\frac{3}{2}$ is the mixed number of $1\frac{1}{2}$.

I was impressed with student reasoning, placing $4^{1.5}$ between 4^1 and 4^2. I wanted to honor their thinking while still applying the more accurate procedure of finding the root before raising to the exponent, as well as the procedure of raising to the power first then taking the root. What a treat though! The lesson was dragging on with the number of cards, so I spontaneously had the last three placed simultaneously and had other students explain the potential reasoning behind it. It turned out to be a productive engagement technique that will be good to keep in mind for the future.

The cards for this lesson are available in the Digital Resources (lesson20.pdf).

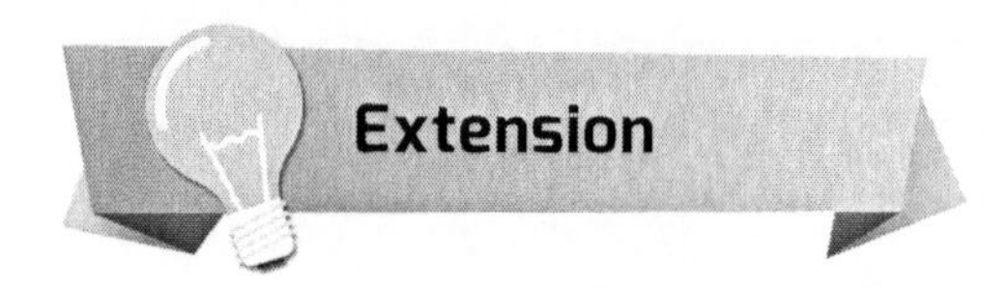

Place 2^3 and $\frac{1}{2}^{-3}$ on the number line, and explain why they are equivalent.

Lesson 21: Rational Expressions

x	$\frac{1}{2}x$	$\frac{x}{2}$	$\frac{1}{2x}$

Objective: Discern the difference between multiplying $\frac{1}{2}$ by x and dividing 1 by $2x$.

Teacher: **Good afternoon, class. I hope that you are ready to do some thinking today. I noticed that some of you have been writing $\frac{1}{2}x$ as $\frac{1}{2x}$. You have figured out that these two expressions graph differently when we enter them in the graphing app. So, I thought it would be worthwhile today to warm-up with a Clothesline activity involving these values. We will actually investigate the three you see up on the screen. Read them after me.**

One-half x.

Class: One-half x!

Teacher: **x divided by 2.**

Class: x divided by 2!

Teacher: **One divided by $2x$.**

Class: One divided by $2x$!

Teacher: **Let's investigate what the order of operations looks like for each. I'm going to write it on the board** (see Figure 8.43).

Figure 8.43 Teacher Equations

$$\frac{1}{2}x = (1 \div 2) \cdot x$$
$$\frac{x}{2} = x \div 2$$
$$\frac{1}{2x} = 1 \div (2 \cdot x)$$

Teacher: **I want to draw your particular attention to the first and last equations, here. What is the difference in the order of operations of these two expressions? Philippe.**

Philippe: The first one says that you divide first. The third one says that you multiply first.

Teacher: Interesting. Let's have a thumb vote to Philippe's statement.

Students vote thumbs-up in support of Philippe's answer.

Philippe is correct. One expression asks us to divide before we multiply; the other asks us to multiply before we divide. I wonder whether that makes a difference. If you think that will give the same value, raise your hand.

A few students raise their hands.

Thank you, hands down. All of those who think that changing the order of operations with the parentheses will yield two different values, raise your hand.

Many more students raise their hands.

Okay, these are the results of our voting.

The teacher writes the following on the board (see Figure 8.44):

Figure 8.44 Voting Results

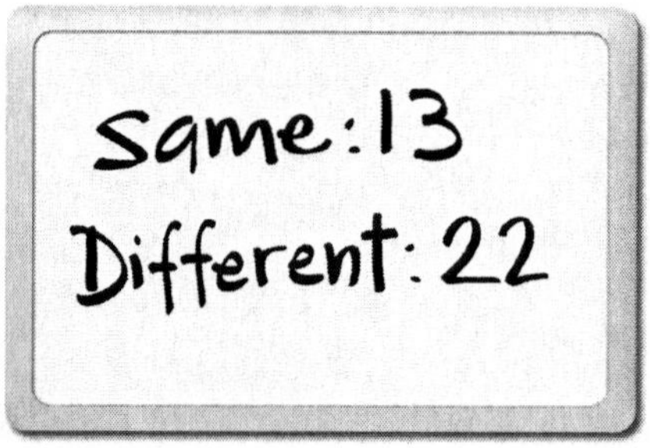

same: 13
Different: 22

Now, for x over 2. I wonder if it gives us the same value as the first one, the third one, or both. In other words, are all three all the same, or neither? Let's vote by a show of hands.

Students vote, and the teacher tallies the following totals on the board (see Figure 8.45):

Figure 8.45 Tallies

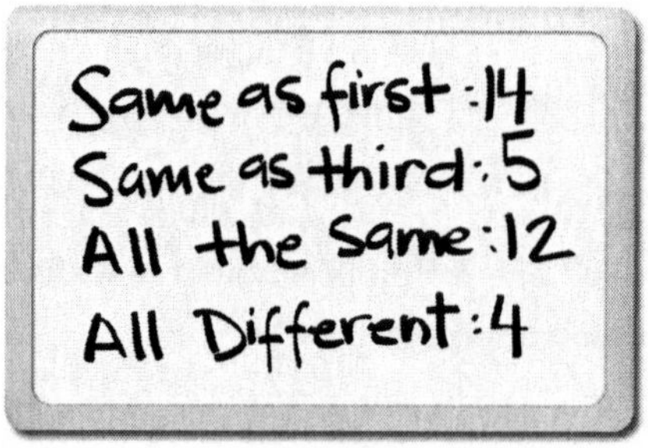

Same as first: 14
Same as third: 5
All the Same: 12
All Different: 4

Teacher: Let's let the Clothesline settle the argument. I'll start by placing x here at 1. I would like Haley's group to show us where our three expressions go (see Figure 8.46). The rest of you will show placement on your lapboards with an elbow partner.

Figure 8.46 Haley's Placement

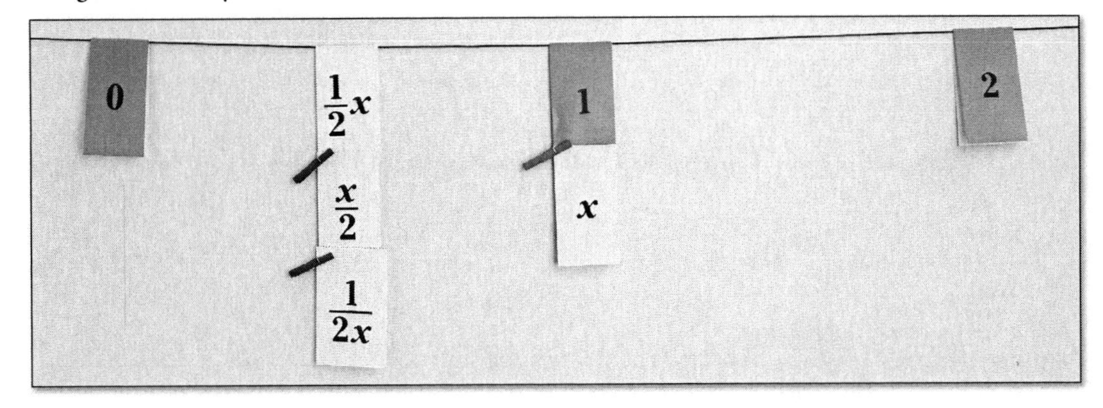

Haley's group is claiming these are all equivalent. Any challengers? None? So far, the votes for all three being the same look strong. Haley, who gets to place the same three expressions for another value of x?

Haley chooses Chin.

Chin, your group gets these same three, but now I'm going to pin x at 2. Your group, up here. Everyone else, work on your boards.

Chin's group pins three cards to 1 (see Figure 8.47).

Figure 8.47 Chin's Placement

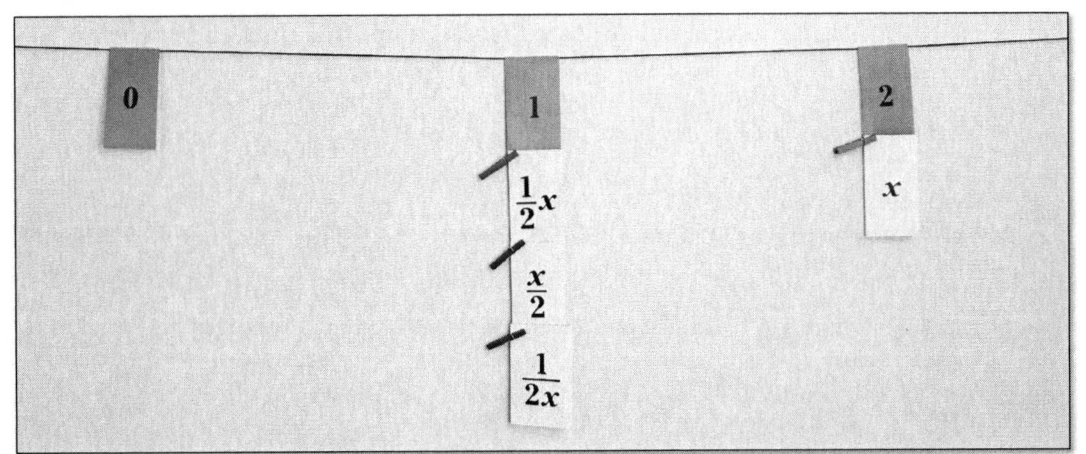

Chin's group is claiming these are all equivalent to 1. Any challengers?

Irene raises her hand.

Irene.

Irene: The 1 over $2x$ should be at $\frac{1}{4}$.

Teacher: Explain why you think 1 divided by $2x$ is different from $\frac{1}{2}x$.

Irene: Like you showed us up there, we need to multiply 2 times x first. Then, divide 1 by that.

Teacher: Let's see whether Chin agrees with your challenge. Chin?

Chin: Yes, I agree. $\frac{1}{4}$.

Teacher: Class, if you are in favor of moving this to $\frac{1}{4}$, please raise your hand.

Students all raise their hands, and the teacher moves the card (see Figure 8.48).

Figure 8.48 Teacher's Placement

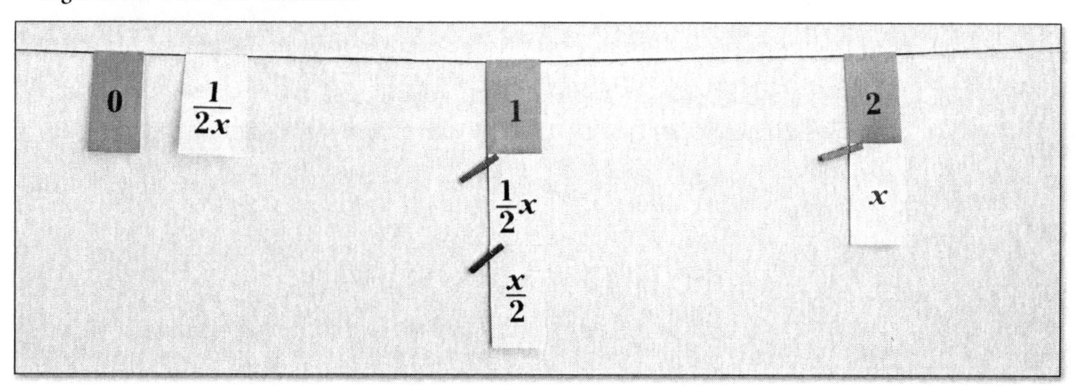

Chin, you get to pick the next group.

Chin chooses Irene's group.

Nice. Give it right back to Irene's group. This time, the value of x is $\frac{1}{2}$. Everyone else, work on your lapboards.

Irene places the card, as shown in Figure 8.49.

Figure 8.49 Irene's Placement

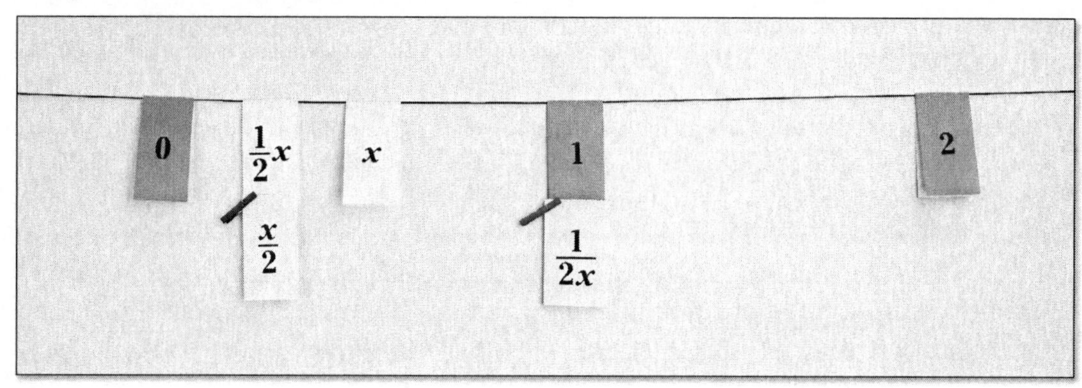

Teacher: **Whoa, Irene, your group just changed everything on us. You switched the places. Any challengers?**

Dustin raises his hand.

Dustin.

Dustin: Not a challenge. I just don't get it.

Teacher: **Get what?**

Dustin: Why is $\frac{1}{2x}$ at 1 instead of $\frac{1}{4}$?

Teacher: **Great question. Let's substitute and evaluate. Remember that in this expression, we are being asked to do the operations in the denominator first. $2(\frac{1}{2})$.**

Dustin: Oh…I see it now. Thank you.

Teacher: **See what?**

Dustin: That will be 1 over 1.

Teacher: **Let's vote again.**

Students vote again, and the teacher writes the totals on the board as follows (see Figure 8.50):

Figure 8.50 New Votes

Wow. We have consensus in less than 10 minutes. Fine work, crew. I want to record these three values for x = 1 on one number line, $x = 2$ on another, and $x = \frac{1}{2}$ on a third. Then, pick a fourth value on a fourth number line, and place the three expressions. Once both are done, have your partner check your answers.

Why These Numbers?

The values of $\frac{1}{2}x$ and $\frac{1}{2x}$ and similar choices are often misinterpreted as equivalent expressions. Students also have a difficult time understanding that $\frac{1}{2}x$ is equivalent to $\frac{x}{2}$. So, we are really discerning between having x in the numerator versus x in the denominator. This lesson can be used in an Algebra 1 course when students are miswriting $\frac{1}{2}x$ for a slope of $\frac{1}{2}$, or in Algebra 2 class, when students are discerning the difference between linear and rational functions but need to properly evaluate these expressions before plotting the values.

The Key Questions

- What is the difference in the order of operations of these two expressions?
- Does the order of operations matter?
- Will these yield the same values?
- Which of these expressions are equivalent?
- Are there any challengers?

Analysis

One of the noticeable differences between this activity and how we typically conduct the Clothesline is the fact that I did not ask students who placed the cards to explain their placements. I simply asked for challengers. This was for the sake of simplicity and time. The Clothesline lesson was meant to be a warm-up activity, so I did not want to bog down the discussion with a triple explanation each time.

I also took a conjecture vote. This is a very effective engagement technique in which the teacher has students vote on the outcome, then tests the conjecture publicly. Most of the time, students couldn't care less about what the person with the college degree says, but they want to know whether they are correct and whether their friends are wrong. When this is first done with a class, it is depressing how no one substitutes a number before they vote. They simply guess. But, with consistent use of this technique, students begin to test their hypotheses before they declare their opinions.

I also spent some brief, but critical, time demonstrating how to read the notation and how to apply the order of operations. This is the root cause of many misconceptions. The number line activity is just a reinforcement of these two critical ideas.

The cards for this lesson are available in the Digital Resources (lesson21.pdf).

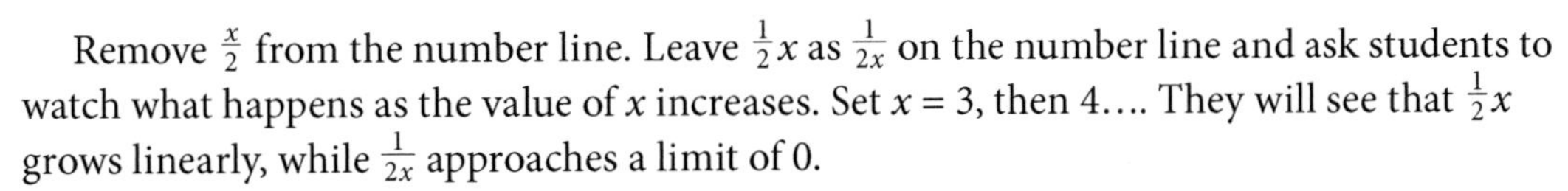

Extension

Remove $\frac{x}{2}$ from the number line. Leave $\frac{1}{2}x$ as $\frac{1}{2x}$ on the number line and ask students to watch what happens as the value of x increases. Set $x = 3$, then 4.... They will see that $\frac{1}{2}x$ grows linearly, while $\frac{1}{2x}$ approaches a limit of 0.

Geometry

Lesson 22: Ratio and Proportion

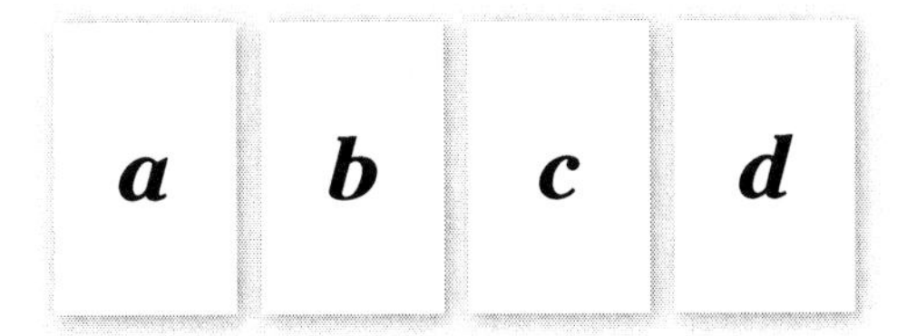

Objective: Determine two values that have the same ratio as two given values.

Teacher: **Hello, all! We have studied ratios and proportions. Today, let's warm-up with an activity that will test your understanding of ratios. Take a look at the number line I have written on the board** (see Figure 9.1).

Figure 9.1 Number Line

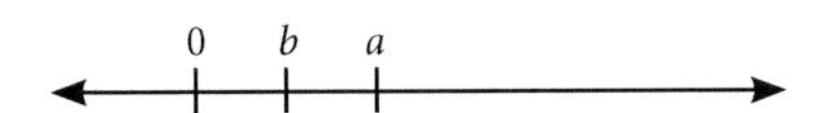

Now, take a look at the ratio I have waiting for you on the clothesline. Your task is to place *c* and *d* on the number line so the ratio of *c* to *d* is equal to the ratio of *a* to *b*. The value of *a* may not equal *c*. Jacob, you choose our first group.

Jacob selects Kenya's group.

Kenya, your group is coming to the front. Everyone else, work on your lapboards.

Kenya's group places the variable cards on the clothesline, as shown in Figure 9.2.

Figure 9.2 Kenya's Placement

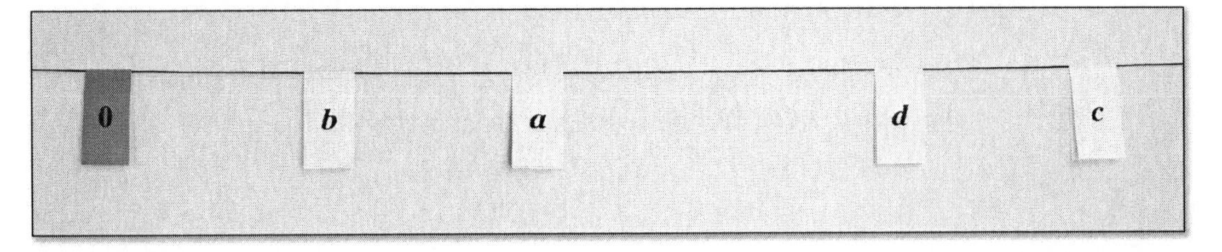

Teacher: Declare your boards.

Students hold up boards, and most do not match the clothesline.

Kenya, we have a great deal of disagreement with your group's response. Please explain the reasoning behind your placement of your values.

Kenya: Since the ratio has to be the same, we made the distance between c and d the same as the distance between a and b. Is that wrong?

Teacher: Let's hear some other arguments and let the class decide. Victor, do you and your partner disagree with this?

Victor: Yes, we estimated that a is twice as large as b, so c should be twice as large as d.

Teacher: So, what ratio are you seeing?

Victor: Two to one?

Teacher: Two to one. Will you show us what you mean?

Victor's group adjusts the cards on the clothesline (see Figure 9.3).

Figure 9.3 Victor's Placement

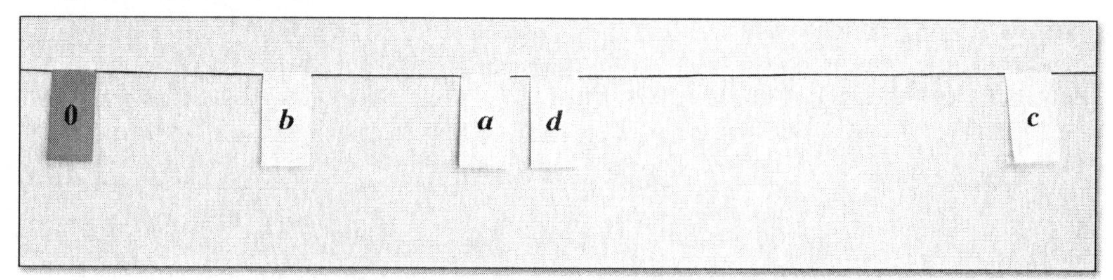

Kenya, what do you think of Victor's response?

Kenya: I agree.

Teacher: What changed your mind?

Kenya: His shows a same ratio. Ours shows a same distance.

Teacher: Class, if you agree with this response, thumbs-up. If you agree with the first response, thumbs-down.

The class gives a unanimous thumbs-up.

Let's check everyone's understanding of this. What if I moved a and b to here, and placed c here (see Figure 9.4).

Figure 9.4 Teacher Adjustment

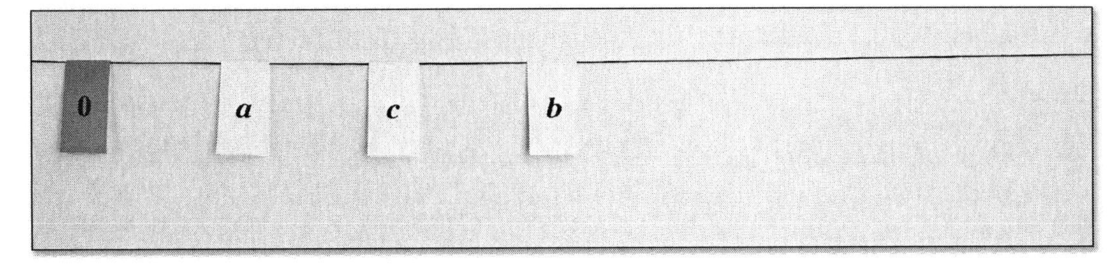

Teacher: **Where should we place *d*? Victor, please choose our next group.**

Victor selects Heather's group.

Heather, your crew is next. Class, work on your lapboards, please.

Heather's group adds d *to the clothesline (see Figure 9.5).*

Figure 9.5 Heather's Placement

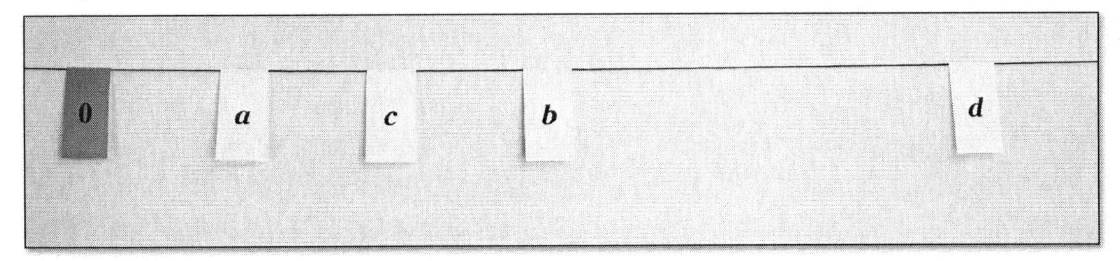

Class, declare your boards.

Students all agree.

Heather, the class agrees with you. Joseph, can you explain why you agree?

Joseph: It looks like *a* is about one third of *b*, so *d* has to be about three times as big as *c*.

Teacher: **Class, that was a very brave conversation. Well done on sharing so openly today. Let's record both our examples on the activity sheet. Be sure to make any notes about ratio that you may have learned.**

Why These Numbers?

The strength of this lesson is the opportunity it provides for students to see a proportion. A traditional proportion such as $\frac{1}{2} = \frac{2}{4}$ is a far different visual than the following image in Figure 9.6.

Figure 9.6 Ratio Clothesline—Why

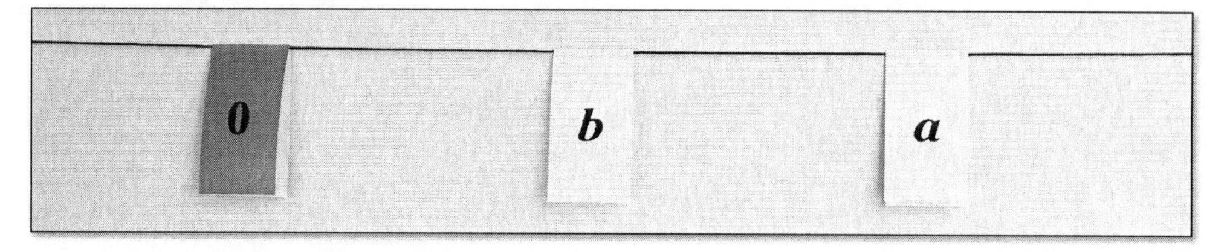

Of course, the algebraic representation is more easily manipulatable and, thus, more useful than the visual number line when applying proportions. However, when learning proportions the visual is superior. To start, the ratio of 2 to 1 was intentionally chosen, because it is the easiest proportion to visually estimate. The second example was planned to follow a ratio larger than 1 with a ratio less than 1.

The Key Questions

- What was your reasoning?
- What ratio are you seeing?
- Why do you disagree?
- Why do you agree?
- What changed your mind?

Analysis

Fortunately, for the purpose of class discourse, the anticipated incorrect response occurred first. Had the first group answered correctly, I would have called up the student with the distance response to contrast the answers and have the class vote. Having students choose between these two lines of thinking helps solidify the concept of ratio.

Since the first ratio was the simplest of all ratios (*2 to 1*), I intentionally posed the second ratio as less than 1 (*1 to 3*) so that students were required to think of scale factors that both enlarge and reduce (*dilations versus contradictions*).

The cards for this lesson are available in the Digital Resources (lesson22.pdf).

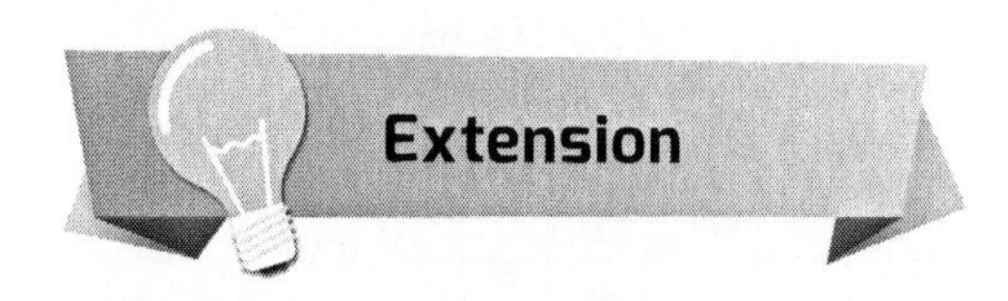

Extension

Place d such that $\frac{a}{b} = \frac{c}{d}$, and $c < a < b$.

Lesson 23: Linear Pairs

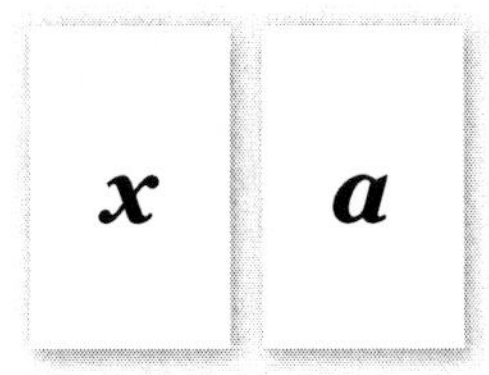

Objective: Place the measures of two angles on the number line to show the supplementary relationship of linear pairs.

Teacher: **Hello, class. Today, we are going to investigate the relationship of linear pairs that we defined yesterday. Take a look at this diagram of a linear pair** (see Figure 9.7).

Figure 9.7 Linear Pair

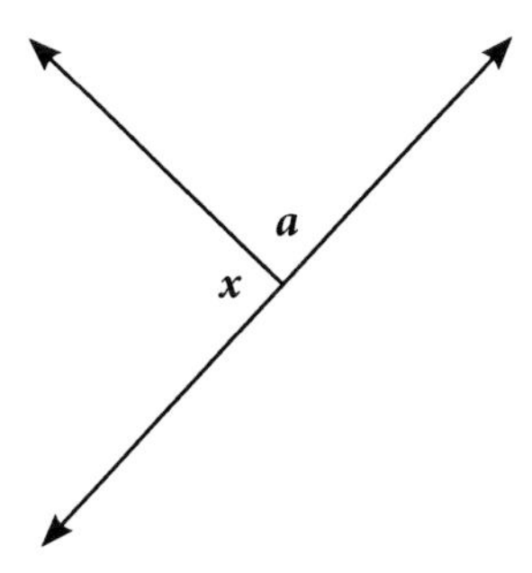

The variable x represents the measure of one angle, while a represents the measure of the other. Notice that I have given you the benchmarks of 0° and 180° on our clothesline. I am going to place x on the number line (see Figure 9.8).

Figure 9.8 Linear Pair on the Clothesline

Now, tell me where you think the value of a goes. If x is the measure of one angle of a linear pair, what is the measure of the other angle? Lamar, your group gets to come up to the clothesline today. Everyone else, draw the number line on a lapboard with your elbow partner.

Lamar's group goes to the clothesline.

Teacher: Place the benchmarks 0°, 180°, and *x* approximately where I did. Then, place *a* on the number line so it represents the relationship these two angle measures have with each other.

Lamar's group places their card, as shown in Figure 9.9.

Figure 9.9 Lamar's Placement

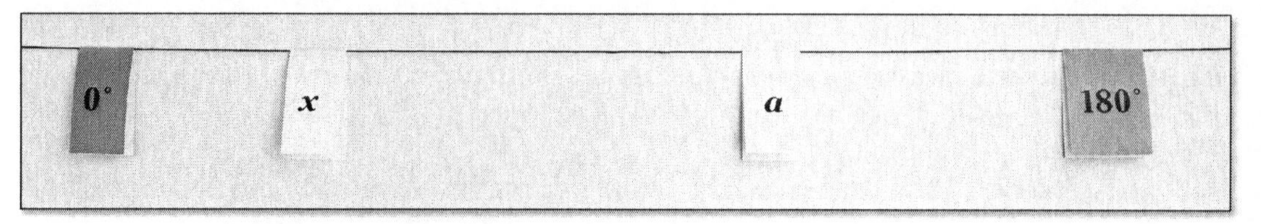

Class, declare your answers.

Student pairs show their lapboards, revealing that the entire class placed a as an obtuse angle. Several, but not all, had it also correctly spaced so that x *and* a *showed a supplementary relationship.*

Karla and Hank, you two agree with what is shown on the clothesline. Please tell us why.

Karla: The x° angle is less than 90°, so *a* has to be more than 90°.

Teacher: Can anyone tell us the mathematical terms for angles that measure less than 90° and more than 90°? Denise?

Denise: *Acute* is "less than 90°," and *obtuse* is "more than 90°."

Teacher: Thumb vote, class. I placed *x* as an acute angle. Do you agree with Karla and Hank that *a* must be obtuse?

Students give a unanimous thumbs-up.

Everyone agrees with you that *a* must be obtuse. Freda and Gabriel, you have placed *a* as an obtuse angle, but you spaced it differently than what is shown here. Will one of you please go to the clothesline and move the value of *a* to reflect your response? Class, watch carefully to see where they move it.

Gabriel moves the value of a *closer to 180°. With his hands, he visibly measures the distance from 0° to* x*, and copies that distance from 180° backward to* a *(see Figure 9.10).*

Figure 9.10 Gabriel's Placement

Teacher: Freda, please tell us why you and Gabriel spaced a there.

Freda: We know that x and a are 180°…

Teacher: They are each 180°?

Freda: Together they are 180°.

Teacher: Together how?

Freda: x and a have a sum of 180°.

Teacher: Thank you! So, then why does that mean a has to go where you placed it?

Freda: Because 180° minus x will tell you what a is.

Teacher: Taylor, did you notice the hand reasoning Gabriel used?

Taylor: Yes.

Teacher: Why was he doing that?

Taylor: I don't know.

Teacher: Gabriel, will you show us your hand reasoning again? Taylor, follow along with finger reasoning from your seat.

Gabriel repeats his hand reasoning, as shown in Figure 9.11. He shows the distance from 0° to x *is the same as from* a *to 180°.*

Figure 9.11 Gabriel's Hand Reasoning

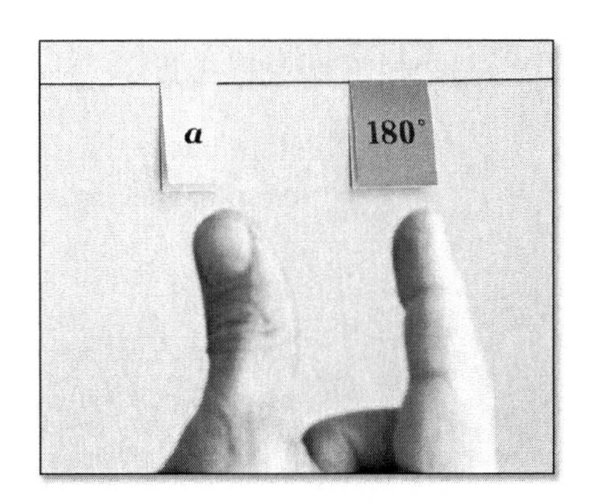

Can you now tell us why Gabriel did that?

Taylor: He is making sure that from 0° to x is the same as from a to 180°.

Teacher: Why?

Taylor: I don't know.

Teacher: Nadine, can you explain why Gabriel was so particular about making sure the distance from 0° to x is equal to the distance from a to 180°?

Nadine: Because 180° minus x equals a.

Teacher: Back to you, Taylor. Why is it important to make sure the distance from 0° to x is equal to the distance from a to 180°?

Taylor: Because 180° minus x equals a.

Teacher: What is the term for two angles that have a sum of 180°, Petra?

Petra: Supplementary.

Teacher: Again, what is the term for two angles that have a sum of 180°, Annie?

Annie: Supplementary.

Teacher: One last time, what is the term for two angles that have a sum of 180°, all of you?

Class: Supplementary!

Teacher: Nice. Did anyone else place $a°$ here but thought about it differently than subtracting x from 180°? Yes, Gina.

Gina: We measured from x all the way to 180°, then measured that distance from 0°.

Teacher: So, instead of $180 - x = a$, you thought of it as $x + a = 180$.

Both of these equations are written on the board.

Thumb vote. All in favor that these two equations are equivalent forms of the same relationship, thumbs-up.

The vote is unanimously thumbs-up.

Time to record on our *Clothesline Math* activity sheet. Be sure to write a note to yourself using the terms *linear pair* and *supplementary*.

Why These Numbers?

The key to this lesson is having students "see" a supplementary relationship on the number line. Since this lesson is an introduction to the relationships of linear pairs, the benchmarks of 0° and 180° are offered, as is the placement of x, which is intentionally made acute for the first placement.

The Key Questions

- What is the relationship between the angles of a linear pair?
- If one angle is acute, then must the other be obtuse?
- Is the spacing correct?
- Is showing that the angle is obtuse enough?
- Why did you space a that specific distance from 0° or from 180°?
- Can you explain why your classmate was so particular about the spacing of a?
- Did anyone else get the same answer in another way?
- If one angle is a right angle, what do we know about the other angle?
- Do you agree or disagree?

Analysis

The use of the mathematical process *Attend to Precision* pertains to more than just the accuracy of the placement of values on the number line, which is an important part of being precise, of course. It also applies to the use of mathematical language. Therefore, there was a strong push in this lesson to have students use mathematical terms accurately.

The lesson, however, focuses mostly on the visual tool of seeing that x and a have a sum of 180°. Therefore, in the first case, I pressed students to not just claim that a was obtuse but to space it accurately to display the supplementary relationship between the angles. That is why I emphasized the hand and finger reasoning with Gabriel and Taylor as well as intentionally using specific phrases, such as "place a on the number line so that it *represents the relationship* that these two angle measures have with each other."

The cards for this lesson are available in the Digital Resources (lesson23.pdf).

Extension

Start by placing x as an obtuse angle or as a right angle (see Figures 9.12–13).

Figure 9.12 Sample Vertical Angle

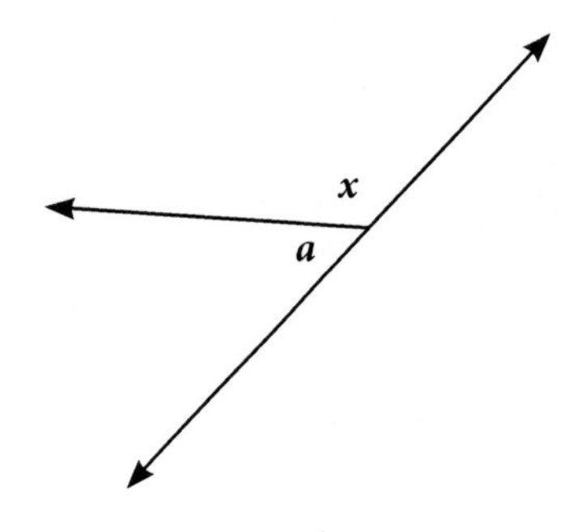

Figure 9.13 Sample Right Angle

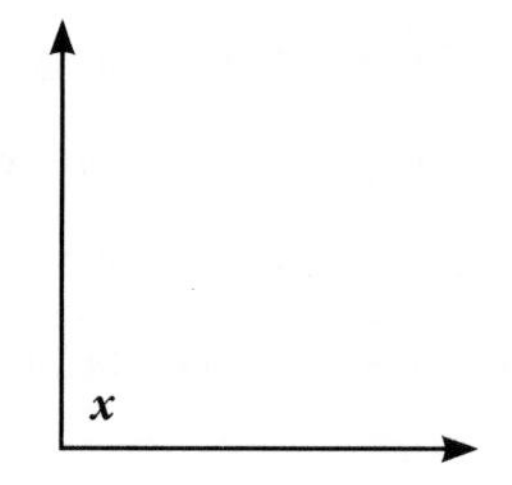

Lesson 24: Vertical Angles

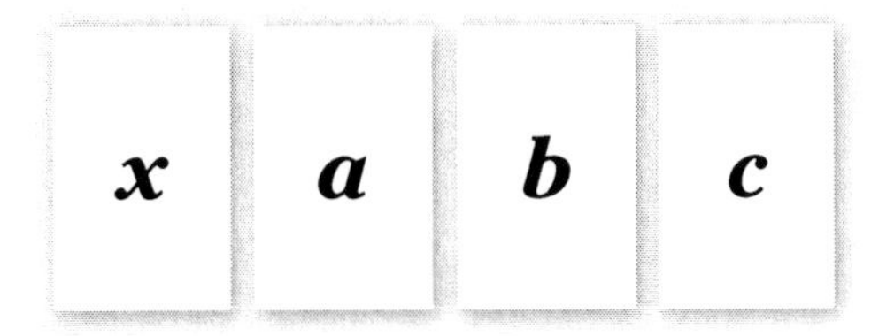

Objective: Place the measure of two angles on the number line to show the equivalent relationship of vertical angles and its logical extension from the supplementary relationship of linear pairs.

Teacher: **Good day, class. Today, we are going to take the relationships we have studied about linear pairs and vertical angles and explore them on an open number line. Take a look at the diagram of two intersecting lines** (see Figure 9.14).

Figure 9.14 Sample Vertical Angles

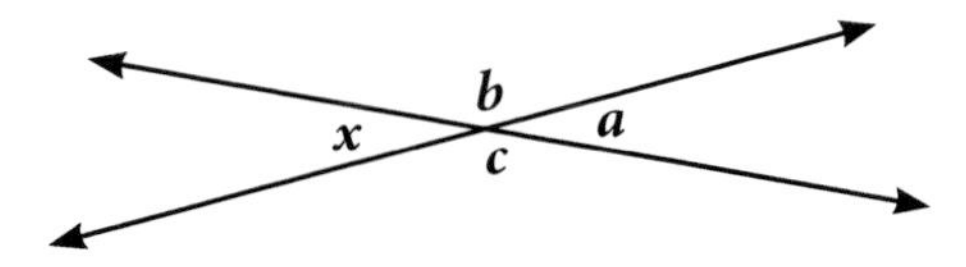

The variable x represents the measure of one angle, while a represents the measure of its vertical angle. You have the benchmarks of 0° and 180° on our clothesline. I am going to place x on the number line as well (see Figure 9.15).

Figure 9.15 Teacher Placement

Now, tell me where you think the value a goes. If x is the measure of one vertical angle, what is the measure of the other vertical angle? Ali, your group gets to come to the clothesline first. Elbow partners, draw your number line with benchmarks of 0° and 180° on a lapboard with the variables placed approximately where I have them. Then, decide where to place a on the number line so it represents the relationship these two angle measures have with each other.

Ali's group accurately pins a *to* x *(see Figure 9.16).*

Figure 9.16 Ali's Placement

Teacher: Everyone, declare your boards.

Students show their boards, but not everyone matches with what is on the clothesline.

Most everyone agrees, but not all. Ali, why did your group pin *a* to *x*?

Ali: Because they are equal.

Teacher: Why?

Ali: All vertical angles are equal.

Teacher: Shawn and Mike, you disagree with what is displayed here. Instead of being equal, you placed *a* closer to 180°. Why?

Shawn: We thought they were supplementary, but now we agree with Ali. They are equal, and *b* and *c* are supplementary to *x* instead.

Teacher: Since you changed your mind, your group gets to come to the clothesline next. Show us where you would place *b* and *c*. Everyone else, correct what you had before, if necessary, and place *b* and *c* on your number line.

Shawn's group at the clothesline accurately pins b *and* c *and is conscientious in the spacing, showing they understand these new angles must be supplementary to* a *and* x *(see Figure 9.17).*

Figure 9.17 Shawn's Placement

Teacher: **All in favor, two claps.**

The class gives a resounding two claps.

Class, what you see here is proof that vertical angles are equal. After all, if *b* is supplementary to *x*, then *b* must be placed here. If *c* is also supplementary to *x*, then it, too, must go here. Therefore, they must be equal. Does that work for all values of *x*? Let's see. What if I move *x* and *a* over here? Where must *b* and *c* go? Everyone, point in which direction I should move *b*. Clap once when I get there.

The class accurately directs the teacher to the new placement of b. *The teacher keeps* c *pinned to it (see Figure 9.18).*

Figure 9.18 Teacher Adjustment

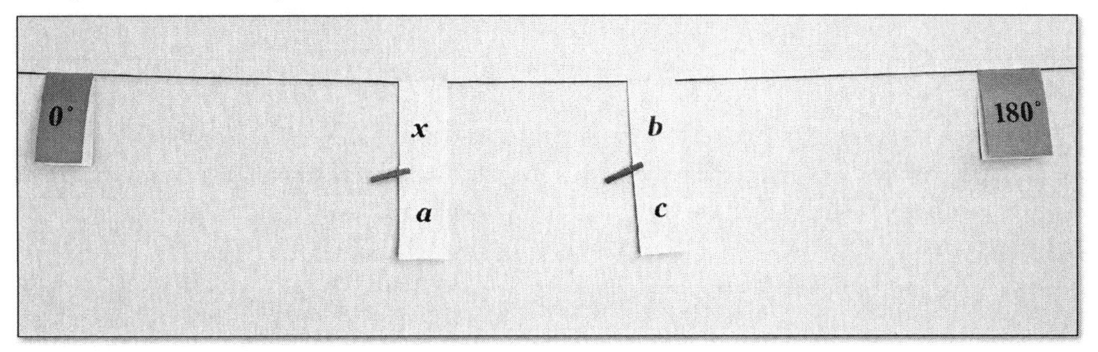

Should I keep *c* equal to *b*?

Class: Yes!

The teacher plays with a few more new placements of x. *Students are accurate with all of them.*

Teacher: **Okay, now let's record. Draw the diagram. Write *x*, *a*, *b*, and *c* as the value set. Then, record two different places on your sheet as 1 and 2. Be sure to note the relationship of vertical angles on your paper.**

Why These Numbers?

This lesson would be much shorter and simpler if only *x* and *a* were presented. That may serve as a preliminary lesson, but the linear pairs are intentionally included here to illuminate the proof that all vertical angles are equal given that linear pairs are supplementary.

- What is the relationship between any two vertical angles?
- How do you know the two angles are equal?
- What is the relationship between the angles of the linear pairs?
- Why do you disagree?
- Why are we pinning these values together?
- Would these relationships be true for any value of x?

This lesson was taught simulating some investigative work on vertical angles that was conducted previously. I expected students to know vertical angles are congruent. Therefore, I started with the placement of the vertical angle and corrected the few students who had misconceptions.

From there, I tied the relationship of the vertical angle to both of its adjacent angles, which also happen to be adjacent angles of x. This is the basis for the vertical angle theorem. Since $x + b = 180$ and $a + b = 180$, then $x + b = a + b$. Therefore, $x = a$.

The cards for this lesson are available in the Digital Resources (lesson24.pdf).

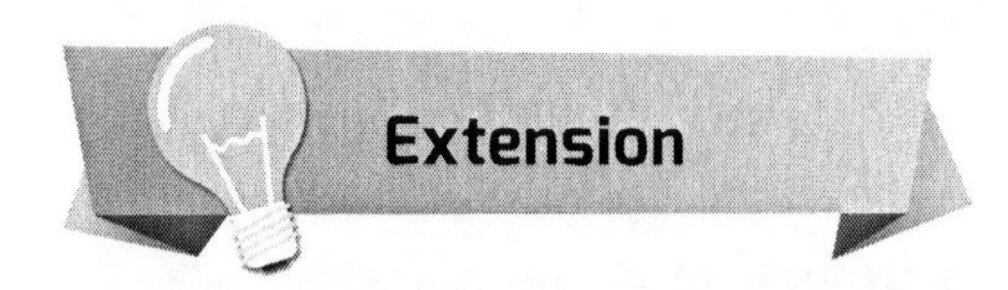

Challenge students to find the location for which the measure of all four angles are equal. Alternatively, write proofs for the vertical angle theorem, which states, "Two angles supplementary to the same angle are congruent."

Lesson 25: Transversals

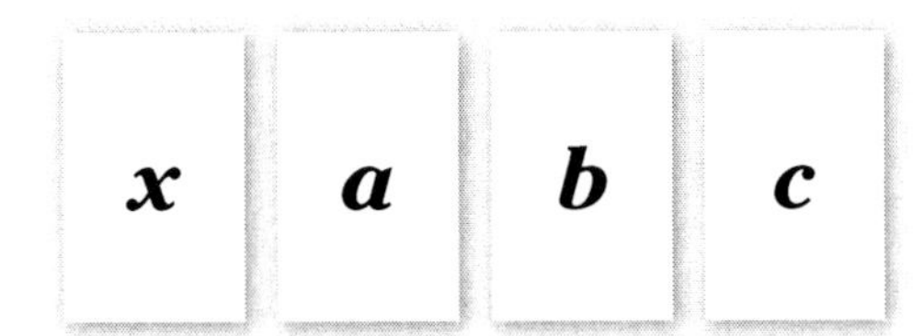

Objective: Place the measure of three angles on the number line to show the relationship of corresponding angles, consecutive angles, and alternate interior angles when two parallel lines are cut by a transversal.

Teacher: **Good day, class. During the last few days, we investigated the relationships of angles involved in parallel lines cut by a transversal and formalized our thinking with postulates and theorems** (see Figure 9.19).

Figure 9.19 Transversal

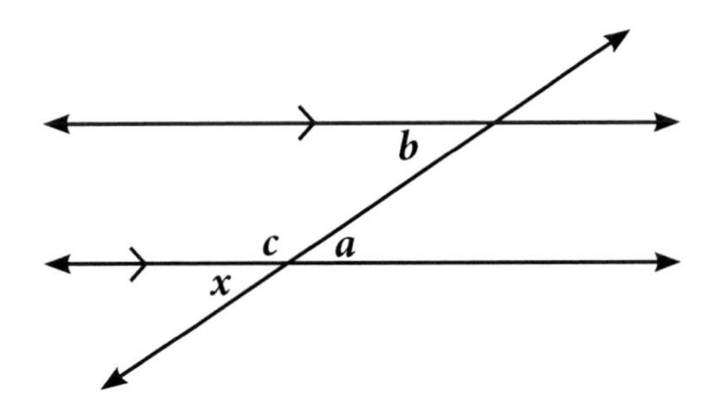

Today, we will take those postulates and theorems to the clothesline. I already placed x on the clothesline (see Figure 9.20).

Figure 9.20 Teacher's Placement

In reference to the given diagram, I would like our first group to place $b°$. The person to choose that group is the person with the most letters in their first name. Charlotte, that's you. Who do you pick?

Charlotte selects Austin's group.

Teacher: Austin, your group is at the clothesline. Everyone else, work with your partners on your lapboards.

Austin's group places b*, as shown in Figure 9.21.*

Figure 9.21 Austin's Placement

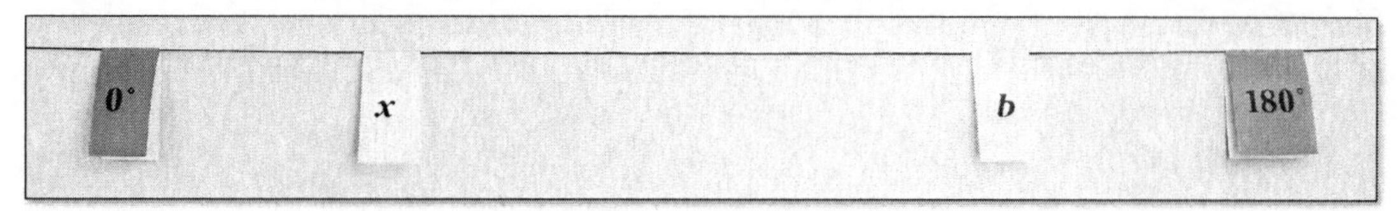

Everyone, declare your responses.

Students present their boards, showing two different answers: either x = b*, or* x *is supplementary to* b*.*

Interesting—you are split into two camps. Half agree with Austin, and the other half all gave the same alternate answer. Austin, let's start with your group's thinking.

Austin: Well, since *x* and *b* are on the same line, they are supplementary.

Teacher: Nice use of mathematical vocabulary. Thumbs-up if you agree with Austin's group, thumbs-down if you disagree.

The class is still split on the decision.

Sharlene, you and your partner disagree. Please show the class where you two think *b* should go.

Sharlene's group pins b *to* x*, as shown in Figure 9.22.*

Figure 9.22 Sharlene's Placement

A great deal of discussion erupts between partners.

Okay everyone, back together. I love when student thinking causes this type of disturbance. Thumb vote for this response.

Many voters now agree with the current solution, but some are still holding out.

Teacher: **Interesting again—many of you switched your vote. Karla, you originally agreed with Austin, but now you are thinking *x* and *b* are equal. Why?**

Karla: Corresponding angles are always equal.

Teacher: **Always?**

Karla: When lines are parallel.

Teacher: **How do you know? Have we proven this?**

Karla: Yes, the other day.

Teacher: **Brandon, you seem eager to say something.**

Brandon: We didn't prove that one; we assumed it.

Teacher: **And what do we call our formal assumptions in this class? Everyone.**

Class: Postulates!

Teacher: **Good. So, we don't have to prove it, but we do need to have reasoning to support it. I believe that is what Karla was thinking. Karla, back to you. Can you give some support for why we agreed to assume that corresponding angles are equal when lines are parallel?**

Karla: If you drag that whole top intersection down, it makes sense that the angles in the same parts of the two intersections lying on top of each other will be equal, but only if the lines are parallel. If you start moving one of the lines so they are not parallel, the angles with no longer be equal.

Teacher: **Is there anyone who wants to challenge that argument? No? All in favor, two claps.**

A unanimous two claps.

Karla, the class agrees with you, but I wonder how we know these lines are parallel.

Karla: The arrows.

Teacher: These at the end?

Karla: No, those extra arrows mean the lines are parallel.

Teacher: I see. Thank you. Now, I would like our next group to place a°. Austin, who is it going to be?

Austin selects Michael.

Michael, your group is up. Everyone else, adjust your answer for b, if necessary, then place a° on your lapboard's number line.

Michael's group pins a, *as shown in Figure 9.23.*

Figure 9.23 Michael's Placement

Class, what do you think? Now, this is looking weird. Michael's group says all three of those values are equal. Heidi, can you support this answer for us?

Heidi: Vertical angles are equal.

Teacher: Always? Or, only when lines are parallel?

Heidi: Always. Vertical angles are formed when lines intersect. It has nothing to do with being parallel.

Teacher: So, if b is equal to x, and x is equal to a, then b is also equal to a. What do we call this property, class?

Class: Transitive!

Teacher: What name do we have for this pair of angles, a and b? Discuss with your partners, and write the term on your board. What is the name of these angles on opposite sides of the transversal and inside the parallel lines? Go.

Students are predominantly correct.

Teacher: **So, you are claiming alternate interior angles are equal when lines are parallel. Now, for our last one. Michael, who is going to place *c* for us?**

Michael selects Greg, who places c on the number line (see Figure 9.24).

Figure 9.24 Greg's Placement

Look at Greg's group breaking the trend and placing *c* way over here. Greg, what's with that?

Greg: We put it over there because *x* and *c* are supplementary.

Teacher: **Partners, on your boards. Ready? These angles are supplementary because they share a vertex and a ray and thus form a line. The name for these types of angles is what?**

Most, but not all, write "linear pair" on their boards.

We use *pair* to refer to two angles and *linear* to imply a line, so we call these a linear pair. And yes, since the sum of their measures will equal 180°, we claim that linear pairs are supplementary.

Now, according to what you placed up here, *b* and *c* are also supplementary. Do we have a name for these types of angles? Brandon.

Brandon: I can't remember.

Teacher: **Since they are both on the inside, we use the word *interior.* *Consecutive* means "in a row," or "one after the other," so we call these consecutive interior angles. Partner on the right, hold the marker and write, but do not speak. Partner on the left, tell your partner how to spell *consecutive interior angles.***

Students declare their boards and show a widespread need for the spelling of the phrase.

Teacher: I'm going to write it on the board but only briefly. Take a quick look. Now, while I erase it, go back to your lapboards—the opposite partner spells this time.

Only minor errors remain.

Angles *b* and *c* are examples of what types of angles?

Class: Consecutive interior!

Teacher: So, linear pairs are supplementary always, and consecutive interior angles are supplementary when lines are parallel. Thumb vote.

Students show a unanimous thumbs-up.

Let's record the results of our discussion. Be sure to note the three relationships we have discussed. If you need to, use the sentence frames I have on the board (see Figure 9.25).

Figure 9.25 Sentence Frames

angles x and b are called ___.
angles a and b are called ___.
These angles are ___ when the lines are parallel.
angles c and b are called ___.
These angles are ___ when lines are parallel.

Why These Numbers?

Corresponding, alternate interior, and consecutive interior angles are the three most important pairs of angles when dealing with parallel lines cut by a transversal because they apply to many other topics, such as congruent triangles and parallelograms. They also involve some of the earliest proofs in geometry. Academic language is key here as well as understanding that the names of angles are consistent, but their properties may depend on conditions. In the sample above, *x* and *b* are always corresponding, but they are only congruent when lines are parallel.

The Key Questions

- Why did you change your mind?
- Why are these angles congruent?
- What has to be true for these angles to be equivalent?
- Will these measures always be equal?
- What is the name for these types of angles?
- What do we call our formal assumptions?
- What is the name of that property?
- Are these angles also supplementary?
- What has to be true for these angles to be supplementary?

Analysis

We can start with any of types angles, but for this example I intentionally choose b to highlight the idea that we start with our postulates and use those to prove theorems. Our lesson objective emphasizes the necessary condition that lines must be parallel for corresponding and alternate interior angles to be congruent and for consecutive interior angles to be supplementary. Therefore, I pressed students to state this condition as each type of angle was discussed and contrasted. I reminded students that linear pairs are always supplementary and vertical angles are always equal. I used this activity to hammer mathematical vocabulary, as it is new for many students. It is important to note that I did not have students spell the words for accuracy in spelling—I simply wanted them to focus their attention on the names of these pairs of angles and their meanings by requiring them to read and write the terms. The technique of having one student silently write what the other one says is called the Robot and works very well when you want students to practice articulating their thinking or use academic vocabulary.

As with all Clothesline activities, we want students to summarize the main points of the lesson. Since there are quite a few relationships and a couple of conditions ruling them, I chose to offer sentence frames. This is not always necessary but can help students.

The cards for this lesson are available in the Digital Resources (lesson25.pdf).

Extension

Place x at 90°.

Lesson 26: Special Right Triangles

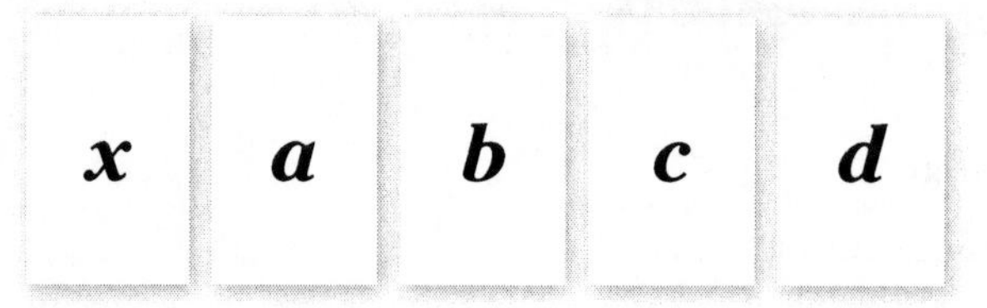

Objective: Place approximate decimal equivalents of the sides *a*, *b*, *c*, and *d* of 45°-45° and 30°-60° right triangles.

Teacher: **Good day, class. We have studied the ratios of special right triangles like this diagram** (see Figure 9.26).

Figure 9.26 Sample Special Right Triangle

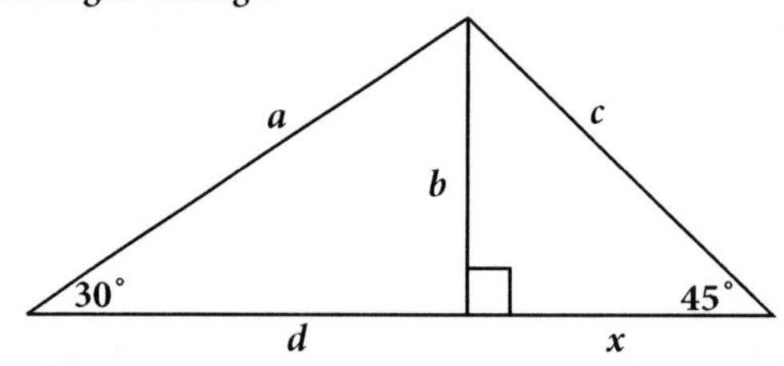

Today, we are going to apply those ratios from the diagram on the open number line. Before we do, let's warm-up by stating our ratios on our lapboards. Individually, please complete these diagrams, finding the length of the two missing sides (see Figures 9.27–28).

Figure 9.27 45°-45° with *x*

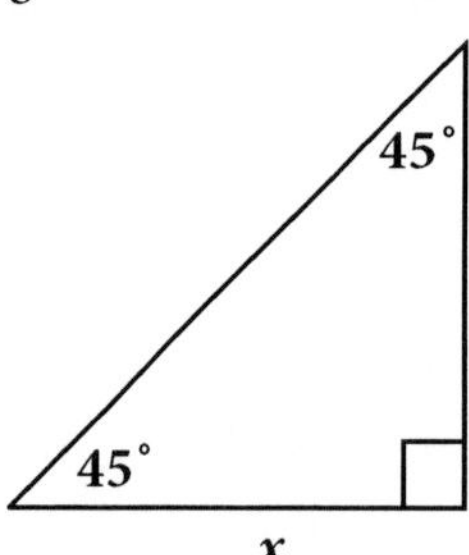

30°-60° with *x*

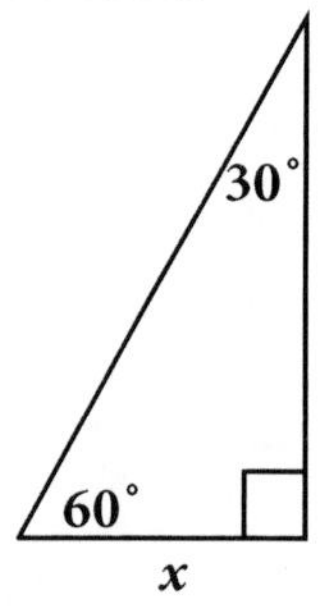

Students show their boards, and the majority are correct.

Figure 9.28 45°-45° with Ratios

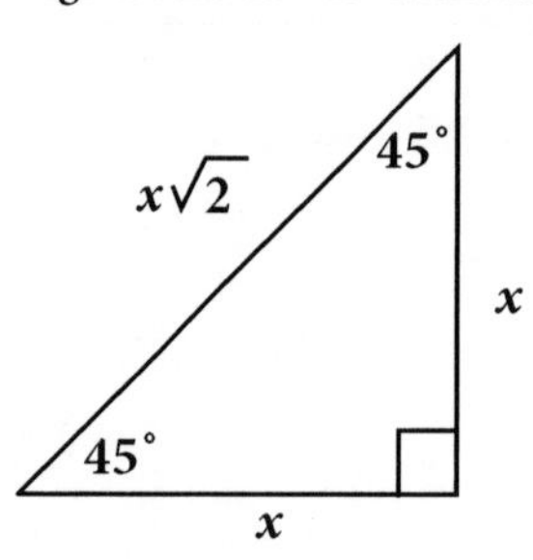

30°-60° with Ratios

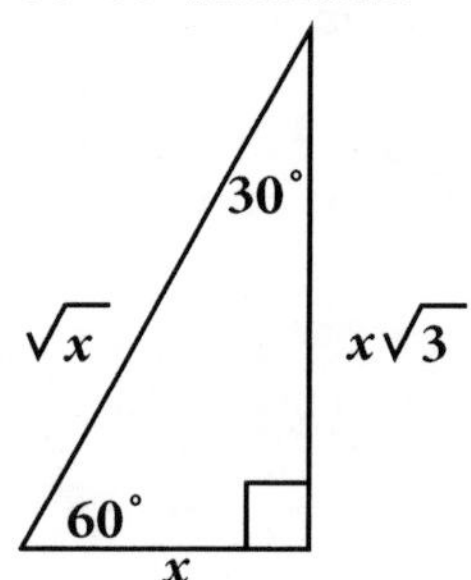

Declare your boards, please.

PART III – Chapter 9

Teacher: Since you seem to know your special right triangle ratios, let's see how you do with placing them on the clothesline. I will place the benchmarks of 0, 1, and *x* here (see Figure 9.29).

Figure 9.29 Teacher Placement

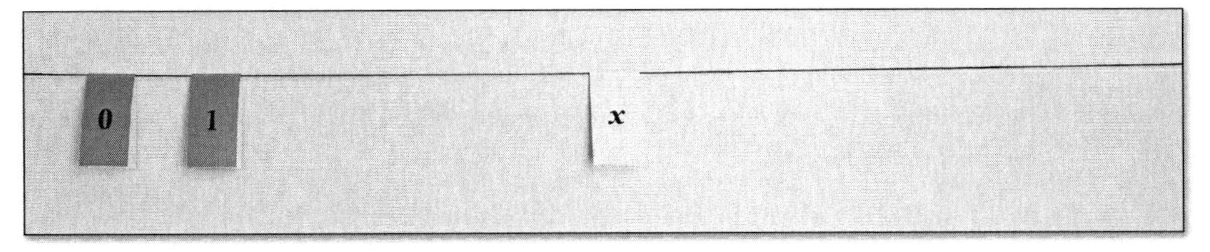

Hank, here is *b*. Place it on the clothesline according to the diagram shown.

Hank pins b *to the clothesline, as shown in Figure 9.30.*

Figure 9.30 Hank's Placement

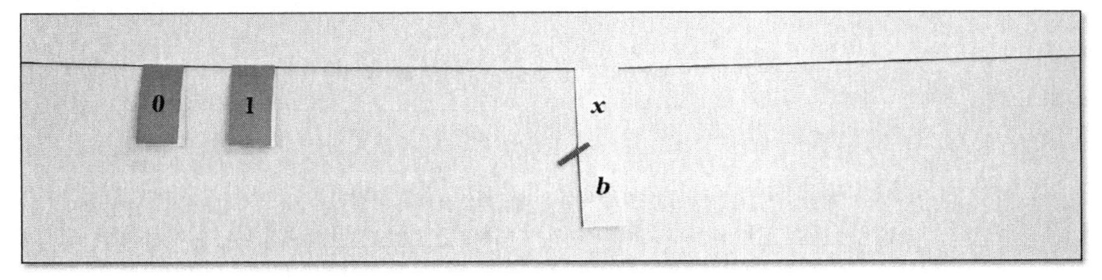

Hank, please explain your thinking.

Hank: Since the triangle on the right is 45°-45°, then *b* has to be the same length as *x*.

Teacher: Thank you. Gracie, here is *c*. If you agree with everything currently on the clothesline, then you may place this one. If you disagree with anything, then you may change it and hand this card to another student.

Gracie pins c *to 1 on the clothesline, as shown in Figure 9.31.*

Figure 9.31 Gracie's Placement

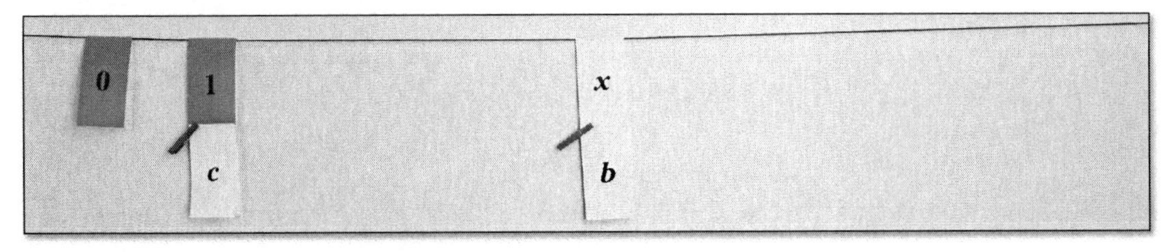

Gracie, please explain your thinking.

Gracie: We are still talking about 45°-45°, so the hypotenuse is the square root of 2, which is 1.

Teacher: Thank you. Greg, here is *a*. If you agree with everything currently on the clothesline, then you may place this one. If you disagree with anything, then you may change it and hand this card to another student.

Greg changes Gracie's response to slightly greater than 1 for c *(see Figure 9.32) and gives his card to Flynn.*

Figure 9.32 Greg's Placement

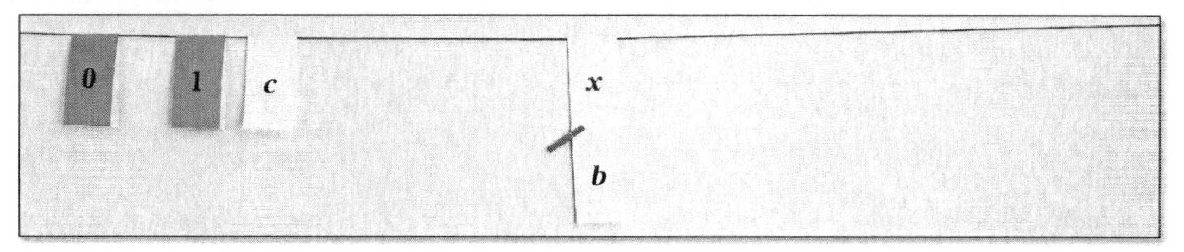

Greg, why did you change the placement of *c*?

Greg: Because the square root of 2 is about 1.4, not 1.

Teacher: Thank you. Flynn, Greg gave you *a*. If you agree with everything currently on the clothesline, then you may place this one. If you disagree with anything, then you may change it and hand this card to another student.

Flynn: I'm stuck.

Teacher: How so?

Flynn: Because *c* is not 1 or the square root of 2. It is the square of 2 times *x*, but I don't know how to show the square of 2 times *x* on the number line.

Teacher: Outstanding. You did a great of job articulating your struggle. I'm sure a lot of others in the room are confused as well. So, let me ask you this—if we multiply *x* by 1, what does that equal?

Flynn: Just *x*.

Teacher: Correct. If I multiply *x* by 2, what does that equal?

Flynn: 2*x*.

Teacher: So, point to where 2 times *x* would be placed on the number line.

*Flynn accurately points to 2*x *on the number line.*

Teacher: Okay, so we know the approximate value of root 2 is about 1.4, so 1.4 times x is somewhere between $1x$ and $2x$. Where do you think it would go? Place c there.

Flynn places c *(see Figure 9.33).*

Figure 9.33 Flynn's Placement

Why there?

Flynn: Because 1.4 is a little less than one and a half, so I chose a little less than halfway between $1x$ and $2x$.

Teacher: Excellent explanation. Class, let's applaud Flynn's bravery for sharing his confusion so we could all learn from it.

The class enthusiastically applauds.

Thank you, Flynn. Since you adjusted c, please hand a to another student.

Flynn hands a *to Michelle, who places it on the number line (see Figure 9.34).*

Figure 9.34 Michelle's Placement

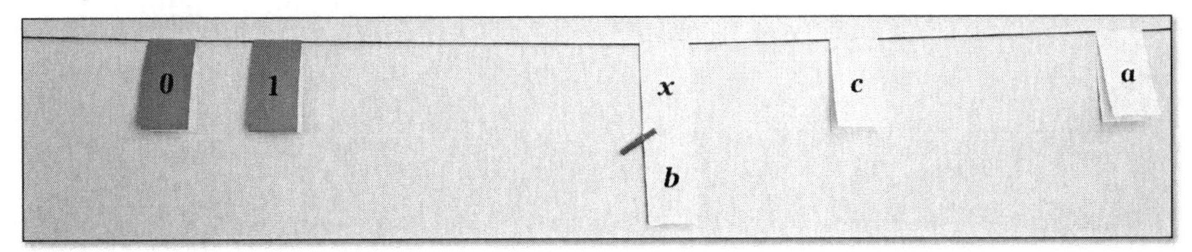

Michelle, please explain.

Michelle: The hypotenuse of a 30°-60° triangle is twice as long as the shortest side. So a is $2b$ or $2x$.

Teacher: How do you know that b is the shorter leg?

Michelle: The b is across from the 30° angle, which is the smallest angle, so it is the shortest side.

Teacher: Thank you, Michelle. Jay, Here is *d*. If you agree with everything currently on the clothesline, then you may place this one. If you disagree with anything, then you may change it and hand this card to another student.

Jay places d, *as shown in Figure 9.35.*

Figure 9.35 Jay's Placement

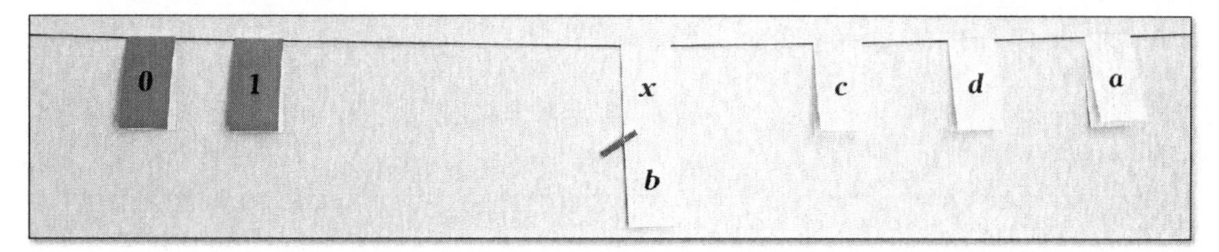

Jay, please explain your thinking for *d*.

Jay: This side is the square of 3 times b and root 3 is about 1.7, so like what we did with the 1.4, I put d a little less than one and three quarters between x and $2x$.

Teacher: Thumb vote for what we have up here.

There is a unanimous thumbs-up.

Since we are in agreement, let's recap. We have the legs of a 45°-45° triangle being equal and the hypotenuse being 1.4 or about one and a half times as long. In the 30°-60°, the hypotenuse is twice as long as the shorter leg, while the longer leg is about 1.7 times as big. Let's record this as well as the ratio diagrams we started with.

Why These Numbers?

There are many students in geometry who, with a traditional problem on special right triangles, can properly write $\sqrt{2}$ and $\sqrt{3}$ along the correct sides of triangles but cannot accurately place the value of these roots on a number line. Therefore, x is placed intentionally without an assigned value. The benchmark of 1 is used to contrast the value of $\sqrt{2}$ and $x\sqrt{2}$.

The Key Questions

- What was your reasoning for the placement of the card?
- Why did you change the placement of that card?
- What is the approximate value of $\sqrt{2}$ and $\sqrt{3}$?
- How do you know that is the shorter leg?

Analysis

The diagrams in the warm-up are an important launching point for the activity because the objective is for students to make sense of the ratios $1:1:\sqrt{2}$ and $1:\sqrt{3}:2$. The process of the student chain was utilized because it is a better tool than the group chain when giving students the option of changing previous answers.

Placements of the equivalent leg of the isosceles triangle and the hypotenuse of the 30°-60° are easier values to deal with because of their whole number ratios. The crucial discussion surrounds the meaning of the irrational ratios and the meaning of the square roots themselves. The common error of thinking $\sqrt{2} = 1$ occurred in this lesson. The error was corrected by another student rather than the teacher. This drove conversation toward the understanding of the difference between $\sqrt{2}$ and $x\sqrt{2}$.

Having a student explain his or her confusion was as useful as having a student explain his or her reasoning behind an answer. The public conversation on this led to the understanding of $x\sqrt{3}$.

The cards for this lesson are available in the Digital Resources (lesson26.pdf).

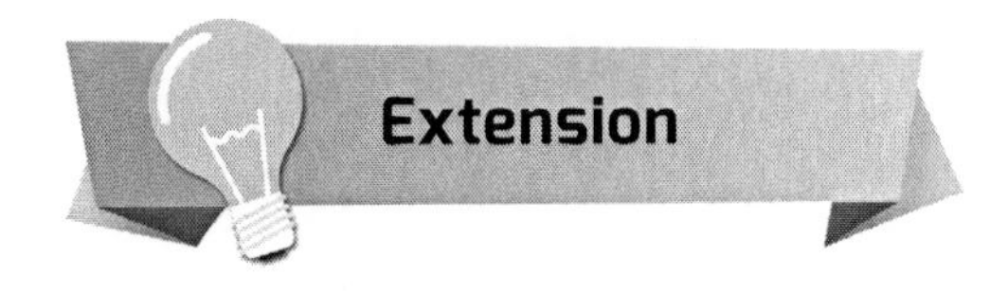

Extension

Attach a value, such as 5, to the value of x and ask students to attach the approximate values on blank cards for the other three values of c, a, and d.

Lesson 27: Polygon Angle Properties

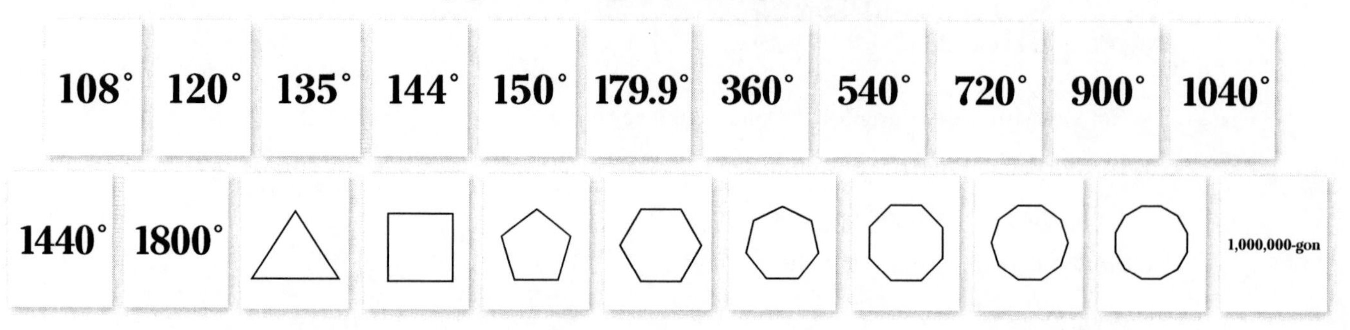

Objectives: Determine interior and exterior angle measures of regular polygons and their angle sums. See the linear or rational (asymptotic) relationships for the formulas.

Teacher: **Hello, class. You have a group warm-up today. Your goal is to match each of our four formulas for the angles of regular polygons to its corresponding diagram** (see Figure 9.36). **Your choices: the measure of an interior angle, the sum of interior angles, the measure of an exterior angle, and the sum of exterior angles.**

Figure 9.36 Polygon Angles

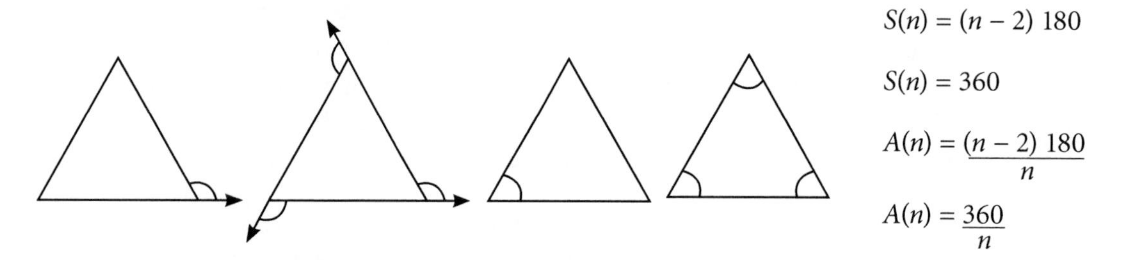

Have each person in your group draw one of these diagrams on a lapboard and write the corresponding formula underneath. Discuss as a group, and decide who will display which diagram and formula.

Most groups correctly match the formulas to the diagrams. Some need help, which the teacher offers in a group conversation.

JoJo, explain why you matched the formula $S(n) = (n - 2)180$ to this diagram (see Figure 9.37) **for the sum of the interior angles.**

Figure 9.37 Polygon Interior Angle Sum

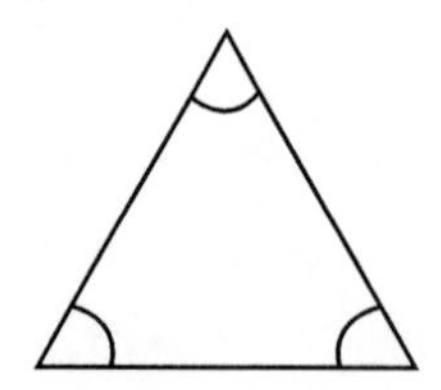

JoJo: If you cut the polygon into triangles, you have three triangles, which is two less than the five sides. Each triangle has 180°.

Teacher: **Will we always have 3 triangles for every polygon, or 2 fewer than the number of sides?**

JoJo: Two fewer.

Teacher: **What in the formula tells you that?**

JoJo: $n - 2$.

Teacher: **Thank you, JoJo. Miguel, Why did you match the formula** $A(n) = \frac{(n-2)180}{n}$ (see Figure 9.38) **for the measure of a regular interior angle?**

Figure 9.38 Polygon Interior Angle

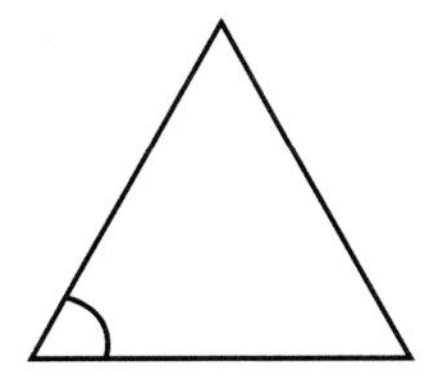

Miguel: If you take the total JoJo just talked about and divide by the number of the angles, it equals 1 angle.

Teacher: **That is true if all angles are congruent, so this formula is for regular polygons. Thank you, Miguel. Shawna, why did you match the formula** $S(n) = 360$ (see Figure 9.39) **for the sum of the exterior angles?**

Figure 9.39 Polygon Exterior Angle Sum

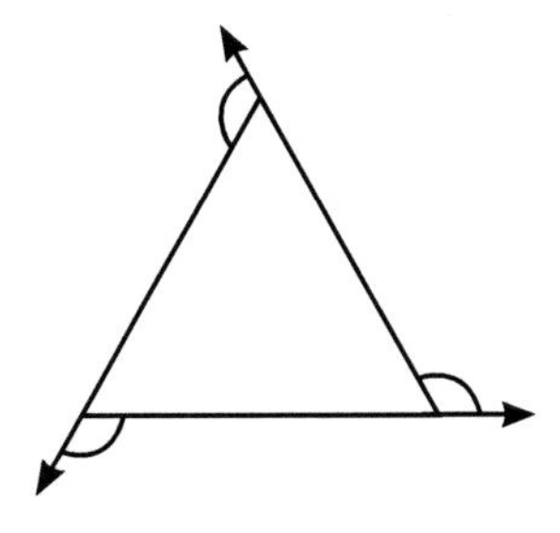

Shawna: If you connect the outside angles, they surround a point that is 360°.

Teacher: **Thank you, Shawna. Kevin, why did you match** $A(n) = \frac{(360)}{n}$

(see Figure 9.40) **for the measure of a regular exterior angle?**

PART III – Chapter 9

Figure 9.40 Polygon Exterior Angle

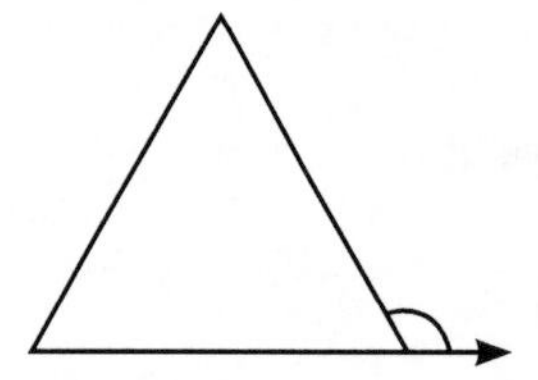

Kevin: If you take 360° for all the exterior angles and divide by how many you have, then you get just one of them.

Teacher: **Let's give a big hand to those who just explained those formulas.**

Students enthusiastically applaud.

Now, let's take these polygons to the clothesline. We are going to start with the polygon sum theorem. I am going to start with the first two polygons and place the triangle and square on the clothesline according to the sum of their interior angles (see Figure 9.41).

Figure 9.41 Teacher's Placement

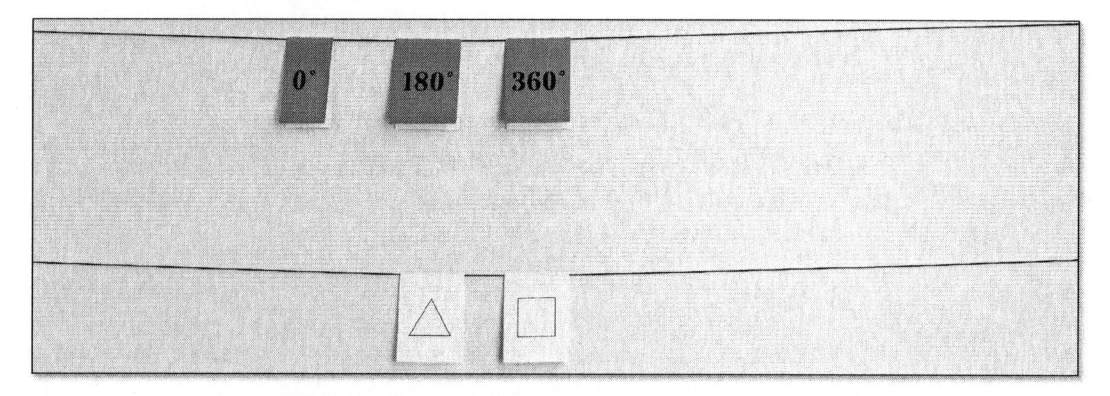

Joelle, what do these numbers represent?

Joelle: All the inside angles added up.

Teacher: **Or, the sum of the interior angles. You try that.**

Joelle: The sum of the interior angles.

Teacher: **Good. Now everyone with your partner, place the pentagon, hexagon, octagon, decagon, and dodecagon on the number line with corresponding sums of the interior angles. On your lapboards. Go.**

Students work on their lapboards.

Joelle, back to you. Please choose the group that will place the pentagon on the number line showing the sum of its interior angles.

Joelle selects Miguel's group.

Teacher: Miguel, your group's turn.

Miguel's group places the pentagon on the bottom line, as shown in Figure 9.42.

Figure 9.42 Miguel's Placement

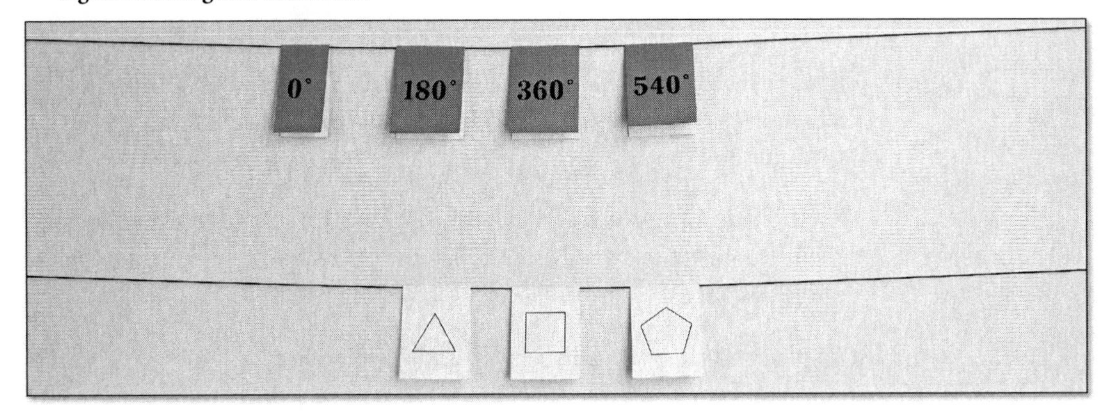

Miguel, how did you determine where to place the pentagon?

Miguel: We used the formula to get the sum of the angles.

Teacher: Did anyone else get the same answer as Miguel's group in a different way? Lynn.

Lynn: I just spaced it the same distance between the triangle and the square.

Teacher: And what is the distance on the number line?

Lynn: 180°.

Teacher: Paul, your group had similar thinking to Lynn's. Please show where you placed the hexagon and octagon.

Paul's group places the hexagon and octagon, as shown in Figure 9.43.

Figure 9.43 Paul's Placement

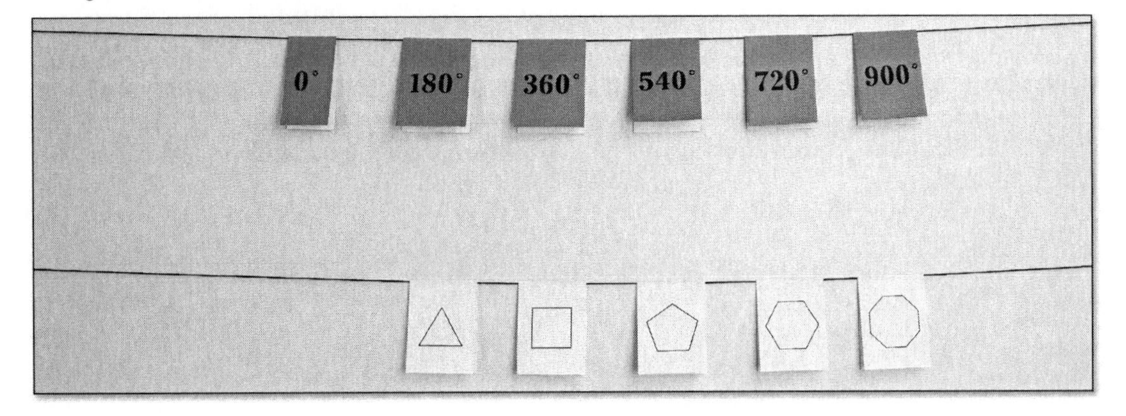

Teacher: JR, you are shaking your head like you disagree.

JR: The octagon has to move over another 180°. We are skipping the 7-sided one.

Teacher: Paul, does that make sense?

Paul: Yes.

Teacher: JR is agreeing with Paul's thinking that for every angle we need to add another 180°. Since we added two angles, we need to add two 180°s, or 360°.

Rylie, I see your group agrees with JR as well. Let's have your group place the last two while everyone else adjusts what's on their lapboards.

Rylie's group places the last two polygons, as shown in Figure 9.44.

Figure 9.44 Rylie's Placement

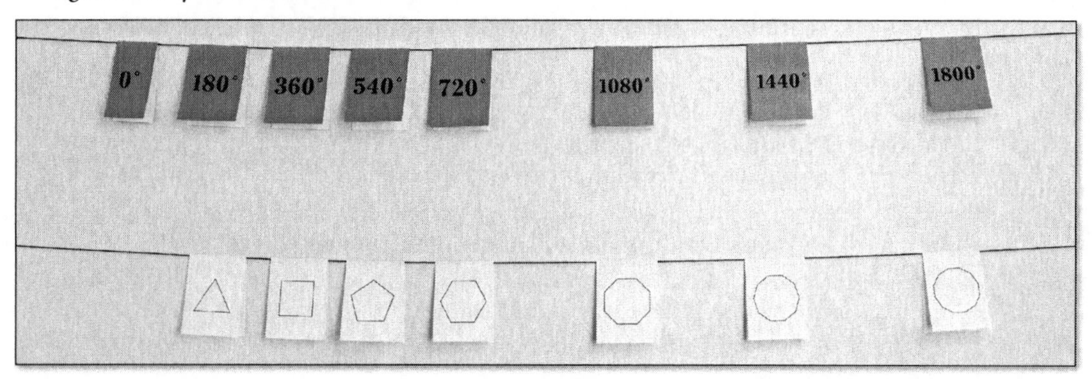

I'm curious about this pattern you have been discussing. So, from what you are showing me, I'm not sure I want to place the 1,000,000-gon on the number line.

JR: What the heck is a one-million-gon!

Teacher: That is a polygon with a million sides. We have names for the most common polygons, but not for all the infinitely many, so we can call it by its number of sides with *-gon* at the end. So, why might I be reluctant to try to place this on the number line?

Rylie: It would be way off the end of the number line.

Teacher: How far?

Rylie: 999,998 times 180. It would go outside the room.

Teacher: Would it have to?

JR: Or, we could scrunch everything else together.

Teacher: **Both of those sound fun, but for the sake of time, are you all okay with simply recognizing this would be a very large number?**

Many students give affirming nods.

Great, let's record this on our activity sheet. Be sure to place the formula next to the number line and write the values of the angle sums with each polygon. Now, let's investigate our next formula for the angle measure. I will start again with the triangle and square. Take a look (see Figure 9.45).

Figure 9.45 Teacher's Adjustment

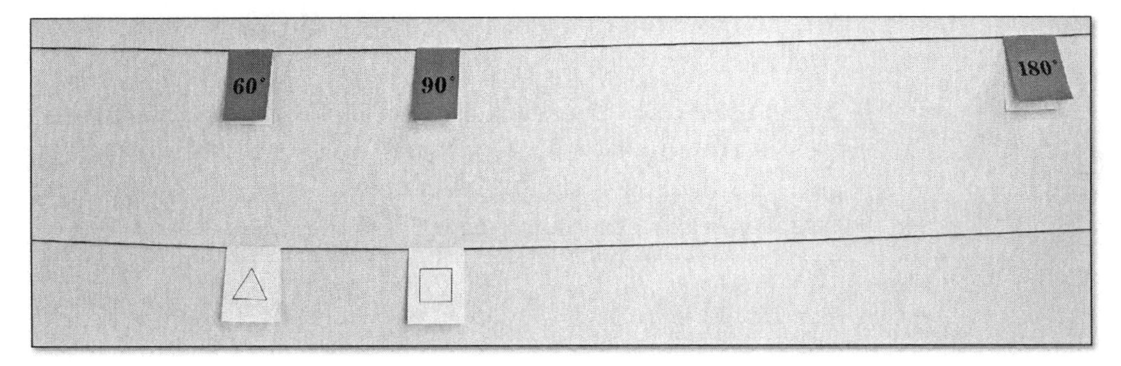

Preston, what do you notice is already different in this pattern for the individual angle measures versus the sum of the interior angles?

Preston: The distance between the polygons doesn't seem to be the same.

Teacher: **Let's see whether Preston's conjecture is correct. Work with your partners on the lapboards, and place the rest of the polygons except the 1,000,000-*gon*. Let's save that one for a special conversation. Place the others according to the measure of their interior angles. Be sure to show each interior angle measure, and check whether the spacing between the polygons is all different.**

Students work on their lapboards.

Jake, your group's response looks interesting. Please show it on the clothesline while everyone else finishes his or her work.

Jake's group places the cards on the clothesline, as shown in Figure 9.46.

Figure 9.46 Jake's Placement

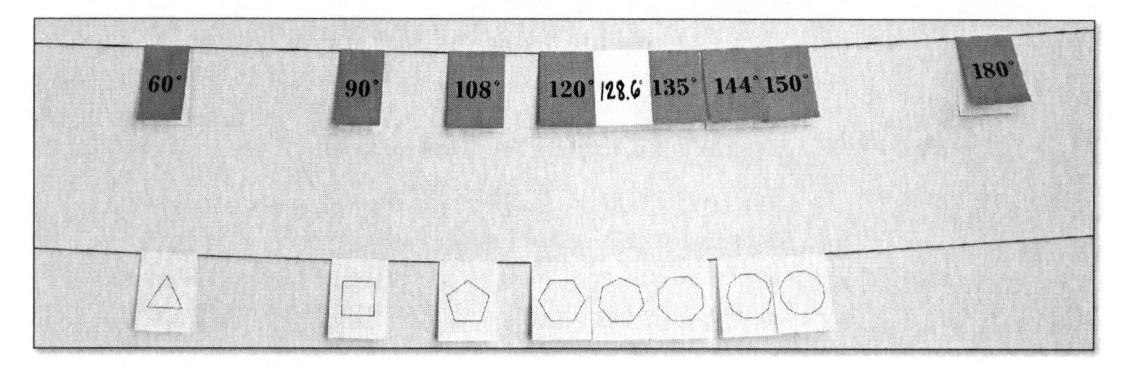

Teacher: Declare your boards.

Students show their boards which closely match Jake's work on the clothesline. Jake, however, added an extra self-drawn polygon.

Looks like everyone agrees with Jake's group. You are supporting Preston's conjecture that the spacing between the polygons is different. I like how you added in a missing polygon to make the pattern more obvious. Is there a pattern to these different gaps? Julie.

Julie: They are getting smaller.

Teacher: Why is that? Luis.

Luis: When you add an angle to the triangle to get the square, you add 30°. But, when you add an angle to a decagon, it barely makes a difference.

Teacher: Thumb vote on Luis's explanation.

There is a unanimous thumbs-up.

Well Luis, now you have me curious. Where will the 1,000,000-gon go? Will it be way past the end of our clothesline and outside the room, like with the sum formula?

Luis: I don't think so because the gaps keep getting smaller.

Teacher: Bring your group to the clothesline and test your idea. Everyone, use your formula to determine the measure of one interior angle of a million-gon.

Luis's group places the million-gon, as shown in Figure 9.47.

Figure 9.47 Luis's Placement

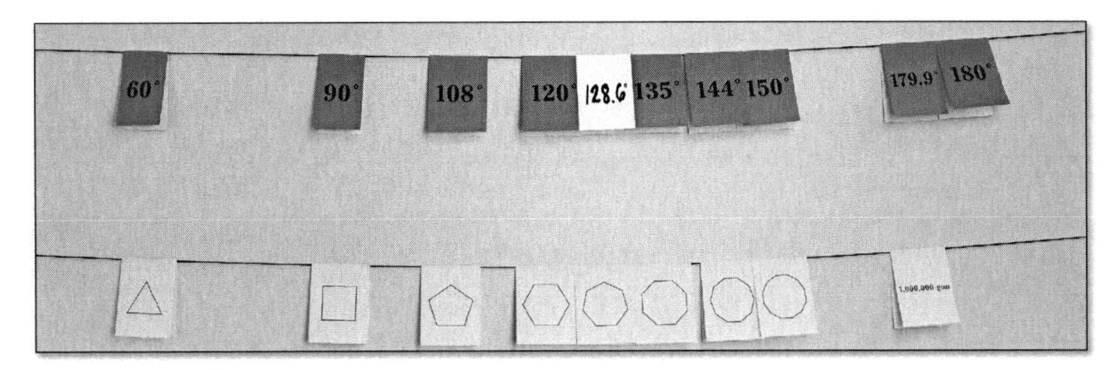

Teacher: Luis, what did your group discover?

Luis: It's really close to 180°.

Teacher: Interesting. Raise your hand if you got the same result.

All students raise their hands.

Without using the formula, I wonder—what if we had a billion-gon? Would the angle be more than 180°? Sonya.

Sonya: No.

Teacher: Why not?

Sonya: Because then each angle would be more than a straight line and wouldn't make the inside of a polygon anymore.

Teacher: Sonya is talking about a limit. That is a value we get closer and closer to without ever reaching. So, what if we had a trillion-gon, or even a google-gon? A google is a 1 with a hundred zeros. Do you agree with Sonya that the measure of the angle would get really, really close to 180° but not actually reach it? Thumb vote.

The class unanimously agrees with Sonya.

Let's record this one with the appropriate formula as well. We have just seen two different types of functions. The first one was a linear function; the rate of change was constant at 180° per side. The second one was a rational function; we divided by a variable. This particular rational function had a limit of 180°. Let's see whether these linear and rational relationships occur with the exterior angle formulas.

We will start with the exterior angle theorem. I have placed the triangle and square on our number line (see Figure 9.48).

Figure 9.48 Teacher's Second Adjustment

Teacher: **Please place the others including the 1,000,000-gon. Sonya, which group is going next?**

Sonya selects Becca's group.

Becca, your group.

Becca's group pins all the polygons together in a vertical line, as shown in Figure 9.49.

Figure 9.49 Becca's Placement

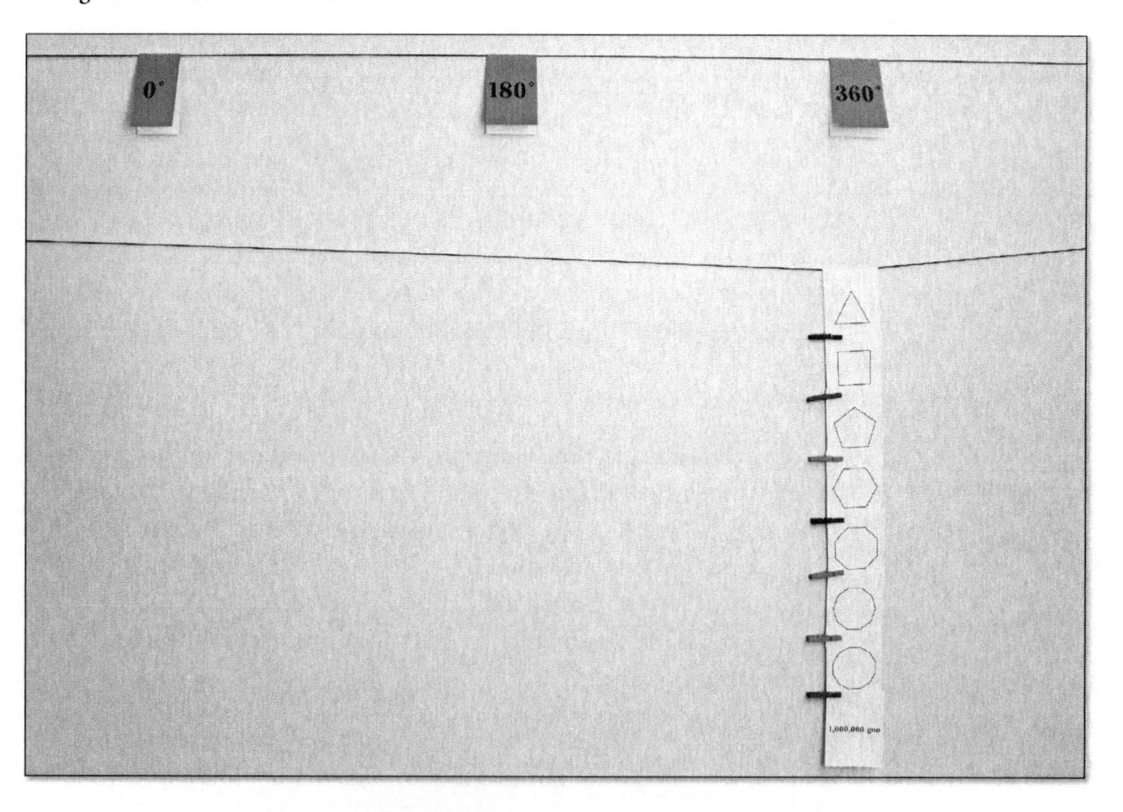

Declare your boards.

Students show their boards, which all match.

Teacher: **Wow, it's unanimous. Is this a linear, a rational, or a different kind of function? On your boards, write which function you believe this is.**

The boards vary among linear, rational, and different.

We have a variety of responses. Let me rewrite the formula: $S = 0n + 360$. This function has rate of change of zero. We don't add or subtract anything to the angle sum as we add sides. Also, there is no limit because we are not approaching any value for angle measure as we add sides. With this in mind, change your answers if necessary. Linear, rational, or something different?

The boards universally display linear.

Now, we all agree this function is linear. Let's record it with the formula for the sum of the exterior angles. How about our last formula for the measure of an individual angle of a regular polygon? I already have the triangle and square and their corresponding angle measures displayed on the clothesline. Place the other polygons, including the 1,000,000-gon. Be sure to display the angle measures for each. I will take a volunteer group for our last one.

Fred raises his hand.

Fred's group, thank you.

Fred's group places the cards on the bottom clothesline, as shown in Figure 9.50.

Figure 9.50 Fred's Placement

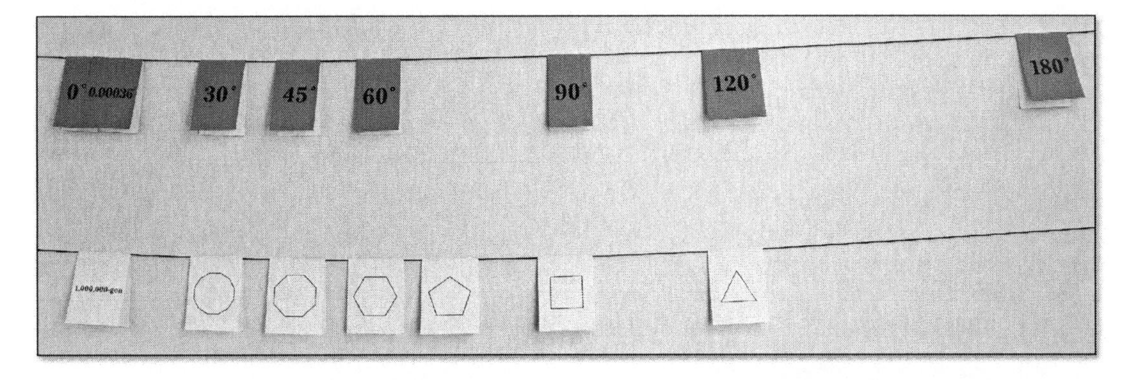

So, does this function look linear or asymptotic? If you think linear, stretch your arms out to form a straight line. If you think it is asymptotic, face your palms really close to each other without touching.

The entire class claims asymptotic.

Teacher: Jillian, why do you think this is asymptotic?

Jillian: Because we are getting smaller and smaller without touching zero.

Teacher: Anyone have another reason?

Jillian: In the picture of the polygon, you can see that as the interior angle gets bigger, the exterior gets small. They are supplementary. So, if one gets really close to 180°, then the other should get really close to zero.

Teacher: Both of those are solid explanations. How can we tell from the formula this a rational function? Annie.

Annie: I'm not sure.

Teacher: Take at a look at the formula. What do you notice about the variable?

Annie: Oh yeah, it is in the denominator.

Teacher: Excellent. Okay crew, let's record this last one with our formula. You bent your brains really hard today. Well done.

Why These Numbers?

Or maybe I should ask, why these shapes? Or maybe, why these formulas? The answer is that geometry students are often asked to find an answer using these formulas without really understanding what the formulas represent. Comparing the various measures for polygons helps correlate the formula to the aspect of the polygon it is used to calculate. The Clothesline also forces students to analyze each formula, which helps them understand formula types in general (e.g., linear vs. rational) as well as reinforcing the long-term memory of the formula.

The Key Questions

- Why did you match this formula to that diagram?
- What in the formula makes you think that?
- What do these numbers represent?
- What do you notice about the variable?
- Is this function linear or asymptotic?
- Does anyone have a different reason?
- Do you see any kind of pattern?
- How is this pattern different than the previous one?

Analysis

Before I had students begin the analysis, I did an extensive warm-up to revisit the formulas as a whole and to tie each equation to a visual. I knew some students were still struggling with determining which of the four formulas to apply in a given situation. I was betting the connection between the symbolic (equation), the visual (polygon diagrams), and the numeric (clothesline) representation would make students more proficient in using the formulas.

My questions dealing with the algebraic properties may have more relevance in an integrated course, but I felt it was healthy and reasonable to have students analyze the behavior of the functions. Note that this could also be an Algebra 2 or Math 3 lesson in which the geometry formulas are simply reviewed and the focus is on asymptotic behavior.

For this lesson, though, my questions kept drawing attention back to the conceptual understanding of what the formulas represented. Was it yielding the sum of the angle measures or an individual angle measure? Why is one sum constant while the other constantly grows? This idea should be derived from the diagrams as well as from the equations.

The cards for this lesson are available in the Digital Resources (lesson27.pdf).

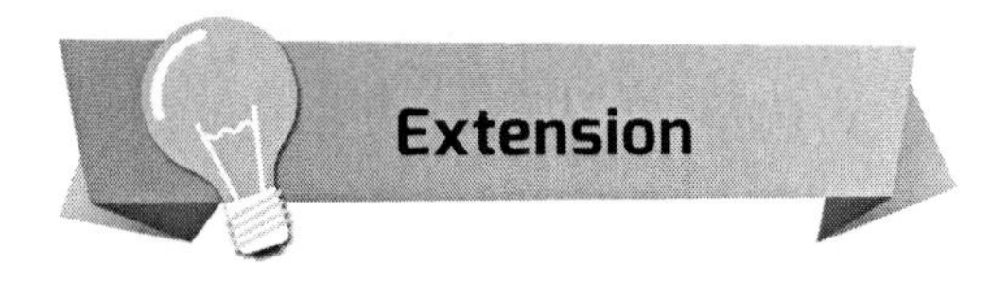

Extension

Graph each of the four formulas, and determine the shared characteristics of the two linear functions and those of the two rational functions. Write the equations of the limits (asymptotes).

Statistics

Lesson 28: Average

μ

Objective: Compute the average of various data sets and understand that *average* is a central measure for a set of variable values.

Teacher: Hello, class. I am curious about something—how tall are the boys in this class? Anybody want to venture a guess?

Benny: We are all different heights.

Teacher: That's correct. So, we need a statistical measure to measure or describe a set of numbers that varies. Let's start with the tallest person in the class. Jesse, will you please stand in front of the clothesline for a moment. How tall are you?

Jesse: Five foot ten.

Teacher: Okay, we are going to convert to inches. 12 inches times 5 feet is 60 inches. Add another 10 inches, and you are 70 inches tall. I am going to write your height on this card and have you place it on the right end of the clothesline.

Jesse places the card on the number line (see Figure 10.1).

Figure 10.1 Jesse's Placement

PART III – Chapter 10

Teacher: **So, are all the boys in this class 70 inches tall?**

Class: No!

Teacher: **Okay, so we can't have Jesse's height represent all the heights. I get that, so I will call up another boy. Before I do, let's have everyone in the class convert their height to inches. This means boys and girls. Now, we need our shortest boy. Kyle, please write your height on this card and place it at the left end of the clothesline.**

Kyle writes his height on a card and places it on the clothesline (see Figure 10.2).

Figure 10.2 Kyle's Placement

Does Kyle represent the height of all boys in our class?

Class: No.

Teacher: **Okay, then. Please calculate the average of Jesse's and Kyle's heights. Write it on your lapboard and hold it up.**

Students offer two answers: 11 and 59.

Eleven what? Fifty-nine what? Units, people, units!

Students add inches after their numerical answers.

Much better. So, we have two answers offered. Chandra, please explain why you say 11".

Chandra: I subtracted 70 minus 48 and divided by 2.

Teacher: **Dustin, how did you get 59"?**

Dustin: I added and divided by 2.

Teacher: **So, which one would better represent the heights of the boys in the class: 59" about here on the number line, or 11" way over there somewhere? If you think 59" is the better representation, point to the right. If you think 11", point to the left.**

All students point to the right.

That's how we will remember to add the values, rather than subtract before dividing. It gives us a more reasonable representation of the data set.

Teacher: **I am going to place 59" on our number line and pin this symbol to it** (see Figure 10.3).

Figure 10.3 Adding Mu

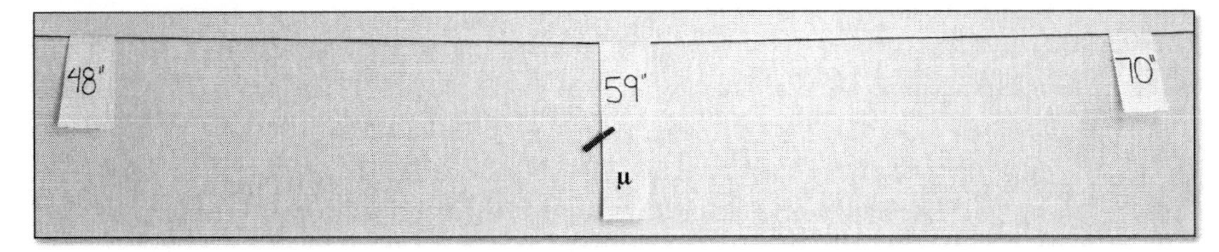

This symbol is actually pronounced *mu*. It is a Greek letter used in mathematics to represent average. I need a volunteer who is a few inches below the average of 59". Evan, place your height on the clothesline.

Evan adds his height to the number line (see Figure 10.4).

Figure 10.4 Evan's Placement

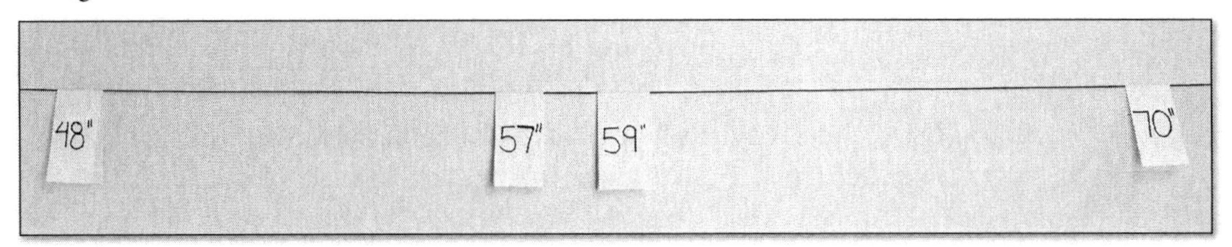

My question is this: Is Evan's height powerful enough to affect the average? If you think Evan's height will not affect the average, give a thumbs-down. If you think it will, point in the direction you think the average will shift.

A handful of students vote no. The rest of the class is split between Evan's height lowering and increasing the average.

As a class, we are voting for all three. Let's calculate and settle the dispute. Everyone, work on your lapboards.

Most students declare either 58.3" or 58". Two students declare 87.5".

Tina and Louisa, does 87.5" look like a representation of the heights of the class?

Tina: No, but I added and divided by 2.

Teacher: **How many data points should you have divided by?**

Tina: Oh yeah, 3.

Teacher: **Make your corrections. So, we all agree our new average is about 58"; therefore, Evan's height lowered the average. I will shift our μ to our new average** (see Figure 10.5).

Figure 10.5 New Average

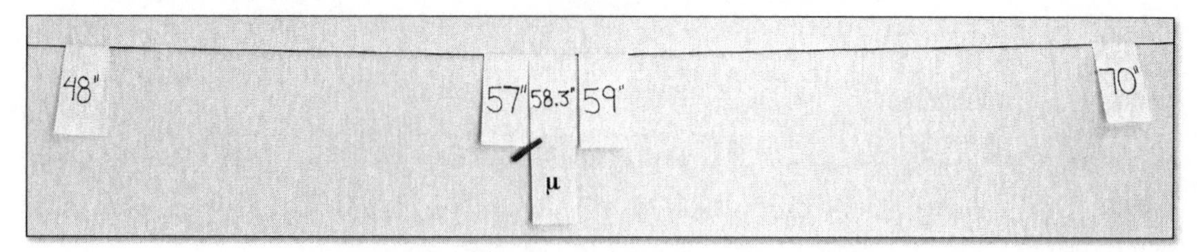

Teacher: Let's have a volunteer who is above average in height. In fact, do we have someone close to Jesse's height? Tyler, come up and place your height on our number line.

Tyler adds his height to the clothesline (see Figure 10.6).

Figure 10.6 Tyler's Placement

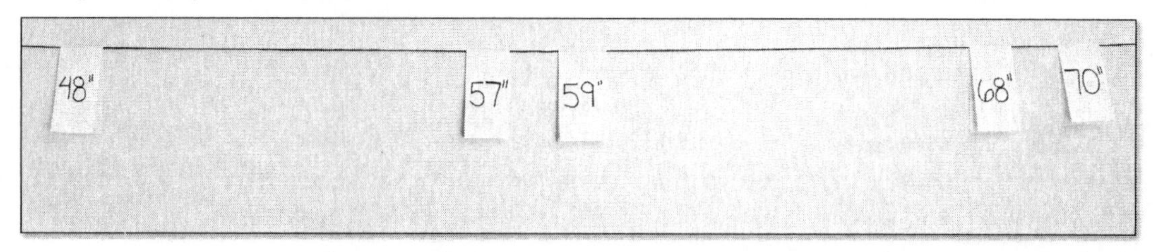

Okay crew, is Tyler's height powerful enough to affect the average? If no, thumb down. If yes, point in which direction.

Everyone points right, implying the average will increase.

You claim Tyler's height will increase the average. Calculate and verify. Show our new average on your lapboards.

Students display answers of 60.75", 60.8", and 61".

You are confirming the average will increase and it will be just under 61". Let me adjust μ to show our new average. Now comes the important question—which height best represents the height of these four boys? Write the best representation of the height of the four boys and declare your boards.

Everyone writes some reasonable approximation of the average.

So, we agree the average is a superior representation of a data set rather than choosing one value. Let's have the rest of the boys place their heights on the clothesline. Now, let's find the average of all the boys' heights in this class.

Students all agree to the correct average.

I will display your unanimous answer on the number line. Record our results on your activity sheet, and make note of the average being a measure of a variable set of data.

Why These Numbers?

This lesson could be conducted with any set of numbers. Choosing to run it with the data of the boys' heights not only enhances the engagement of the lesson, it also highlights the idea the average is a measure for data that displays variability.

The Key Questions

- What is your estimate for the height of the boys in this class?
- So, are all boys in this class 70" tall?
- Is Kyle's height representative of the heights of all boys in the class?
- Will the new height affect the average?
- Will the new average be less than or more than the current average?
- Which of these heights best represents the height of the boys in the class?

Analysis

I intentionally started with extremes of the maximum and minimum heights to emphasize that one data point does not necessarily represent the entire data set. It also establishes a good visual range on the clothesline. The typical error of subtracting instead of adding while computing the average comes early when students are instructed to calculate the average of the first two points. The other typical error of dividing by two for more than two values arrived with the next calculation. The front end of the lesson explained these common procedural issues while focusing on the conceptual understanding that the average describes a data set having variability.

The cards for this lesson are available in the Digital Resources (lesson28.pdf).

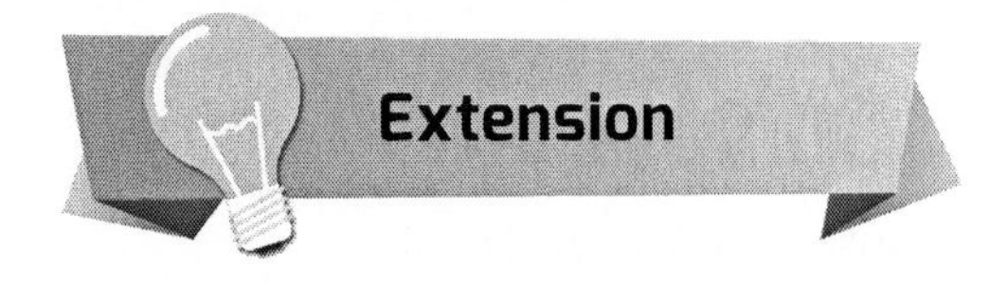

Extension

Pose the following question: If we were to include girls' heights in this data set, would we expect the average to be higher or lower?

Lesson 29: Mean, Median, Mode, and Range

Med **Mean** **Mode** **Range** **5** **9** **10** **11** **13** **14**

Objective: Identify the mean, median, mode, and range of a data set.

Teacher: **Hello, class. We have been studying statistics, specifically mean, median, mode, and range. Today, we are going to find these measures for a given data set on the number line. Take a look at the data set displayed here** (see Figure 10.7), **and draw it on your lapboards with your partner.**

Figure 10.7 Data Set

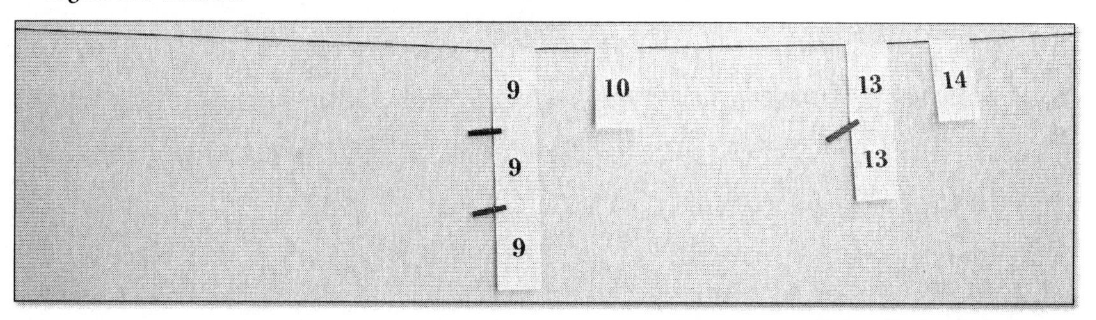

Our first measure is the median. Kendra, your group gets this card that says *median* on it. Discuss it, and send a group member to place it on the number line with this data set.

Kendra's group adds the median to the number line (see Figure 10.8).

Figure 10.8 Kendra's Placement

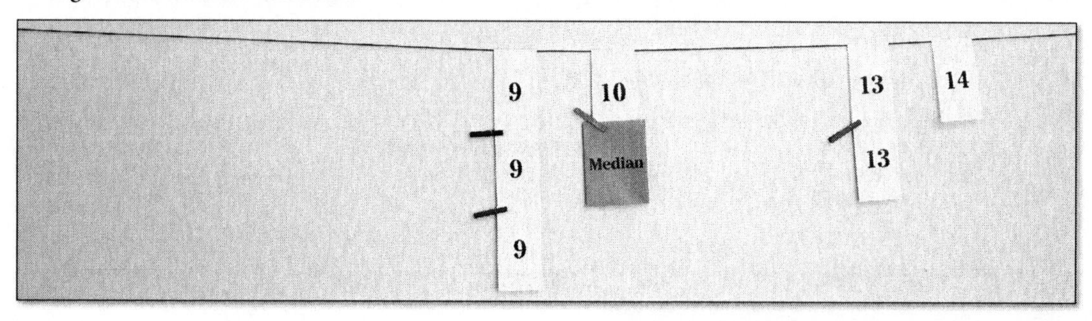

Kendra, why did your group choose 10 as the median?

Kendra: We started from each end and counted evenly to the middle.

Teacher: **Let's see whether the class agrees. Everyone, declare your boards.**

Students show boards with number lines that match Kendra's.

Teacher: Everyone seems to agree with you, Kendra. Jimmy, I heard you and your partner having an interesting discussion. Please explain how you decided that 10 is the median.

Jimmy: There are three numbers to the left and three to the right of 10.

Teacher: Thank you. Kendra, choose the group that will place the mode for us.

Kendra selects Lexi's group.

Lexi, it's your group's turn.

Lexi's group places mode (see Figure 10.9).

Figure 10.9 Lexi's Placement

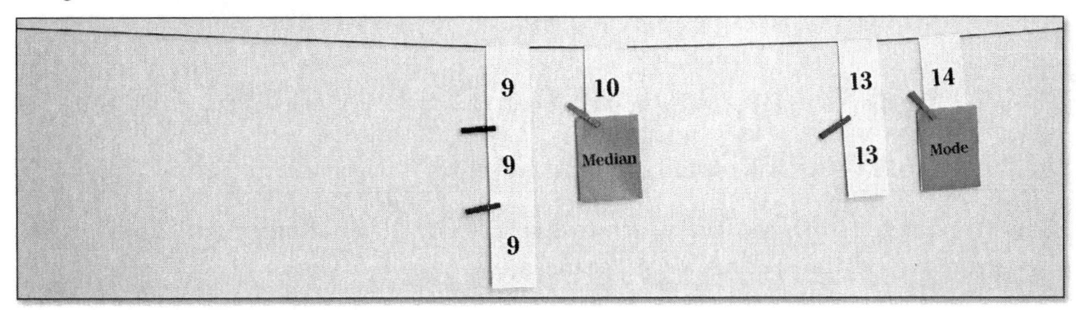

Lexi, please explain why your group chose 14.

Lexi: It's the most.

Teacher: Jimmy, you and your group have something different on your board.

Jimmy: Nine is the mode, because 9 has the most.

Teacher: The most what?

Jimmy: It shows up the most times.

Teacher: I see the class was split between these two answers. I am giving everyone 30 seconds to adjust your boards according to the justification you agree with.

Students adjust their answers.

Declare your boards. Lexi, you changed your mind?

Lexi: Yes, the mode is the one with the most numbers.

Teacher: So, median is the middle number, and mode is the most frequently occurring number. Let's tackle range now. Lexi, who is placing this one?

Lexi selects Linda's group.

Teacher: Linda, your group is up.

Linda's group places range (see Figure 10.10).

Figure 10.10 Linda's Placement

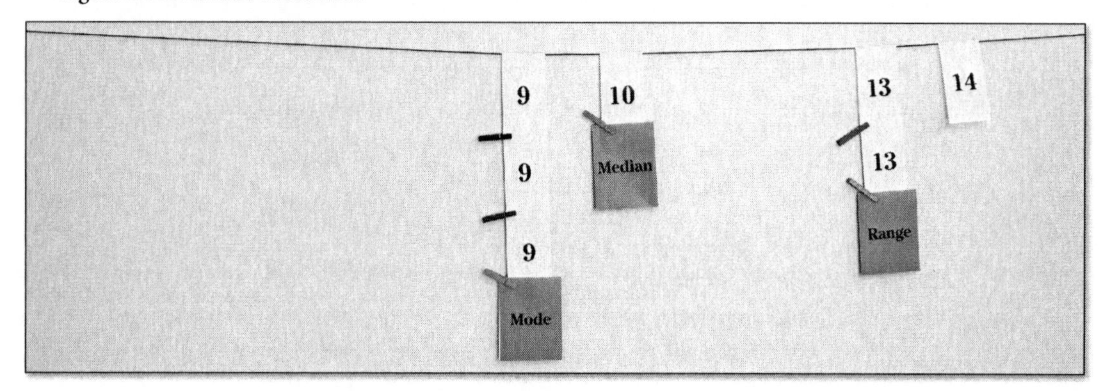

Teacher: Linda, why did you choose 13?

Linda: We had no clue, so we just guessed.

Teacher: Fair enough. At least you made an attempt, thank you. Can anyone help with the definition of range? Paula.

Paula: How far it is from the smallest number to the largest number.

Teacher: Thank you. Yes, the largest number minus the smallest. With that said, Linda, where would your group like me to move the range card?

Linda: On the 9.

The teacher moves range to 9 (see Figure 10.11).

Figure 10.11 Linda's Adjustment

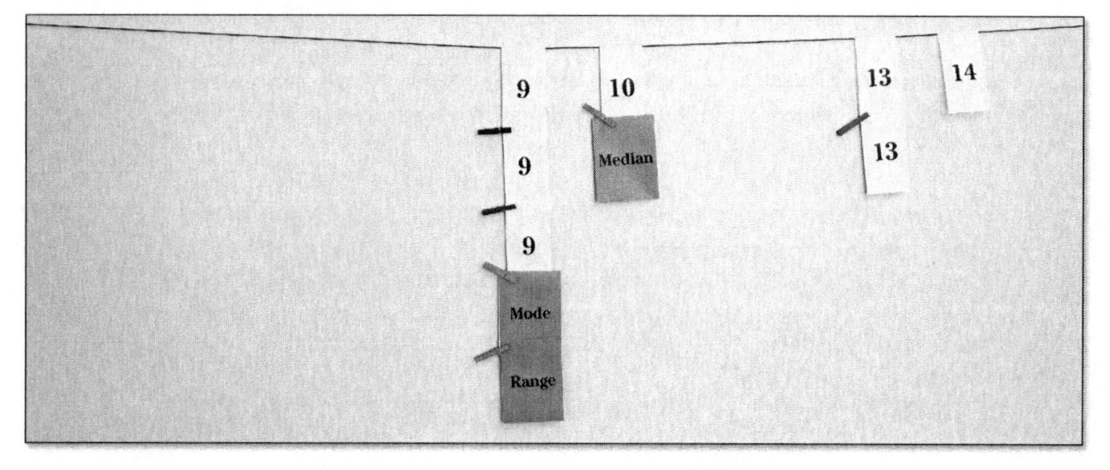

Teacher: Why?

Linda: You said the largest minus the smallest, so we thought it should go on the smallest.

Teacher: Class, thumb vote.

Some students vote thumbs-up, but most vote thumbs-down.

Ben, you disagree.

Ben: It's 5 because 14 – 9 is 5.

Teacher: So, you want it here?

The teacher moves range to 5 (see Figure 10.12).

Figure 10.12 Range Placement —Ben's Response

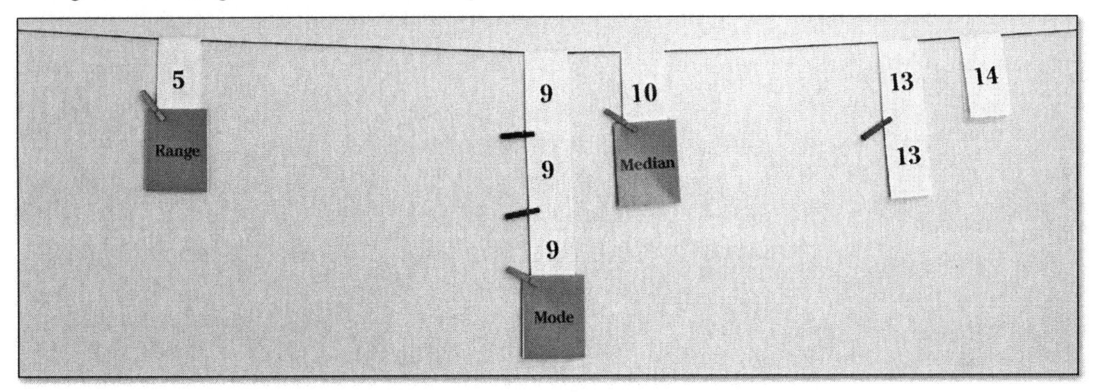

Ben: Yes.

Teacher: Okay, Linda's group. Watch my hands. From 9 to 14 is this much. We want this same amount from 0, which puts us here at 5. Do you agree with Ben?

Linda: Yes.

Teacher: Okay then, time for our last decision. Linda, who is going to place the mean?

Linda selects Andre.

Andre, your group's turn. Everyone else, last time working on your lapboards.

Andre's group places mean, as shown in Figure 10.13.

Figure 10.13 Andre's Placement

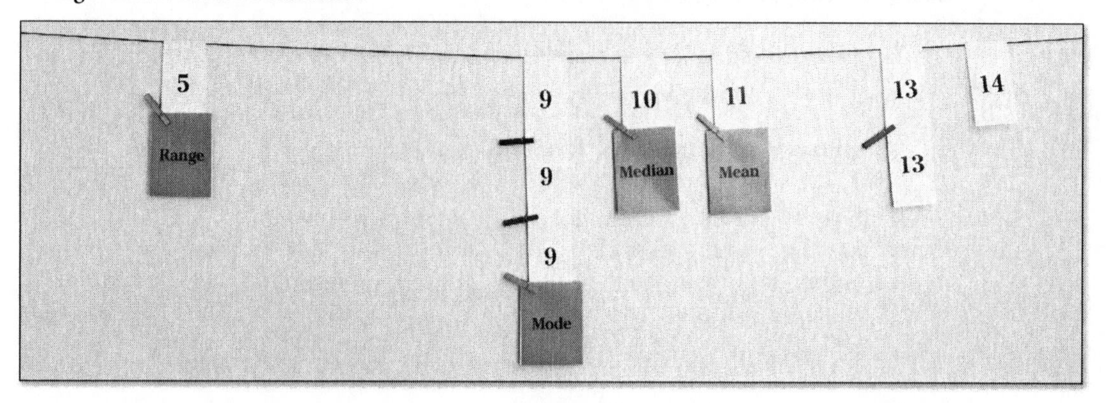

Teacher: Andre, why 11?

Andre: We added the numbers and divided by 7.

Teacher: So the sum of the values divided by the number of values?

Andre: Yes.

Teacher: Isn't that the definition of the average?

Andre: Mean and average are the same thing.

Teacher: All in favor, say "aye."

Class: Aye!

Teacher: Alright, we have a median at the middle, a mode for the most common value, a range as the spread from least to greatest, and an average for the sum of the values divided by the number of values. Please record your fine work on the activity sheet and make note of these four definitions.

Why These Numbers?

I chose a data set that was large enough to demonstrate the four measurements without being too cumbersome. The numbers were relatively small (as opposed to 59, 60, 63), so the range could be shown on the same scale. I chose an odd number of values, so the median would be one of the data points but still allow an extension for adding an extra value to the set. The values were intentionally chosen, so the mean and the median were separate values.

- Why did your group choose that value for the _______?
- What is the definition of _______?
- Where on the number line would you like me to move this?
- What was your alternate strategy?
- ...the most what?
- Why do you disagree?

I started with median and ended with mean, because I wanted to place the two heavy calculations at the beginning and end of the lesson. As it turned out, these were the easiest responses for the class, while the mode and the range were the hardest. Apparently the calculations were not nearly as difficult as the understanding of the academic language.

For example, it was very important when Lexi was struggling with the meaning of *mode* that she and her group members were encouraged to struggle through the sense making. There is a good chance others were having the same trouble discerning between the most frequent number and "the most" as the highest value. I did not anticipate Linda's struggle with *range* either. The visual showing the distance between the minimum and the maximum matching the distance from 0 to 5 was a glorious moment for me and an effective one for Linda.

The cards for this lesson are available in the Digital Resources (lesson29.pdf).

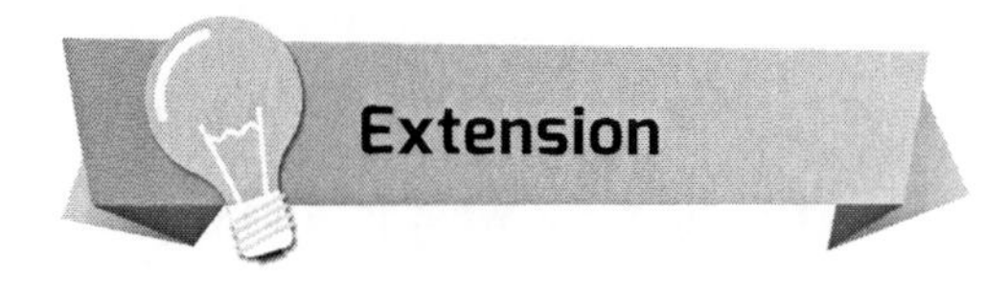

Add the value of 3 to the data set. Ask students to recalibrate the four statistical measures. The new set will have a median of 9.5, a mode of 9, a range of 11, and a mean of 12.

Lesson 30: Five-Number Summary

Min	5	7	Q_1	11	12	Med	22	Q_3	25	Max

Objective: Identify the five-number summary of a data set and understand that each quartile contains 25 percent of the data.

Teacher: **Good day, class. Today, we are going to apply what we learned about the five-number summary to a set of data on the number line. To start, let's remember the names of the five numbers we are looking for. On your lapboards, write as many of the names as you can remember. Share within your groups, and help each other with what you might not know.**

Students work amongst themselves.

Ready? Declare!

Students hold up their boards, showing their work.

Everyone has minimum and maximum. Lola, can you give us another?

Lola: Median.

Teacher: **Good, write that if you don't have it. Audrey, can you give us another name that I see on your group's board?**

Audrey: Q-one.

Teacher: **What does Q stand for?**

Audrey: Quartile.

Teacher: **Good, so Q_1 stands for the first quartile. Sammy, what is our last number?**

Sammy: Third quartile.

Teacher: **Okay, so the five numbers we are going to use to describe this data set today are *minimum, maximum, median, first quartile,* and *third quartile.***

Take a look at the data set I displayed on our clothesline (see Figure 10.14).

Figure 10.14 Data Set

Teacher: Let's start with the minimum and maximum. Ellie, it's your group's turn to place them both.

Ellie's group pins minimum and maximum (see Figure 10.15).

Figure 10.15 Ellie's Placement

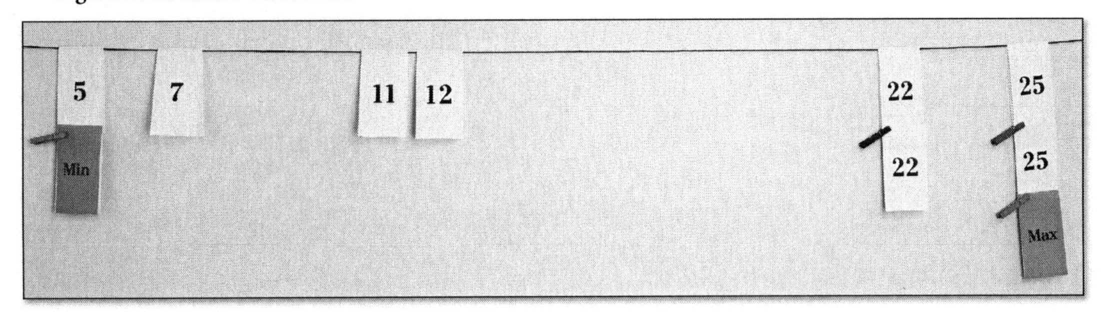

Ellie, why did your group place the minimum and maximum where you did?

Ellie: *Minimum* means "smallest." *Maximum* means "biggest."

Teacher: Thank you. Derek, your group is to place the median.

Derek's group places the median (see Figure 10.16).

Figure 10.16 Derek's Placement

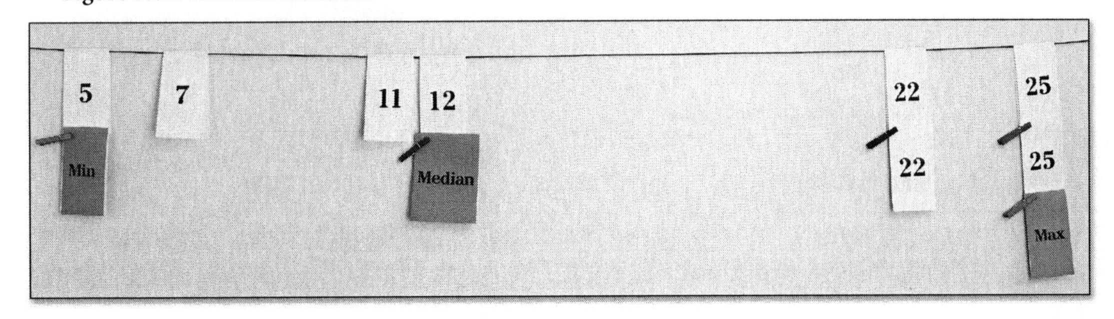

Derek, can you explain why your group claims 12 is the median?

Derek: It's in the middle.

Teacher: In the middle of what?

Derek: The minimum and the maximum.

Teacher: How is 12 the middle of this data set?

Derek: It's about equal on both sides.

Teacher: Yoshi, your group has something different.

Yoshi: We say 15.

Teacher: Please tell us why your group claims 15.

Yoshi: You average the minimum and the maximum.

Teacher: Malik, I see your group has a different answer than 12 or 15. What do you think is the median?

Malik: 17.

Teacher: Why?

Malik: It's the middle, but you have to pick the middle number of all of them. Since there are two middle numbers, 12 and 22, we picked the middle of those two.

Teacher: **So class, we have three cases made for the median: 12, 15, and 17. Time to vote. Write which one you agree with on your lapboards, or you may write a completely different number. Ready? Declare!**

Students hold up their lapboards, showing 17 as their choice.

Everyone seems to agree that the median is 17. I will pin our median card to 17 while Malik chooses the next group.

Malik selects Maria.

Maria, your group is placing our first quartile. Everyone, adjust the median on your lapboards, if necessary, and place Q_1.

Maria's group places Q_1 (see Figure 10.17).

Figure 10.17 Maria's Placement

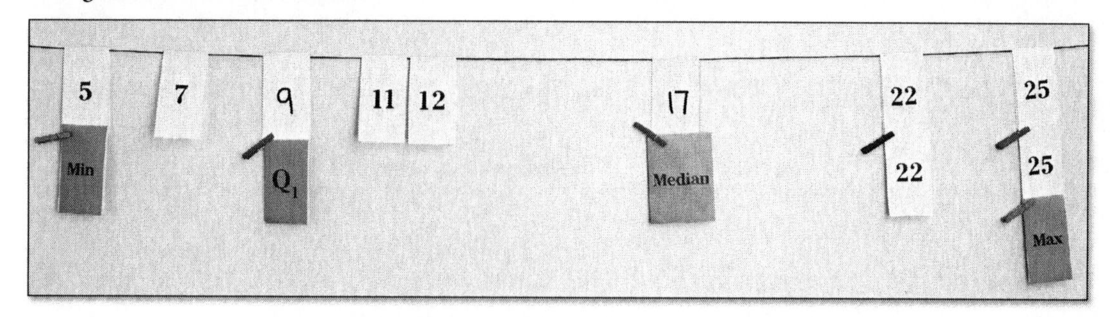

Teacher: Maria, why 9?

Maria: We just did what we all did for the median but to only the lower half.

Teacher: Can you be more specific?

Maria: There are four numbers less than the median. The two middle numbers are 7 and 11. The median between those is 9.

Teacher: Thank you for explaining. Who is going to place Q_3 for us?

Frankie raises his hand.

Frankie, your group has Q_3.

Frankie's group places Q_3 (see Figure 10.18).

Figure 10.18 Frankie's Placement

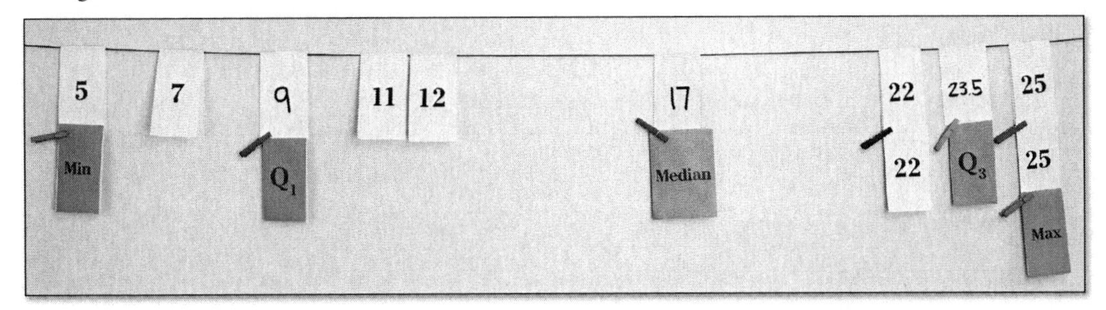

Frankie, well done. Can you tell us what percent of the data points are with any one quartile?

Frankie: I don't know.

Teacher: How many of the values are within the first quartile?

Frankie: Two.

Teacher: Out of how many?

Frankie: Eight.

Teacher: Petra, please tell Frankie the percent that 2 is of 8.

Petra: Hmm...25 percent?

Teacher: Yes, and just like a quarter is 25 percent of a dollar, each quartile is 25 percent of the data. How wide the quartile is tells us the spread of the data in the quartile, but a wide quartile and a narrow quartile have the same number of values. Each has 25 percent of them. Let's record this on our activity sheet including a note about each quartile having 25 percent of the data.

Why These Numbers?

There were eight data values chosen to establish the conversation of quartiles containing one-quarter of the data. This led to a good conversation about calculating the median when its value is not found among the data. I chose values that would also display various spreads of each quartile.

The Key Questions

- What are the five numbers in the Five-Number Summary?
- Can you explain why you claim that is the median?
- What does the *Q* stand for?
- What percentage of the data is within one quartile?
- With which of these answers do you agree?

Analysis

I started with academic language by having students recall the five names in the five-number summary. Then, I started with the easiest of the five numbers—the minimum and maximum—just to start things smoothly. I anticipated the three different responses for the median and relished the opportunity to put all three to a vote. Student defenses of their answers are what convinced other students. Once we cleared that up, quartiles as "medians of the median" were easy for students to calculate.

The five-number cards breaking the data set into four chunks of two values allowed for discussion of the term *quartile*. This also allowed discussion of *spread*. In other words, a wider quartile means a greater spread of the data but not more data. This visual is an example of the power that the Clothesline brings to the conceptual understanding of mathematics.

The cards for this lesson are available in the Digital Resources (lesson30.pdf).

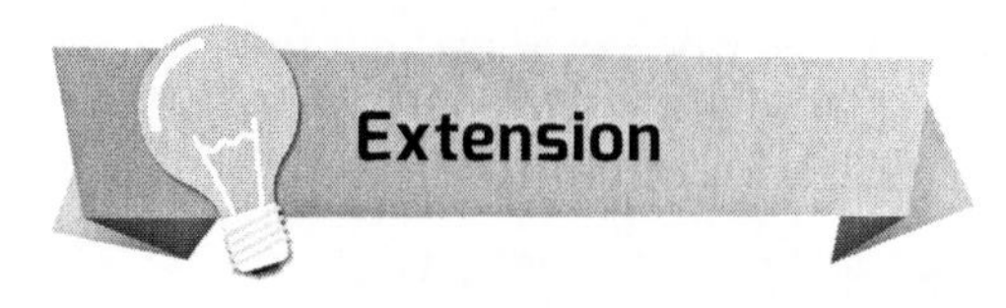

Extension

Why does the summary not include Q_2 or Q_4?

Lesson 31: Standard Deviation

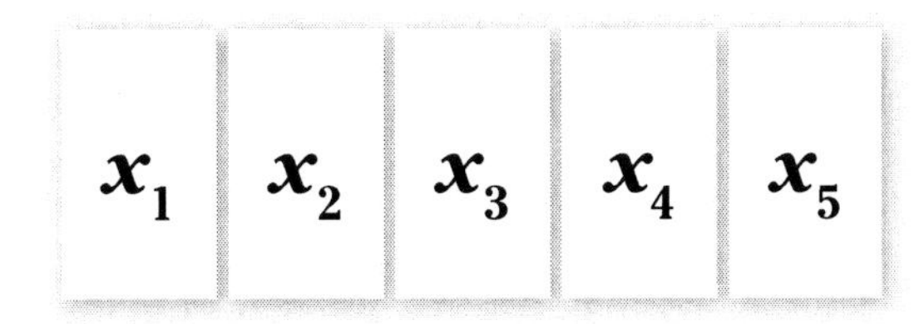

Objective: Compare and analyze the range, mean, and standard deviation of two data sets.

Teacher: **Hello, class. As you can see, we have two data sets** (see Figure 10.19).

Figure 10.19 Data Sets

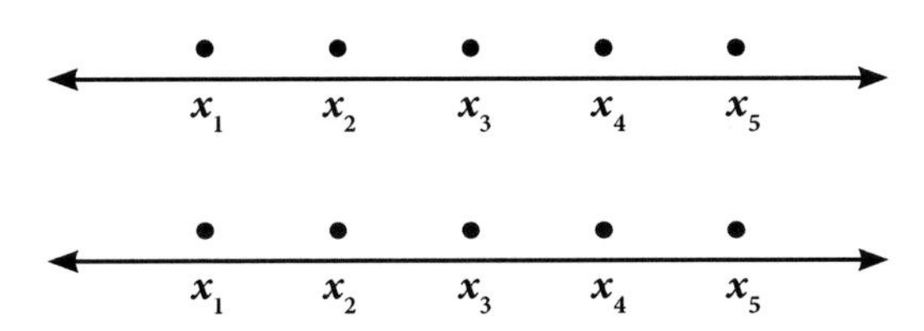

Each has five data points. The values are unknown because they are unimportant for our discussion today. The bottom set will remain the same. You will adjust the top set according to the criteria I give you. I can tell you the criteria will include range, average, and standard deviation. By comparing the sets, you will gain a better understanding of what these words mean as well as an understanding of how they are used to describe the center and the spread of data.

In fact, let's start with understanding what these three measures describe. Let's take the average. Which attribute does the average measure: the center or spread? Cho.

Cho: What do you mean?

Teacher: **For any set of numbers, we like to know what is normal or typical. We usually find that value somewhere in the middle of the set, or the center. We also like to know how much the numbers vary. Do they tend to be crowded around the center? In this case, the spread is small. If the data is literally "spread out," we claim it has a large spread. So, does the average measure the center or the spread?**

Sarah: Well, it would be somewhere in the middle, so center?

Teacher: **Daniel, what did Sarah choose—center or spread for the average?**

Daniel: She said center.

Teacher: **Do you agree?**

Daniel: Yes.

Teacher: **Which does the range measure—center or spread? Jamil?**

Jamil: Spread.

Teacher: **What about standard deviation—does it measure center or spread? Lacey.**

Lacey: Spread.

Teacher: **Well done, everyone. Average measures the center. Range and standard deviation are both used to measure the spread. Let's apply these descriptions of the data to our two data sets on the double clothesline. While leaving the second set in place, please adjust the top data set so it has a smaller range and the same average. Kyle, please choose the group that will go up first, today.**

Kyle selects Jamil's group.

Jamil, have your group adjust the top data set so it has a smaller range but the same average as the bottom set. Everyone else, do the same on your lapboards with your partner. Go.

The teacher puts up the criteria, while Jamil's group adjusts the data set (see Figure 10.20).

Figure 10.20 Jamil's Adjustment for Smaller Range, Same Average

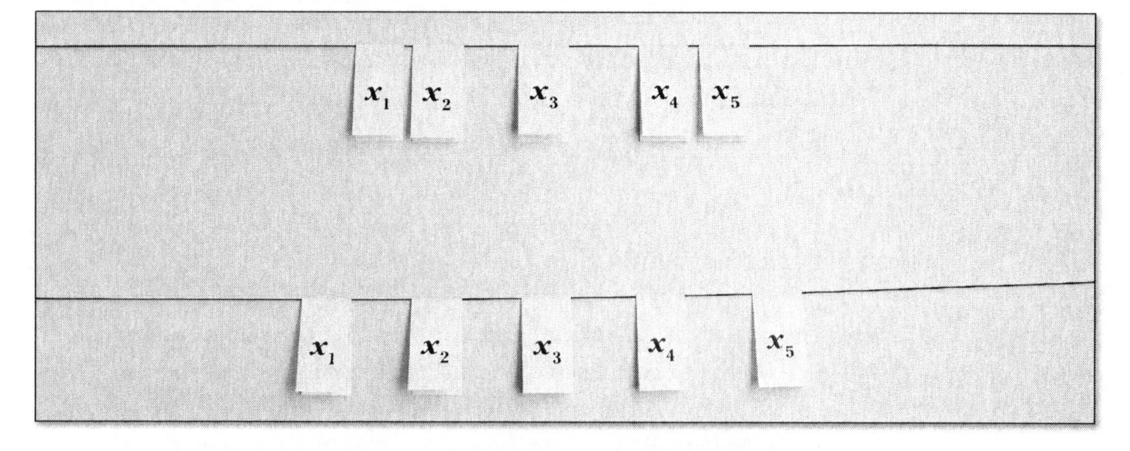

Jamil, please explain your group's response.

Jamil: To make a smaller range, we pinched in the two ends. We thought that as long as we moved them each the same amount, then the average wouldn't change.

Teacher: **What do we call these two at the end?**

Jamil: Minimum and maximum.

Teacher: Let me see whether I understand what you are saying and ask the class whether they agree—as long as I add and subtract the same amount from the total sum, then my average stays the same. Is that what you are claiming?

Jamil: Yes.

Teacher: Class, thumb vote—do you agree with Jamil's group?

The class unanimously agrees.

Sarah, your group had a different response, though. Will one of you please show us what you had?

Sarah's group changes the placement (see Figure 10.21).

Figure 10.21 Sarah's Adjustment for Smaller Range, Same Average

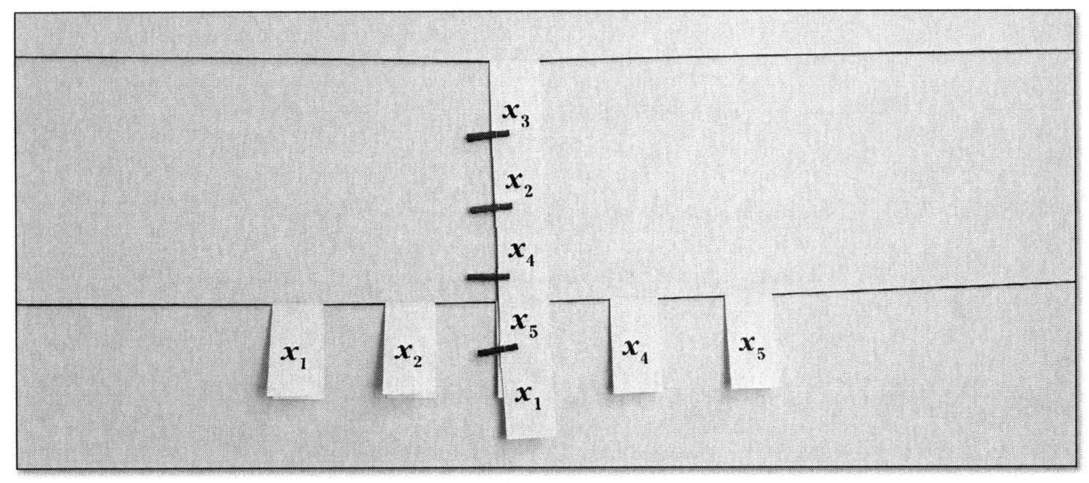

Interesting. You claim that if all the numbers of the top set are the same as x_3 in the bottom set, then the averages will be the same. That means you think x_3 is the average of the other set. What makes you think that?

Sarah: The other values are equally spaced on both sides.

Teacher: Thumb vote, everyone.

There is a unanimous thumbs-up.

Okay, then. Let me put the top data set back to where we started, matching the bottom data set. Now, please adjust the top data set so it has the same range and smaller average. Jamil, please choose our next group.

Jamil selects Peter's group.

Teacher: Peter, that's your group. Everyone else, work with your elbow partner.

Again, the teacher puts up the new criteria. Peter's group adjusts the spacing of the cards (see Figure 10.22).

Figure 10.22 Peter's Adjustment for Same Range, Smaller Average

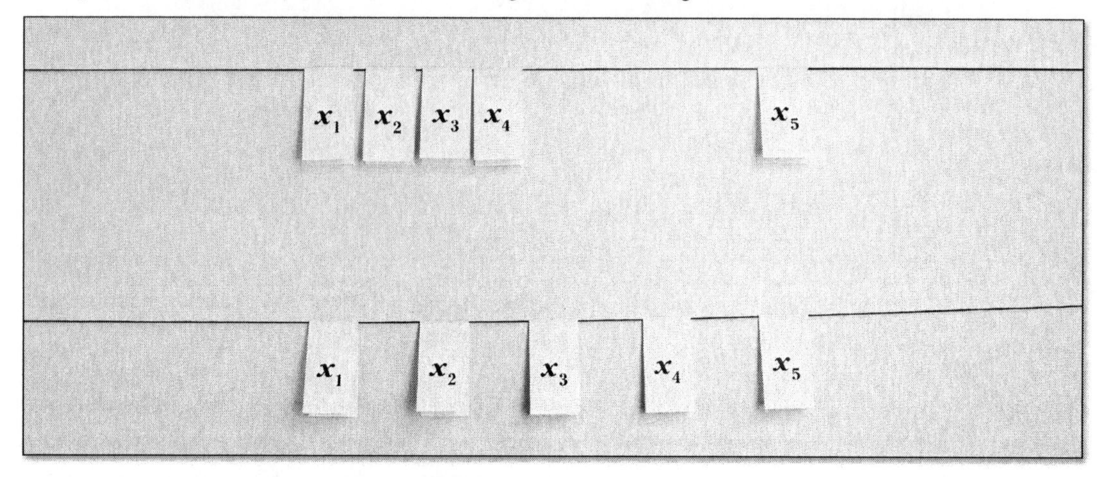

Peter, please explain what your group did here.

Peter: We kept the distance from the minimum to the maximum the same but moved them all lower.

Teacher: Claudia, I notice you and your partner have a different idea.

Claudia: We kept the ends the same but moved the three on the inside lower.

Teacher: Can you show the class what you mean?

Claudia's group changes the spacing of the cards (see Figure 10.23).

Figure 10.23 Claudia's Adjustment for Same Range, Smaller Average

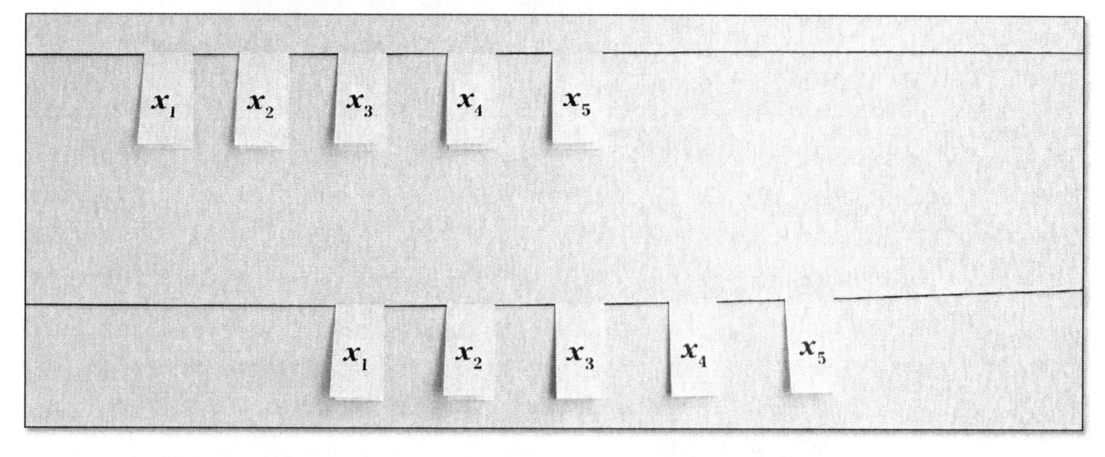

Teacher: If you think these solutions accomplish our objective, clap twice.

There is a resounding two claps.

Teacher: **Terrific. Again, let me put the top data set back where we started, and you adjust the top data set such that it has the same range and a larger standard deviation. Peter, you get to choose our next group.**

Peter selects Claudia's group.

Claudia, back to your group. Everyone else, work with your elbow partner.

Claudia's group returns to the clothesline and adjusts the spacing again, as the teacher displays the criteria (see Figure 10.24).

Figure 10.24 Claudia's Second Adjustment for Same Range, Larger Standard Deviation

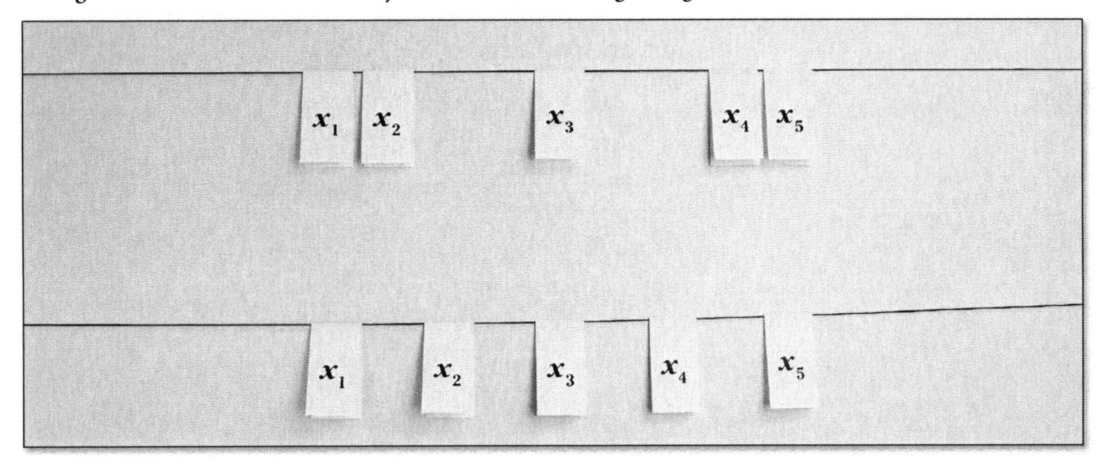

Claudia, what is your group proposing here?

Claudia: We left the minimum and the maximum alone, so range would stay the same. Since we still needed a larger spread, we moved the inside data points wider.

Teacher: **I see most of the class agrees with you, so let's tackle the next one. Back to our original set, adjust the top data set so it has a larger range, smaller standard deviation, and a smaller average. Claudia, choose the next group, please.**

Claudia selects Tanner's group.

Tanner, your group is up. Everyone else, work with your partners on your lapboards.

Tanner's group moves the cards on the top clothesline as the new criteria are posted (see Figure 10.25).

Figure 10.25 Tanner's Adjustment for Larger Range, Smaller Standard Deviation

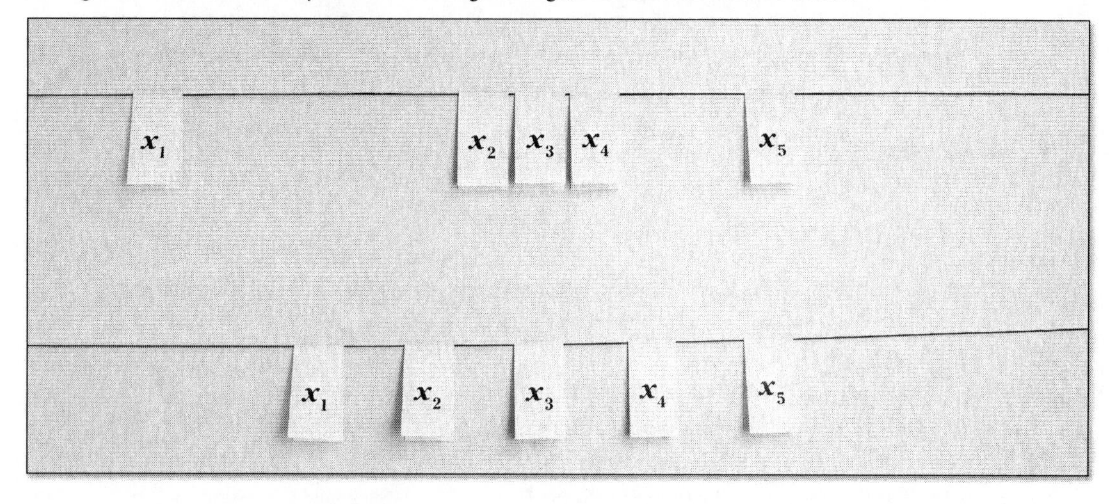

Teacher: Tanner, please have a different person from your group address what you did for each of the three measures.

Ashley: We lowered the minimum. That gave us a larger range. It also lowered the average. Then, we needed to make the standard deviation smaller, so we tightened up the three numbers in the middle.

Teacher: When your group tightened up these numbers, did you raise the average? Maybe even more so than the other data set? Class, I will take a volunteer on this one. Kayne.

Kanye: Not if they moved x_2 and x_4 the same amount like we did before.

Teacher: Thumb vote.

Students vote, but Tanner does not.

Tanner, you did not cast a thumb vote. Are you not sure?

Tanner: No. I'm sure now. I just needed to think for a minute. Yes, I agree with Kayne.

Petra raises her hand.

Teacher: Petra, you have a question?

Petra: May I show something?

Teacher: Of course, please do.

Petra adjusts the spacing of the cards (see Figure 10.26).

Figure 10.26 Petra's Question

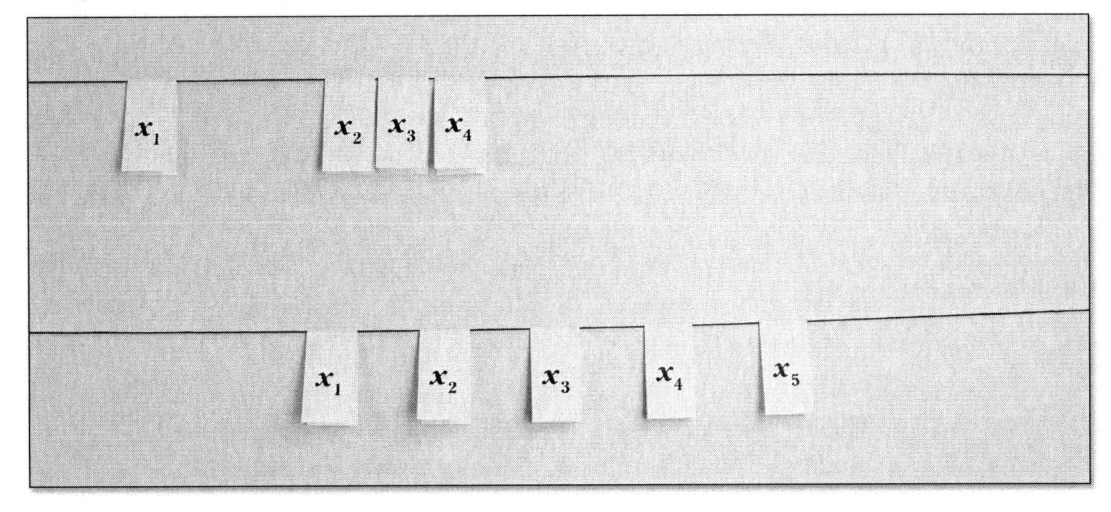

Petra: If we move the maximum higher but then move these three on the inside closer together and lower, are we sure the average would be smaller? Couldn't we move x_5 far enough to the right so that it would make the average bigger?

Teacher: **That is a very interesting question. Let's ask the class. If you think moving the maximum out far enough will eventually make this average larger than the average of the other set, thumbs-up. Thumbs-down if you disagree.**

Students think intently on this one. Eventually, they all agree.

Let's all thank Petra for such a challenging question.

Students applaud enthusiastically.

123 Why These Numbers?

There are no numbers—that is the amazing attribute of this lesson. However, the new twenty-first century standards call for comparing visual displays of data. By not offering numbers, students are not capable of performing any calculations. Therefore, they are required to rely on conceptual understanding of range, standard deviation, and average.

The Key Questions

- What do each of these three measures (i.e., range, standard deviation, average) describe: center or spread?
- What is the meaning of *standard deviation*?
- Is there another way?
- What makes you think the averages are the same?

Analysis

The beauty of this activity is that it is rich in deep conceptual understanding of statistics and heavy on the use of academic language without calling for a single calculation. There is intentionally a great deal of clarification of the three definitions at the front end of the lesson and of the meanings of *center* and *spread*. The lesson started with simple examples and progressively became more complex and challenging.

The second attractive feature of this lesson is its open-ended nature. Each new set of criteria allowed for multiple strategies and deep contemplation, as with Petra's inquiry about how the magnitude of the change affects the measures of central tendency and spread.

The cards for this lesson are available in the Digital Resources (lesson31.pdf).

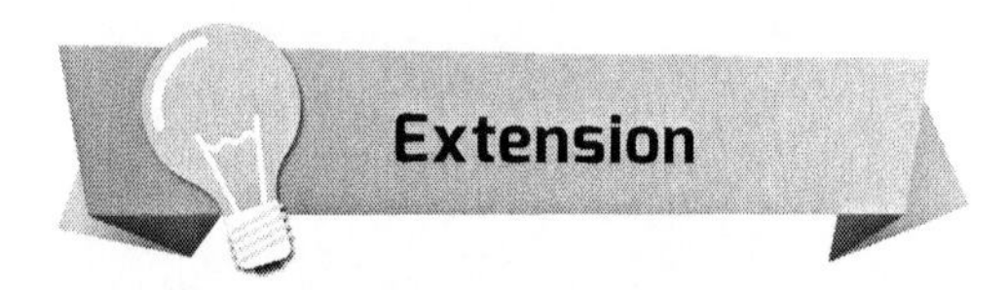

Extension

Partner Challenge: One partner chooses a criterion (e.g., smaller, larger, or same) for each of the three measures and challenges the second partner to create a data set that meets the criteria.

Functions

Lesson 32: Slope

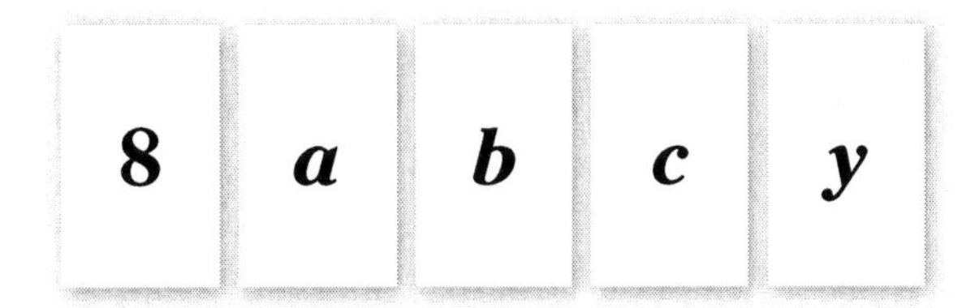

Objective: Determine missing coordinates of collinear points given the origin and one other point on the line, and identify the constant of proportionality.

> Teacher: **Good morning, everyone. Since we have been studying linear equations of proportional relationships, let's see whether the open number line can help us better understand constant of proportionality of these relationships** (see Figure 11.1).

Figure 11.1 Line through the Origin

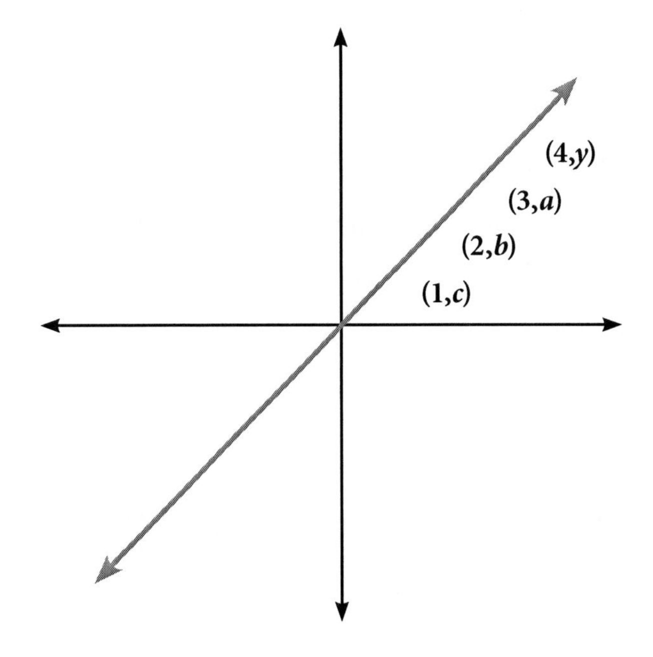

Teacher: **Take a look at this first number line** (see Figure 11.2) **and be ready to share what you notice and wonder.**

Figure 11.2 Slope on the Clothesline

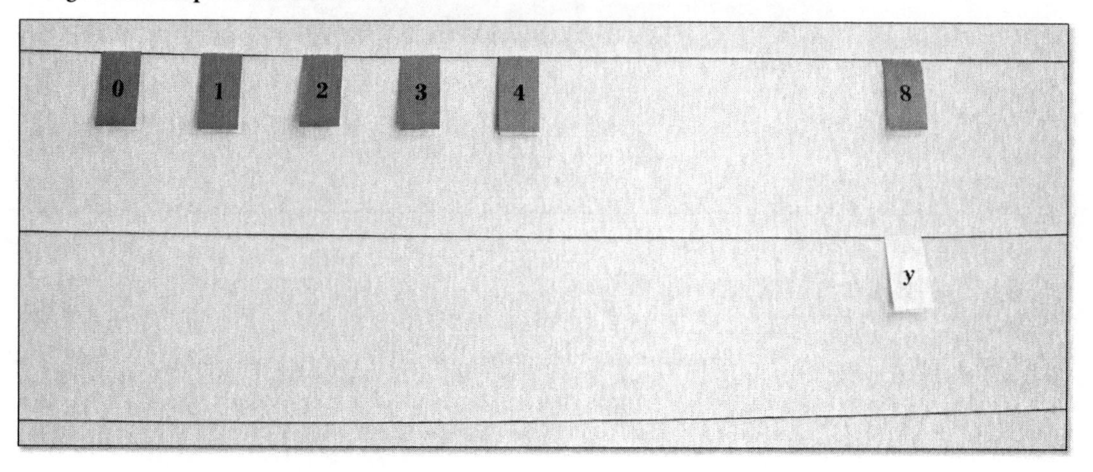

What do you notice, CJ?

CJ: We don't know what the *y* cards are.

Teacher: **Jamie.**

Jaime: The *x* cards are 1, 2, 3, and 4.

Teacher: **Sonia.**

Sonia: It's a straight line.

Teacher: **Anyone else?**

Bena: It goes through (0, 0).

Teacher: **What do you wonder, Oscar?**

Oscar: What are you going to make us do with it?

Teacher: **Cute. Let me ask you—if I place *y* here on the number line, can you accurately place *a, b,* and *c*?**

Students nod.

Oscar, which group is doing this one?

Oscar selects Vanessa's group.

Vanessa, your group is up. Please place *a, b,* and *c* on the clothesline while everyone else works on their lapboards.

Vanessa and her group place the three cards, as shown in Figure 11.3.

Figure 11.3 Vanessa's Placement

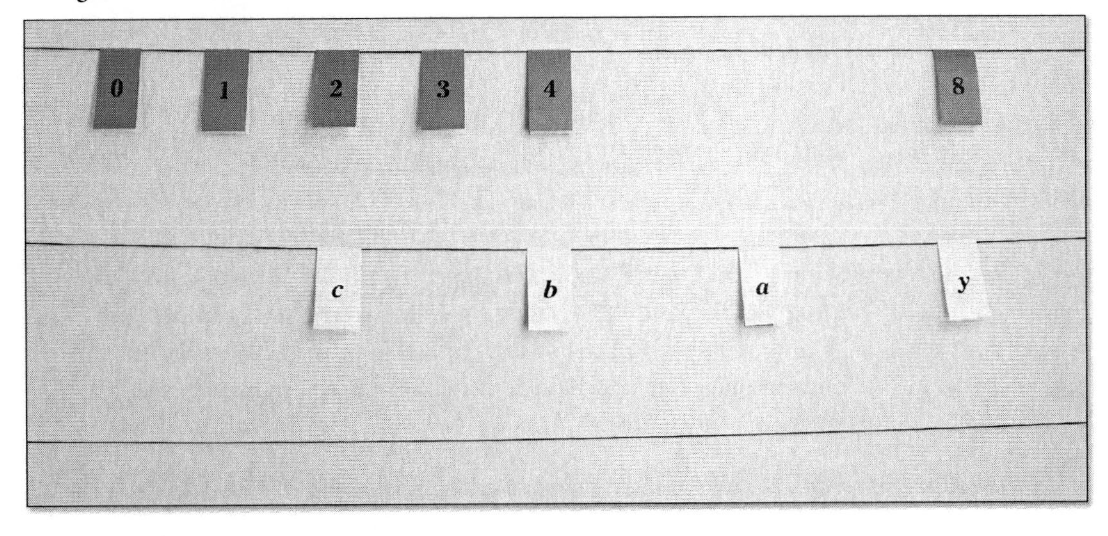

Teacher: **Vanessa, why those values?**

Vanessa: You gave us $y = 8$, which is twice the 4, so we made every y twice the x.

Teacher: **Thank you, but I can see that not everyone here agrees. Greg and Nancy, you two have a different answer on your board which I would like you to come and share on this third clothesline.**

Greg and Nancy add cards to a third clothesline (see Figure 11.4).

Figure 11.4 Greg and Nancy's Placement

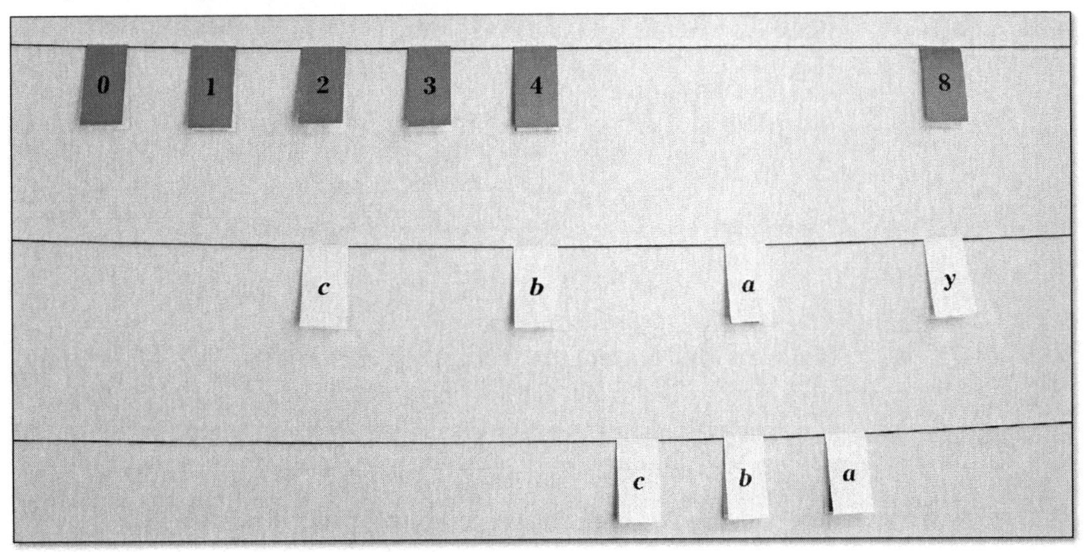

What were you thinking about this?

Greg: Since the x values go up by one each time, we made the y values do the same thing.

Teacher: Well, how are we going to decide which one is correct?

There is a long, silent pause.

I am still wondering, how can we decide which of the two responses is correct?

Mei: Graph the points?

Teacher: Sounds good to me. On graph paper, using 1, 2, 3, and 4 as values for x, plot the four points using the values from our middle line as the y's. Then, plot the four points using the values from the bottom line as the y's. Decide which one, if either, matches the graph in our diagram.

Students graph the plot points. A sample graph is shown in Figure 11.5.

Figure 11.5 Graph of Both Responses

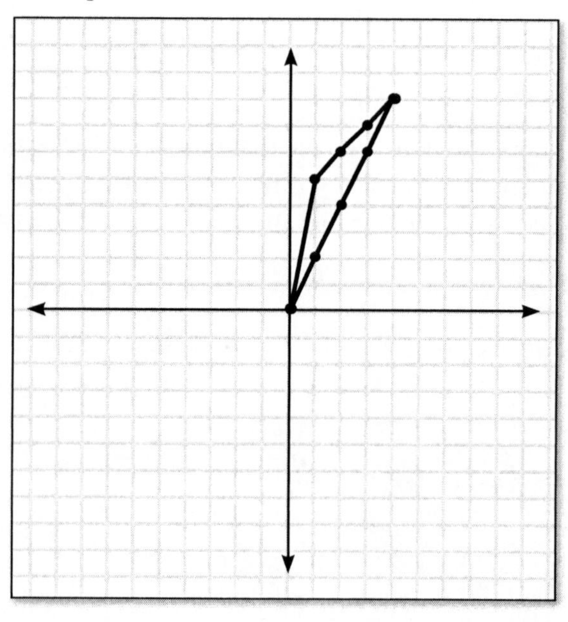

Finger vote. One finger for the middle clothesline answer, two fingers for the bottom line clothesline answer.

The entire class votes for the middle line representing the correct response from Vanessa's group.

Teacher: Greg and Nancy, even you voted against your own answer.

Nancy: Well, they make a straight line, but then you have to bend it to get it to go through the origin.

Teacher: **So, what is the constant of proportionality for the relationship represented by this line? Write it on your boards please.**

There are a variety of answers including 1, 2, and 8.

Hmm, there are a lot of answers on this one. Tell me gang, if proportionality involves a ratio, then what is the ratio of y to x?

Class: Two!

Teacher: **Okay, and what is the y-value when $x = 1$ and our line passes through the origin?**

Class: Two!

Teacher: **Yes. In fact, Vanessa's group explained that all y-values are twice the x-values on this line. She could have said the constant of proportionality is 2 and that would have meant the same thing. Nice work.**

123 Why These Numbers?

One of the new twenty-first century standards is identifying the constant of proportionality from a graph. This involves a graph through the origin showing a constant rate of change. The value of $y = 8$ was chosen to yield a fairly simple constant of proportionality of 2.

The Key Questions

- What do you notice and wonder about the graph?
- Why did you choose those values?
- How do you determine which graph is correct?
- What is the constant of proportionality for the relationship represented by this line?
- What is the ratio of y to x?
- Is this ratio constant for all points?
- Which graph is correct?

I started with the notice and wonder because students do not observe all the details of a graph as quickly as teachers. The notice and wonder requires students to look at the graph in a detailed and critical fashion. I anticipated most students would accurately place the points according to a ratio of 2 and some would follow the consecutive integer strategy (7, 6, 5). Rather than rely on a discussion of the definition of constant of proportionality to settle the debate, I chose to have students graph instead. I also didn't want to give away that strategy, so I intentionally waited out the awkward pause until someone came up with it. Having students graph helped tie the idea of slope to the relationship of the numbers in each ordered pair.

The cards for this lesson are available in the Digital Resources (lesson32.pdf).

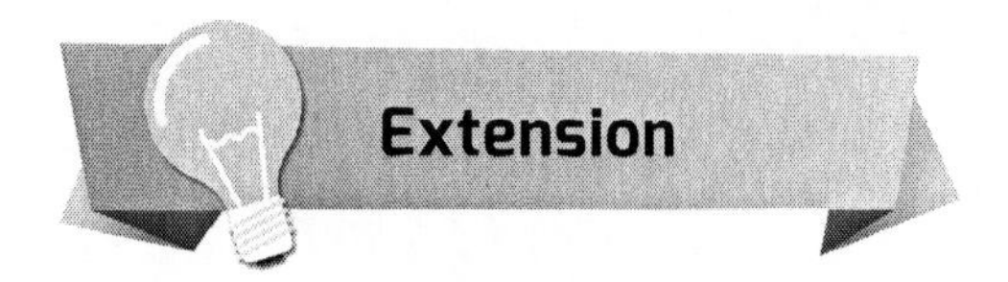

Place y in such a way that it does not represent a whole-number ratio. Alternatively, offer x instead of y, as shown in Figure 11.6 below.

Figure 11.6 Extension Using the *X*-Coordinate

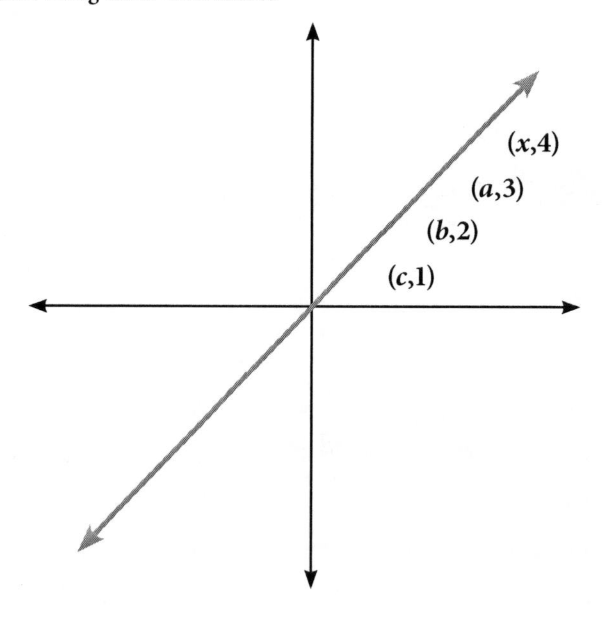

Lesson 33: Slope-Intercept Form

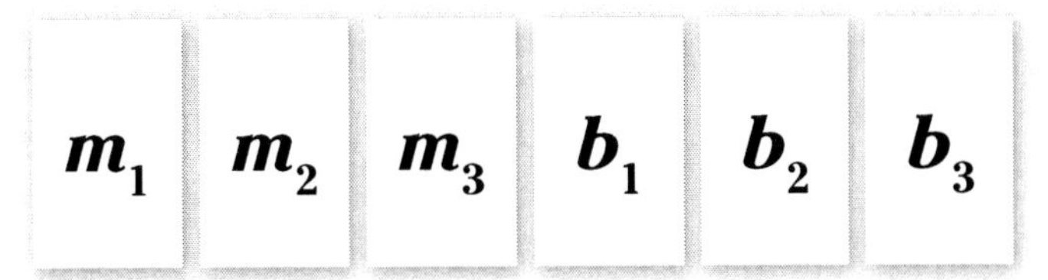

Objective: Determine the approximate slope and y-intercept of the graph of linear equations.

Teacher: **Good day, everyone. Let's warm-up with this diagram today** (see Figure 11.7).

Figure 11.7 Samples of Slope-Intercept Form

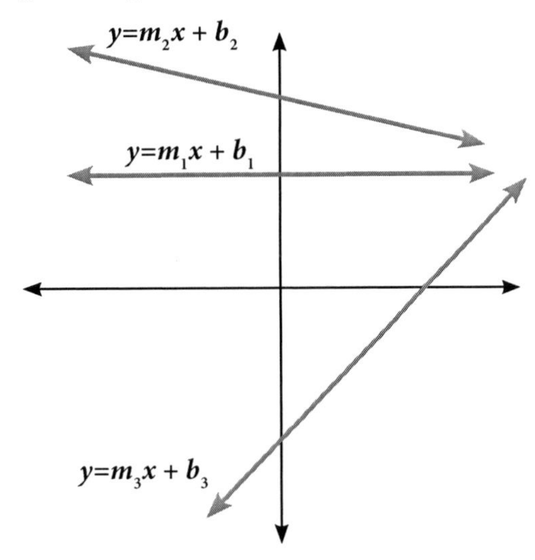

We see the graph of three linear equations. The equations are given to us in slope-intercept form, $y = mx + b$. There is no scale offered on these axes, so let's assume they are both one to one. I would like our first group to come and place m_1, m_2, and m_3 on our number line. Janice, please choose our first group today.

Janice selects Kate.

Kate, your group is going to the clothesline. Everyone else, work with your partner on your lapboards.

Kate and her group place the cards, as shown in Figure 11.8.

Figure 11.8 Kate's Placement

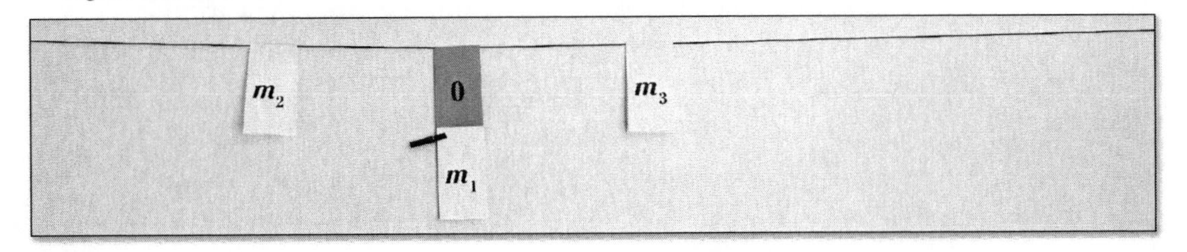

Teacher: Kate, please explain your group's reasoning on this one.

Kate: Well, we knew a flat line has no slope.

Teacher: Has no slope?

Kate: Oh, it has a slope of 0, so $m_1 = 0$. Since the top line is going down, that shows a negative slope. So, we made that slope any negative number and we made the bottom line have a slope of any positive number.

Teacher: First, does everyone agree the *m* in each equation represents the slope of the line? Thumb vote.

Everyone votes yes.

Does everyone agree we have an example of three types of slope: a slope of zero, a slope that is negative, and a slope that is positive? If you agree with Kate, clap twice.

The whole class claps twice.

Did anyone try to get a bit more specific about the slopes? Yes, Jonathan.

Jonathan: We saw the negative slope was not as steep as the positive slope. And, the positive slope looks like it goes up one over one. So, we made m_3 a positive one slope, and m_2 negative one-half slope.

Teacher: Like this?

The teacher moves the benchmark card farther to the left, as shown in Figure 11.9.

Figure 11.9 Teacher Placement

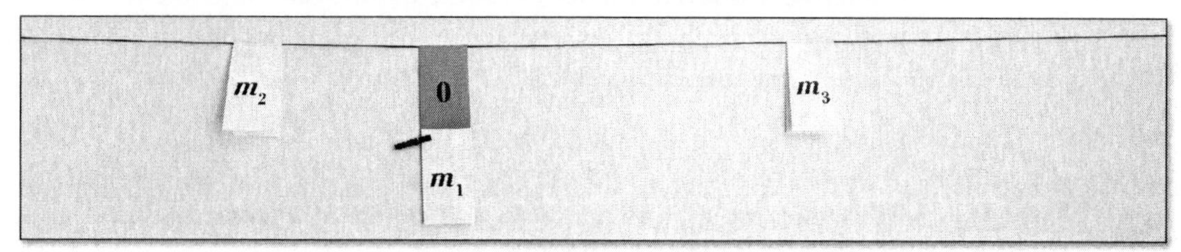

Jonathan: Yes.

Teacher: So, Kate's placement served as a good estimate, which is what I asked for. Jonathan gave us something a bit more specific. Let's now look at the *b* values. Kate, you get to choose which group will place b_1, b_2, and b_3.

Kate selects Travis.

Teacher: Travis, your group is up. Everyone else, erase your last answer, and draw a new number line. Go.

Travis and his group place the cards, as shown in Figure 11.10.

Figure 11.10 Travis's Placement

Interesting, Travis; please explain.

Travis: The b values represent the y-intercepts. We started by choosing any number for b_1. Then, we made b_2 twice as big because it looks about twice as far from zero. Finally, we made b_3 the opposite of b_2 because they both look about the same distance from zero.

Teacher: Let's start with your first statement. Class, do you agree with Travis's group that b represents the y-intercept in our linear equation? Thumb vote.

The class votes, except Kent.

Kent, you are abstaining?

Kent: What does that mean?

Teacher: You didn't vote.

Kent: What was the question again?

Teacher: Does b represent the y-intercept?

Kent: Yes.

Teacher: Do you agree with Travis's group on their logic for the placing and spacing of the b values on the number line?

Kent: What did they say again?

Teacher: Travis, would you like to share again?

Travis: The b_1 is positive, so we picked a positive number. The b_2 looks twice as far from the origin than that, and the b_3 looks like it is the exact opposite of b_2.

Teacher: **Kent, take a close look at the graph before you answer. Do you agree with what Travis said?**

Kent: Yes.

Teacher: **Class, if you agree, clap twice.**

There are two big claps.

Outstanding. Let's record the slopes on one number of the activity sheet and the y-intercepts on another.

There are no numbers in this lesson, and that is part of its strength. There is no calculation of slope and y-intercept, only conceptual understanding of these attributes on the graph and their connection to the symbols and the names.

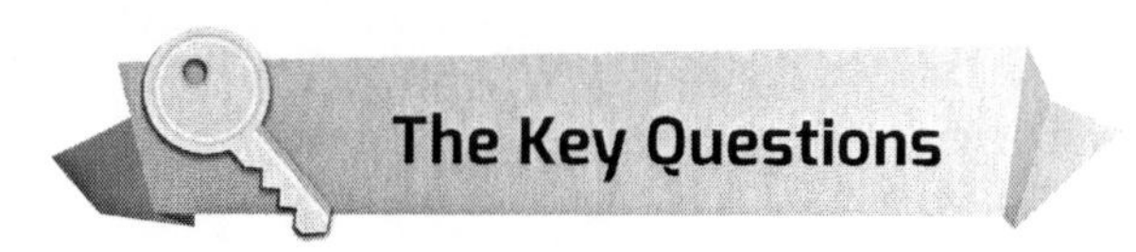

- Does everyone agree we have examples of three different kinds of slopes?
- Does this line have no slope or a slope of 0?
- Can anyone be more specific with the spacing on the clothesline?
- Do you agree with what the other group just said?

I was pleased to see most students recognize m was the slope. There were only a couple pairs of partners who used it as a y-intercept, and they quickly self-corrected after the first explanation. There wasn't much to correct, so I focused on the precision of the spacing to draw students' attention to key concepts, such as 0 slope is different than no slope, and steeper lines have a greater absolute value of the slope.

The cards for this lesson are available in the Digital Resources (lesson33.pdf).

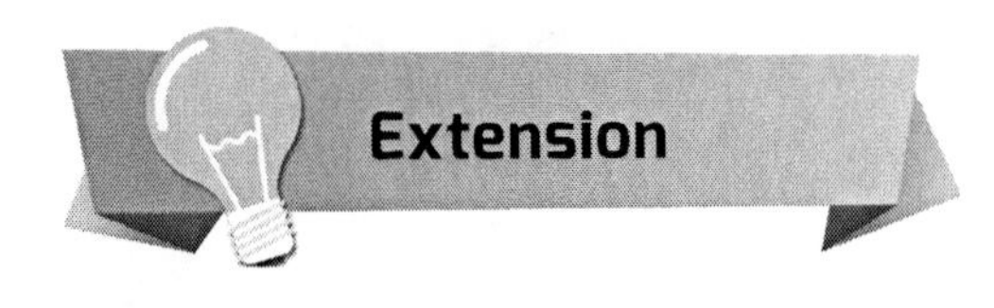

Draw another diagram with three different slopes that are all positive with the same y-intercept.

Figure 11.11 Extension Sample

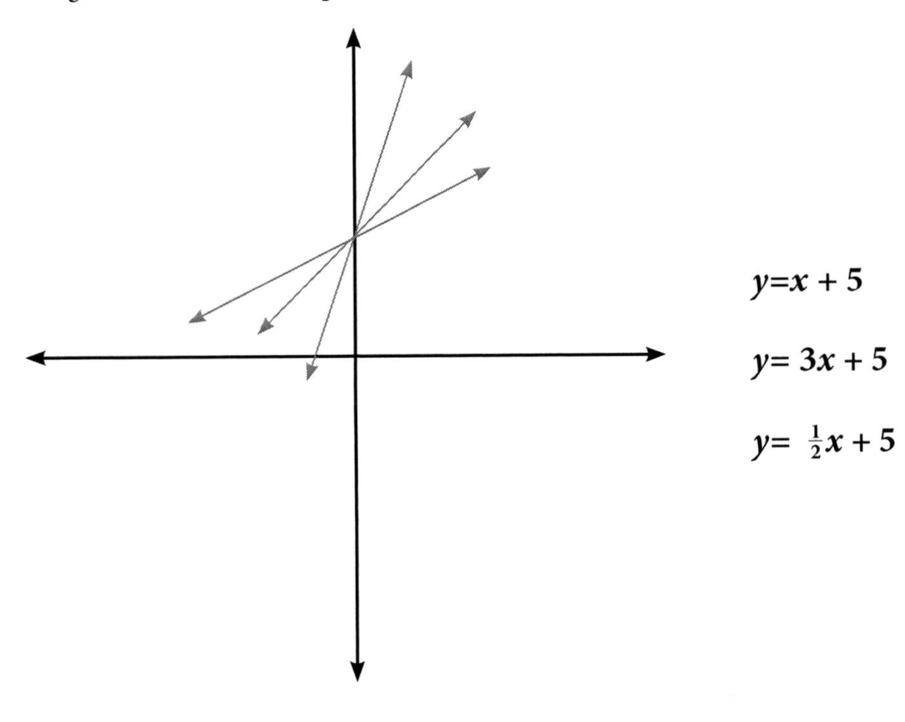

Lesson 34: Exponential Growth and Decay

Exponential Decay $y = r^x$	Exponential Growth $y = r^x$	Neither Growth nor Decay	Neither Increase nor Decrease	10% Decrease	(1 + 0.10)	10% Increase
0.90	90%	100%	110%	1.00	1.10	(1 – 0.10)

Objective: Identify values for r for exponential equations of the form, $f(x) = a(1 + r)^x$ which yield decay versus those which yield growth.

Teacher: **Good morning, class. Before we start our task of applying exponential functions, let's warm-up with an activity to help us determine which exponential functions represent growth and which represent decay. You may notice that we have three benchmarks on our clothesline today as well as a second line** (see Figure 11.12). **Let's draw this on our lapboards with our partners.**

Figure 11.12 Double Clothesline

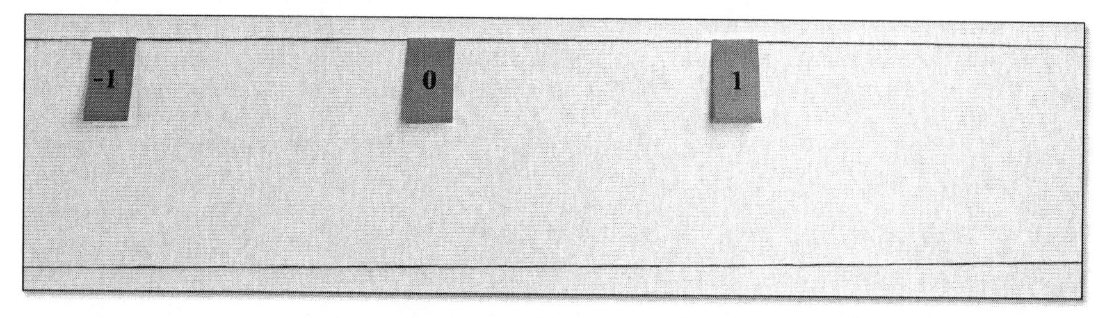

I also have our first two cards: 110 percent and 90 percent. Dalia, will your group please come and show us where these go?

Dalia places the two cards on the clothesline, as shown in Figure 11.13.

Figure 11.13 Dalia's Placement

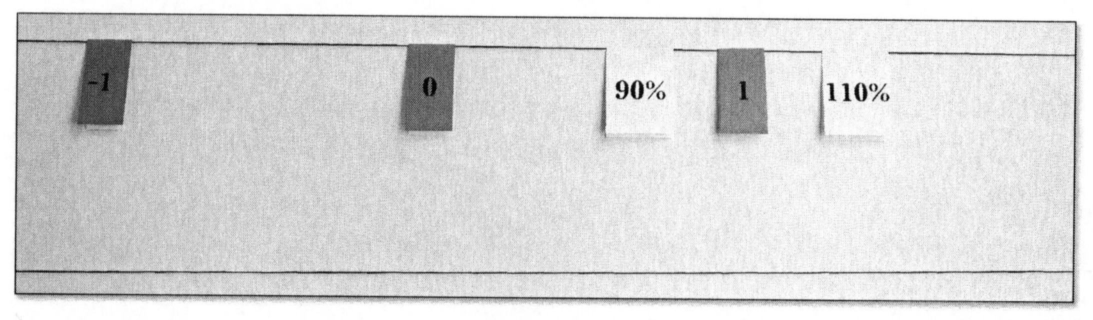

Teacher: **Why there?**

Dalia: Well, 1 is 100 percent, so 110 percent is a little bit more, and 90 percent is a little bit less.

Teacher: **Thank you. Your statement about 1 equaling 100 percent sounds very important to our conversation today, so I think I will add something to our number line** (see Figure 11.14).

Figure 11.14 Teacher's Addition

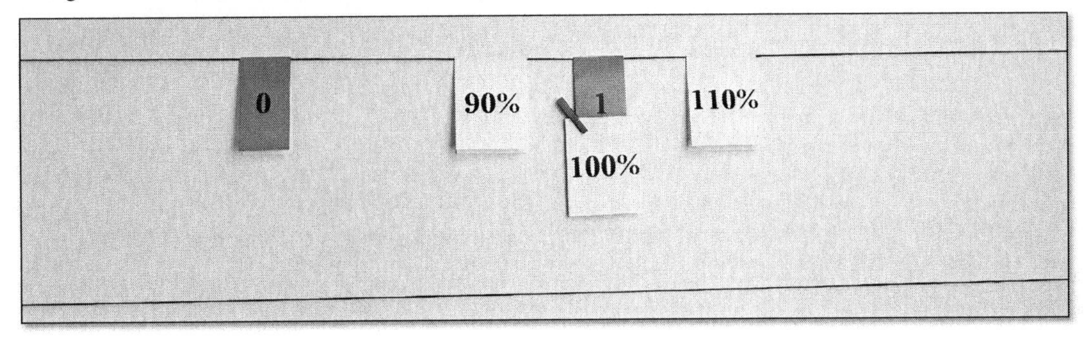

With the idea that 100 percent equals 1, let's now place 0.90 and 1.10. Dalia, who is next?

Dalia selects Odam's group.

Odam, your group is to place the next three cards, 0.90, 1.00, and 1.10. Everyone else, work on your lapboards.

Odam's group places the next two cards (see Figure 11.15).

Figure 11.15 Odam's Placement

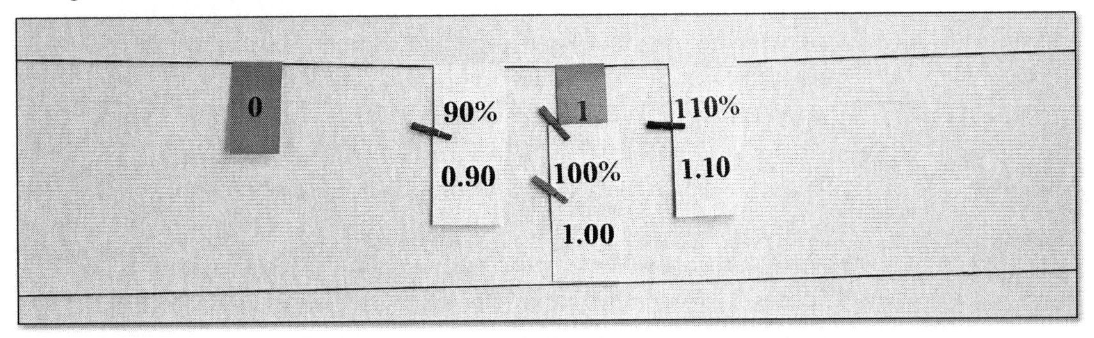

Odam, are you saying 110 percent is the same as 1.10, and 90 percent is the same as 0.90?

Odam: Yep.

Teacher: **Okay, who is next to place the expressions (1 – .10) and (1 + .10)?**

Odam selects Edgar's group.

Teacher: Edgar, it is your group's turn.

Edgar's group adds their cards to the clothesline (see Figure 11.16).

Figure 11.16 Edgar's Placement

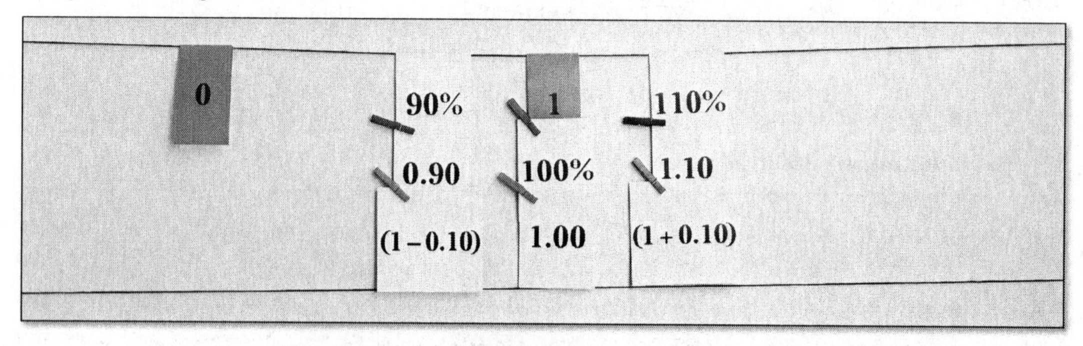

Let's get this straight—a whole minus 10 percent is the same as 90 percent, and a whole plus 10 percent is the same as 110 percent?

Edgar: No?

Teacher: I am not saying that you are wrong.

Edgar: Then, yes.

Teacher: You don't seem so sure. Let's ask what your classmates think. All in favor of what is displayed here, say "aye."

Class: Aye!

Teacher: Edgar, they agree and so do I. So, let's summarize this line before we think about the second one. If we remember that 1.00 and 100 percent represent the same thing because they represent the whole (or, in the case of exponential growth and decay, they represent the initial value), then these others make sense. Ninety percent is 10 percent less than 1, or 0.90, and 110 percent is 10 percent more than 1, or 1.10. Good teamwork! I am now going to move to the second line to place words instead of numbers. But, these words describe the numbers on the top line. I have 3 cards: 10 percent increase, 10 percent decrease, and Neither Increase nor Decrease. You are to place the cards on the bottom line to correspond with the top line. Edgar, which group will place these?

Edgar selects Bonnie's group.

Bonnie, it's your group's turn with the clothesline.

Bonnie's group adds cards to the second clothesline (see Figure 11.17).

Figure 11.17 Bonnie's Placement

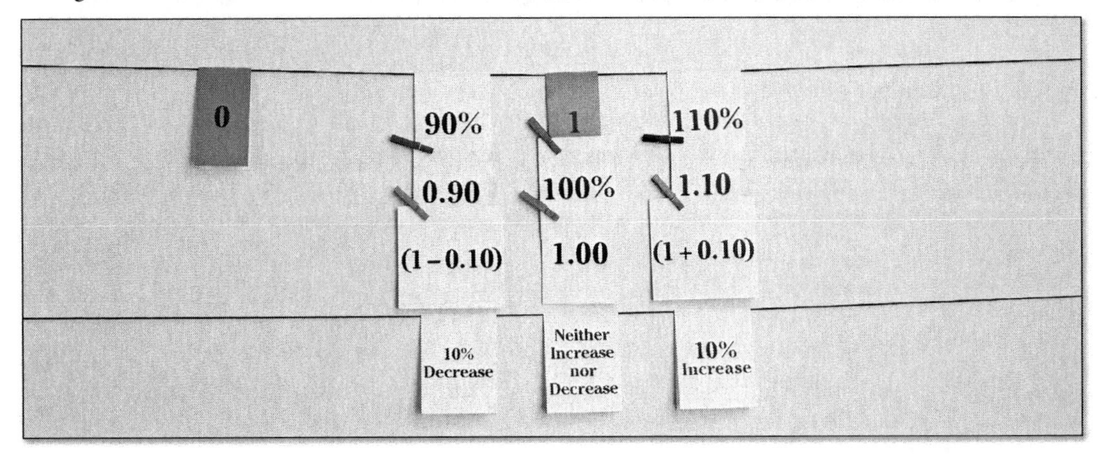

Teacher: Bonnie, please explain your placement.

Bonnie: Well, 1 means you have the same amount because 1 times anything is just the same thing, so that means neither growth nor decay. Growth means you have more than what you started with. Since it says 10 percent growth, we put it at 110 percent. The 10 percent decay means the 1 minus 10 percent.

Teacher: I see a lot of boards in the room stating the 10 percent decay is here at 10 percent. Tia, your group has 10% decay at 10% on the number line. Do you want to defend that idea?

Tia: No, we agree with Bonnie now. It's 10 percent less, not just 10 percent.

Teacher: What if we had a card at 10 percent, how much decay would that be? Anyone?

Rita: Ninety percent decay.

Teacher: Why do you say that, Rita?

Rita: You have to take away 90 percent to have only 10 percent left.

Teacher: That makes sense to me. Class, if that makes sense to you, clap twice.

Students clap twice.

Alright, everyone. Here comes our last warm-up of the day. I have Exponential Growth, Neither Growth nor Decay, and Exponential Decay. For the equation of $y = ar^x$, what value of r would yield these situations? Bonnie, who gets these?

Bonnie selects Roman.

Teacher: Roman, bring your group to the clothesline and place these three cards.

Roman's group pins the final three cards (see Figure 11.18).

Figure 11.18 Roman's Placement

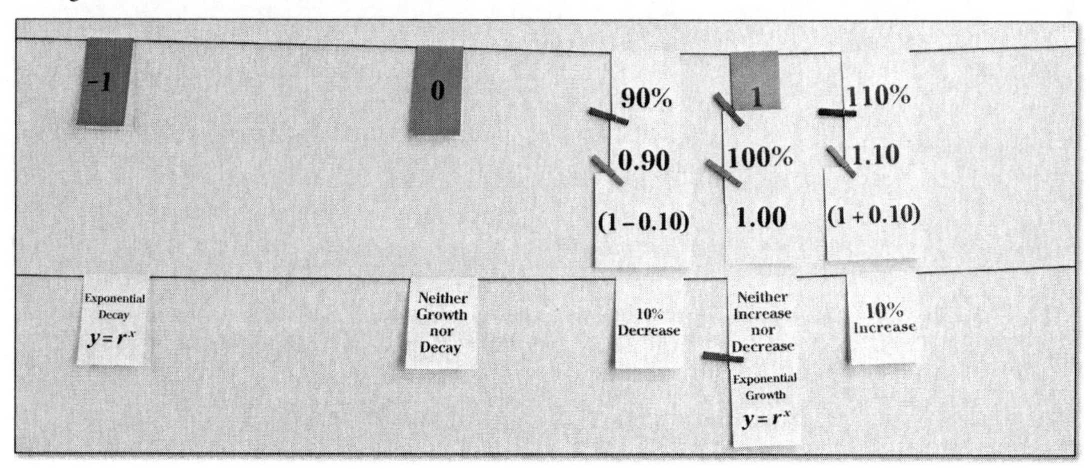

Roman, will you explain your placement?

Roman: Zero means no growth. Negative means it goes down, and positive means it goes up.

Teacher: That sounds logical; let's see if it is correct. I am going to walk us through this. If $r = 0$, that means we need to repeatedly multiply by 0. If you multiply by 0, what does that equal?

Roman: Zero.

Teacher: If you start with some amount and end up with 0, does that mean you had neither decay nor growth?

Roman: Oh, no, no.

Teacher: What would you have to repeatedly multiply by to get the same value?

Roman: One.

Teacher: Correct. So, let's place this card at 1. This means we need to rethink the growth card. What do you have to multiply by to get more than what you started with?

Roman: Something greater than 1.

Teacher: Okay, I'll move the card. Yes, Desiree, question?

Desiree: Can you place it anywhere past 1?

Teacher: Let's think it through. If I repeatedly multiply by 2, or 5.63, or 100, will my amount grow?

Desiree: Yes.

Teacher: There is your answer. For r greater than 1, we will have exponential growth. So, what about $r = -1$? What happens if I repeatedly multiply by -1?

Desiree: You get negative numbers?

Teacher: Do you?

Jameson: Sometimes.

Teacher: Jameson, what do mean?

Jameson: You get negative numbers first. Then, you get positive. Then, you get negative again.

Teacher: Does that sound like a decay pattern?

Jameson: No, that sounds like a crazy bouncing pattern.

Teacher: If we don't want negative or 0, and equal to 1 as well as greater than 1 have already been chosen, what do we have left for our decay? Diane?

Diane: Less than 1.

Teacher: Aren't -1 and 0 less than 1?

Diane: Less than 1, but greater than 0.

Teacher: Somewhere between 0 and 1, Diane says.

The teacher moves the Exponential Decay card (see Figure 11.19).

Figure 11.19 Teacher's Placement

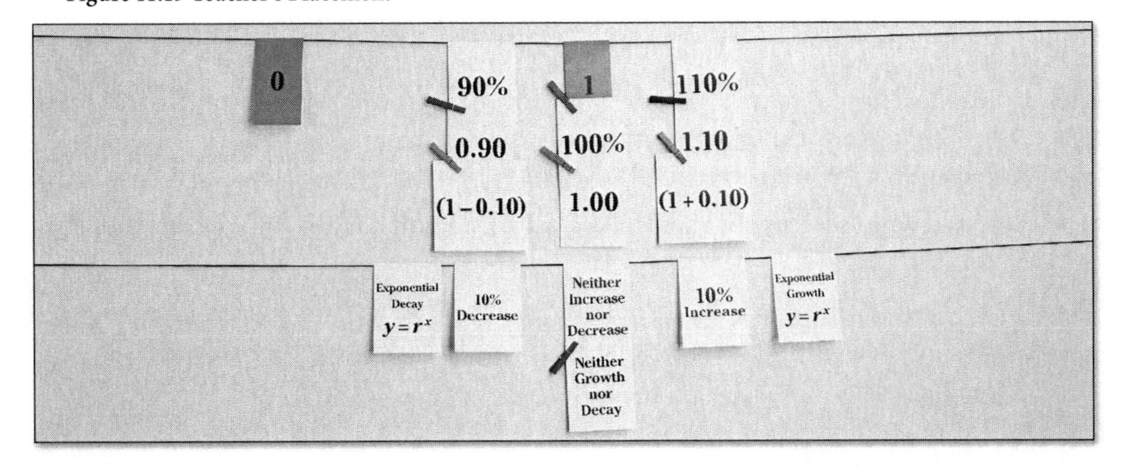

Teacher: All in favor, clap twice.

Students give two resounding claps.

Terrific! After that amazingly intelligent and helpful conversation about exponential growth and decay, let's record our diagram. Be sure to note the values of *r* that generate decay and growth since that was the focus of our big debate today.

Why These Numbers?

A common issue for students in learning exponential decay is the idea that the rate of decay is not the base for our decay equations. In other words, 10 percent decay does not mean 0.10^x, but rather 0.90^x. Juxtaposing 10 percent growth and 10 percent decay on the number line helps address this misconception. The expression $(1 + r)$ is commonly found in growth and decay formulas. This expression evaluates as greater than 1 for growth because *r* will be positive, and it evaluates as less than 1 for decay because *r* will be negative. Tying 10 percent decay, 0.90, and (1 −0.10) together is at the heart of this lesson. The decision to add the verbiage on the second line was intended to drive conversation toward the fact that decay has a positive base between 0 and 1 but never a negative base.

The Key Questions

- Why did you place 10 percent decay there on the number line?
- Are you saying that 110 percent is the same as 1.10, and 90 percent is the same as writing 0.90?
- Are you claiming that a whole minus 10 percent is the same as 90 percent and a whole plus 10 percent is the same as 110 percent?
- What if I placed the card at 10 percent, how much decay would that represent?
- What would you have to repeatedly multiply by to get the same value?
- What do you have to multiply by to get more than what you started with?
- What happens if I repeatedly multiply by -1?
- If you start with some amount and end up with 0, does that mean you had neither decay nor growth?
- If we don't want a negative number or 0, and *equal to 1* as well as *greater than 1* have already been chosen, what type of value do we have left for our decay?
- For the equation of $y = ar^x$, what value of *r* would yield growth? Decay?
- How can you defend your idea?

Students did surprisingly well with placing the values on the first number line. I intentionally paused and summarized the main idea after each card set. I really stressed the difference between 10 percent decay meaning $r = 0.90$ and 90 percent decay meaning $r = 0.10$.

As well as they did with the values, students struggled slightly with verbiage. The -1 benchmark was bait. I placed it on the number line to see whether students would use it. Sure enough, they bit the hook hard. I used this to generate a discussion about values of r that represent decay scenarios. The students' choice of 0 for an r that represents neither growth nor decay continued the rich discussion and cleared up many misconceptions.

Notice my repetitive use of the phrase, "Let's think our way through it." Rather than giving answers in these situations, I led students through the thought process that I go through when thinking about exponential functions and the effects of various values on them.

The cards for this lesson are available in the Digital Resources (lesson34.pdf).

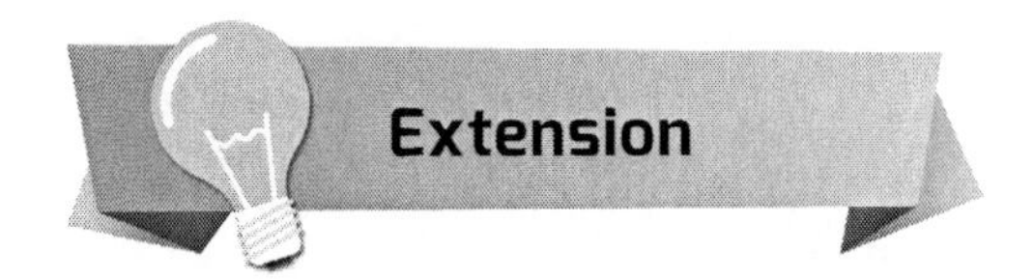

In relation to $(1 - 0.10)$, where would $(1 - \frac{0.10}{12})^{12}$ go?

Lesson 35: Exponential Decay

6 | $6(\frac{1}{2})^0$ | $6(\frac{1}{2})^1$ | $6(\frac{1}{2})^2$ | $6(\frac{1}{2})^3$ | $6(\frac{1}{2})^{100}$

Objective: Accurately apply the order of operation to an exponential decay expression and place the outputs to show understanding of the decay approaching a limit.

Teacher: **Good day, everyone. We have been studying exponential growth for functions written with equations of the form, $f(x) = ar^x$, in which r is greater than 1, as in $f(x) = 2x$ and $f(x) = 3x$. Next, we are going to explore situations in which a quantity decays exponentially instead of grows. In such cases, r is between 0 and 1, as you see with today's function: $f(x) = 6(\frac{1}{2})^x$.**

Let's start by placing two instances of our function, in which $x = 0$ and $x = 1$. Tanya, your group is up. Please place these two while everyone else does so with their partner on the lapboard (see Figure 11.20).

Figure 11.20 Tanya's Placement

Teacher: **Tanya, please explain your group's thinking.**

Tanya: Anything to the 0 power is 1, and six and a half raised to the first power is just $6\frac{1}{2}$.

Teacher: **Do you all agree, or does anyone want to challenge what is up here? Nick.**

Nick: It is not $6\frac{1}{2}$. It is 6 times a $\frac{1}{2}$. The $\frac{1}{2}$ is raised to the first power, so the answer is 3.

Teacher: **Tanya, does that make sense, or do you want to defend your original answer?**

Tanya: I agree. Six and a half would not have parentheses.

Teacher: **Thank you for clarifying that because from what I saw on the boards, you were not the only people reading it like that. I will change it for you all. Are we good then with the 0 power? Maddy.**

Maddy: Only half is to the 0 power. You have to do exponents before you multiply, so that should be 6 times 1. So, 6.

Teacher: **Tanya, what does your group say? Do you agree with Maddy's challenge or…**

Tanya: Yes, we were half right. She is all the way right.

Teacher: **Then, let me adjust that one as well and explain. We can tell this is an exponential equation because the variable is the exponent. We can tell this is a decay situation because the rate is less than 1 but still greater than 0. In fact, I can read this simply as "start with 6 and repeatedly cut in half." This expression with the 0 power basically says cut 6 in half 0 times, and this one with the exponent of 1 says cut 6 in half once. Thinking of it this way, do you agree with our values of 6 and 3?**

The teacher moves the cards to reflect his thinking (see Figure 11.21).

Figure 11.21 Teacher's Placement

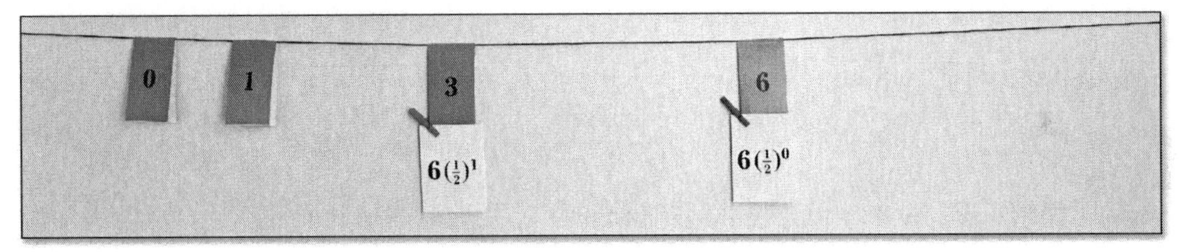

Thumb vote.

Students give a unanimous thumbs-up.

Let's push things a little further. Tanya, which group is up next?

Tanya selects Doug's group.

Doug, your group will place the next two expressions for $x = 2$ and $x = 3$.

Doug's group places their cards (see Figure 11.22).

PART III – Chapter 11

Figure 11.22 Doug's Placement

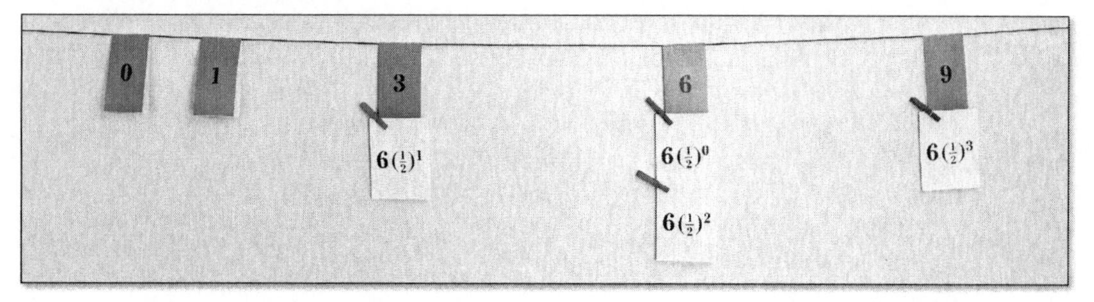

Teacher: Doug, please explain.

Doug: Since the first power was at 3, then 3 times 2 equals 6, and 3 times 3 equals 9.

Teacher: Anyone want to challenge this? Sven.

Sven: You don't multiply by the exponent. The exponent tells you how many times you cut the number in half.

Teacher: Will you come and show us where you think they should go then?

Sven: Sure.

Sven moves the cards to show the correct placement (see Figure 11.23).

Figure 11.23 Sven's Placement

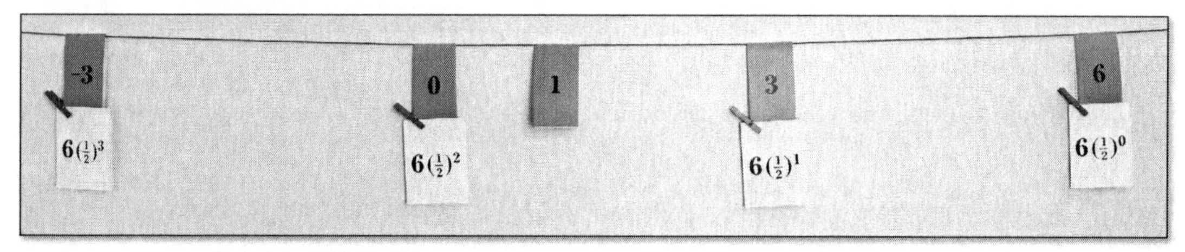

Teacher: Doug, do you agree with how Sven changed your response?

Doug: I agree with what he said, but not what he did.

Teacher: What do you mean?

Doug: I remember what you said after Sven said it—that we start with 6 and cut it in half each time. But 3 cut in half is not 0.

Teacher: Would you like to come up and show us?

Doug moves the cards on the clothesline, as shown in Figure 11.24.

Figure 11.24 Doug's Second Placement

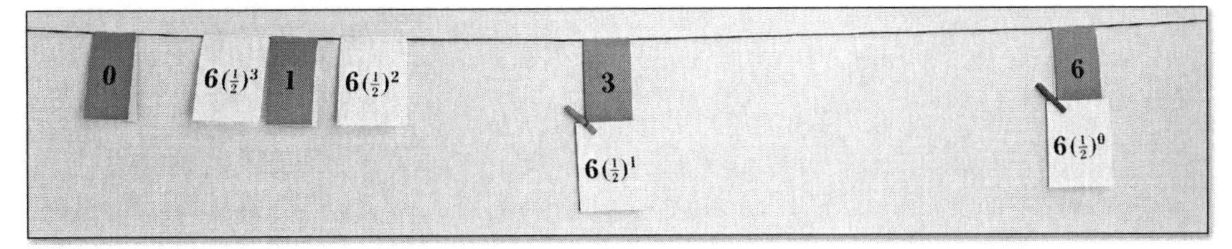

Teacher: **So let me see. Six cut in half is 3, cut in half is 1.5, cut in half is 0.75. All in favor, clap twice.**

Students clap twice.

We could go to $x = 4$ and 5 next, but let's have a little fun…

The teacher writes $6(\frac{1}{2})^{100}$ on a blank number card.

I will take a volunteer for this beast. Shane.

Shane takes the card and places it far to the right of the number line on a box in the classroom, as shown in Figure 11.25.

Figure 11.25 Shane's Placement

Now Shane, that is the most interesting placement of a clothesline card I have ever seen. Why did you place it way off to the right of our number line?

Shane: It's still on the number line, since they go on forever. I'm just saying the 100th power would be way, way out there somewhere.

Teacher: **Interesting thought. I can tell from the show of hands that others might think differently. Lenny.**

Lenny takes the number card and places it far to the left of the number line on the teacher's desk (see Figure 11.26).

Figure 11.26 Lenny's Placement

Teacher: **You two are really getting us thinking here. Shane puts his toward positive infinity, while you say it should be put toward negative infinity.**

Lenny: Yes, if you keep cutting in half 100 times, it gets really small, and the negatives are smaller than the positives.

Teacher: **I understand what you are saying, but Alice is excited to share something. Alice, would you like to show what you are thinking?**

Alice moves the card back onto the number line and close to 0, as shown in Figure 11.27.

Figure 11.27 Alice's Placement

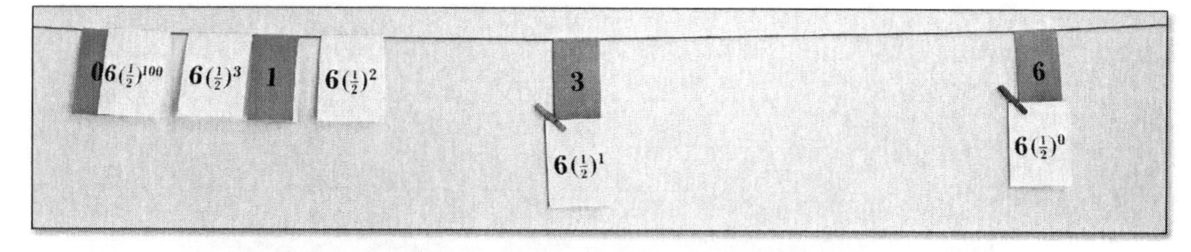

The whole class gives out an audible, collective, "Ooohhh!"

Alice, I think that means a lot of people agree with you. Will you explain?

Alice: If you keep cutting in half, yes it gets smaller, but it gets smaller and smaller closer to 0.

Teacher: **Lenny, I can see from your head nod that you also agree. You don't want those negative values anymore?**

Lenny: No.

Teacher: **Shane, are you good with this or do you want to go back to a very large value?**

Shane: No, I'm good.

Teacher: **All in favor, say "aye."**

Class: Aye!

Teacher: **So, let's recap. *Exponential decay* means "to decrease by the same ratio." In today's case, it was a ratio of $\frac{1}{2}$, so we started at 6 and repeated multiplied by $\frac{1}{2}$. Each successive output kept getting smaller, thus a decay, which meant as x increased, $f(x)$ approached 0. Let's record our fine work on the activity sheet. Be sure to make note that this is a decay function and that this function approaches but never reaches 0.**

Why These Numbers?

A decay function of $\frac{1}{2}$ was chosen because it limited calculations to multiplication by a fraction, keeping student focus on the conceptual understanding of an exponential decay function. Zero was the first value we placed because it sets the groundwork for the y-intercepts of simple exponential functions being equal to 1. The values of the next few exponents being 1, 2, and 3 were to establish the pattern seen on the number line. Then, the leap to the 100th power was intentional. Without a pattern to follow, this value challenges students to think abstractly about what happens when an initial value is recursively multiplied by $\frac{1}{2}$.

The Key Questions

- Does $6(\frac{1}{2})^0$ mean six and a half raised to the 0 power or one-half raised to the 0 power times six?
- Which do we calculate first, exponents or multiplication?
- What does $(\frac{1}{2})^0$ look like on a number line?
- Does anyone want to challenge what is currently on the number line?
- Do you want to change your answer?
- Do you agree with how your classmate changed your answer?
- Why do you think these values approach zero? Positive infinity? Negative infinity?

Analysis

Before students plot points for an exponential decay function, they need to be able to accurately generate the points, which means they must be able to evaluate an exponential decay expression. Therefore, this activity was intended to be the warm-up for a lesson on graphing exponential functions. I did not anticipate the immense issues that were so pervasive in the class. This warm-up took most of the period, but it was time well spent. It offered the opportunity to teach students how to discern the difference between $6\frac{1}{2}$ and $6(\frac{1}{2})$ and the difference between $(6\frac{1}{2})^0$ and $6(\frac{1}{2})^0$. The students obviously need help interpreting the meaning of such expressions as $6(\frac{1}{2})^3$. When I said let's have some fun with $6(\frac{1}{2})^{100}$, I did not realize just how much fun it was actually going to be. I took it as a positive sign of the classroom culture that students felt safe enough to go beyond the clothesline and talk about infinity, even if their thoughts were errant. The debate that occurred over positive or negative infinity brought a hyper-focus onto the concept of exponential decay.

The cards for this lesson are available in the Digital Resources (lesson35.pdf).

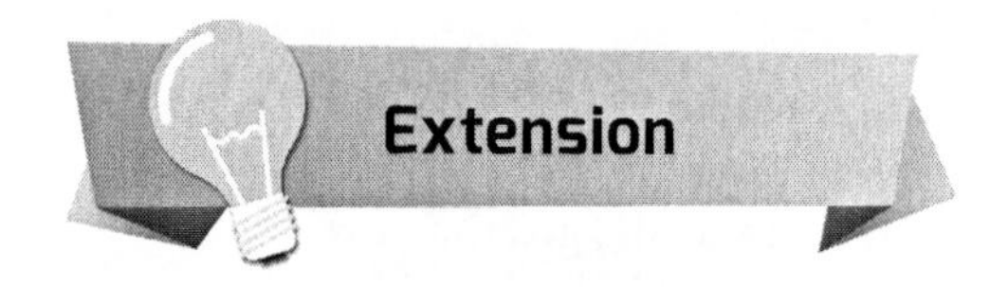

Extension

Have students place for exponent values of $x = -1$, and $x = -2$.

Lesson 36: Three-Equation Models

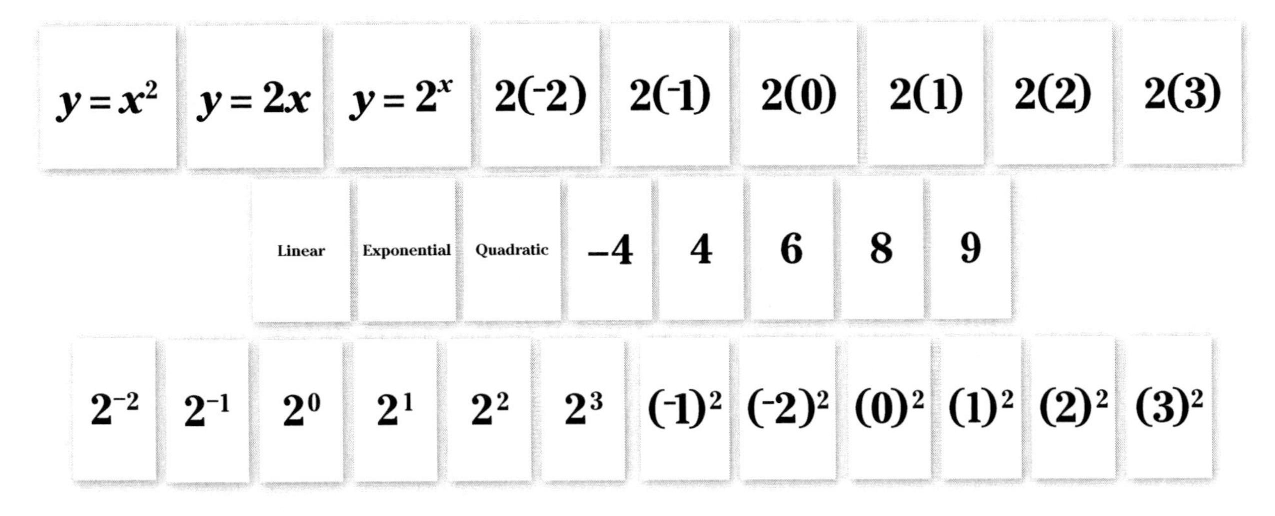

Objective: Compare and contrast the outputs of linear, exponential, and quadratic functions on a number line for the domain of {-2, -1, 0, 1, 2, 3}.

Teacher: **Hello, class. Now that we are approaching the end of the year and have experienced all three of our main types of functions, I would like for us to compare and contrast the outputs of linear, exponential, and quadratic functions. As a warm-up, I would like to first match these three functions to their corresponding names and graphs** (see Figure 11.28).

Figure 11.28 Warm-up Three-Equation Graphs

Linear

Exponential

Quadratic

$y = x^2$

$y = 2x$

$y = 2^x$

On your individual lapboards, sketch the shape of each graph. Then, write the name and equation underneath. Go.

Some students needed correction, but most were solidly competent.

So now that we have a solid agreement of what these graphs look like on a coordinate plane, let's look at their behavior on a number line to get a better understanding of why they have the shape they do on the plane. I have labeled each of these three clotheslines with one of the functions, and now I will also label them with the equations you just showed me.

The teacher adds label cards to the number lines (see Figure 11.29).

Figure 11.29 Triple Clothesline

Teacher: The triple clothesline is a bit different from what we normally work with, so we are going to interact with it differently, too. We are going to sort ourselves into three teams. Groups 1–3, you are our Linear Team. Groups 4–6, you are our Exponential Team. Groups 7–9, you are our Quadratic Team. I'm giving one person from each team an output for their particular function. For example, the first card will have an input of 0. So, we want to see where 2(0), 2^0, and 0^2, respectively, go on the number line. Only one person goes to the clothesline, but the whole team may give feedback. Here you go.

There is enthusiastic instruction shouted by all teams. Students from each team place their cards on the clotheslines (see Figure 11.30).

Figure 11.30 Initial Card Placement

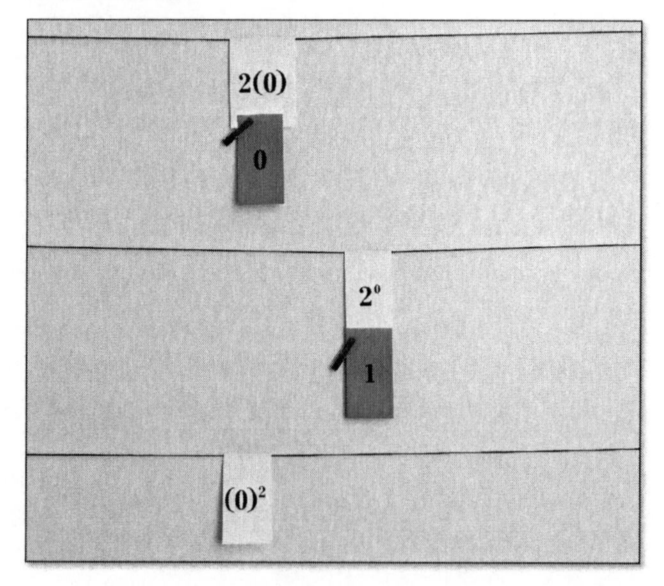

Teacher: **Exponential Team, I noticed you had the most discussion. What was the debate about?**

Exponential Team: Whether 2^0 is 0 or 1.

Teacher: **Is everyone in agreement, or did a voting majority win out? Anybody still think the 2^0 should be at 0 instead of 1? No? Are you sure?**

Class: We are sure.

Teacher: **Okay, my next question is: Will the graphs of linear and quadratic functions look the same for other inputs, since they are the same for an input of 0? Discuss with your teams. I will ask for a vote from all three teams. If you think both will look the same, vote thumbs-up. Otherwise, vote thumbs-down.**

Everyone votes thumbs-down.

Everyone votes no, that linear and quadratic will have different graphs. Let's verify. Send a new person from each team to the clothesline with an input of 1. What do 2 times 1, 2 to the first power, and 1 squared each equal?

Again, there is enthusiastic discussion (see Figure 11.31).

Figure 11.31 Placement for $x = 1$

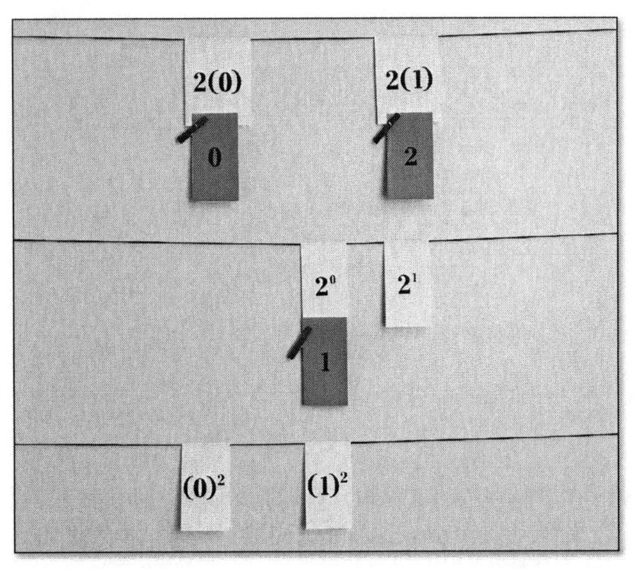

Whoa! Now, the Linear and the Exponential teams are matching outputs. You're getting intense here. I also notice both the Exponential and Quadratic teams have a difference of 1 unit showing between their two outputs. I wonder whether that pattern will continue. Teams, discuss. Will the exponential and quadratic functions continue to increase by one unit? Thumb vote!

Students vote thumbs-down.

Teacher: **Everyone is saying no. Let's verify. A new person from each group comes up for the placement of an input of 2. What is 2 times 2, 2 to the second power, and 2^2?**

The groups place their cards for x = 2 *on the clotheslines (see Figure 11.32).*

Figure 11.32 Placement for $x = 2$

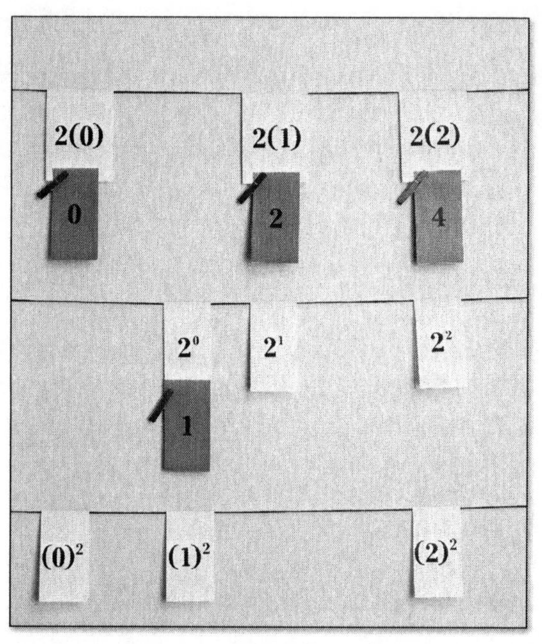

Now, you are all throwing me off. You said these functions would not be the same, but here you are all claiming that an input of 2 yields an output of 4.

Tommy: They won't always be the same.

Teacher: **When will they be the same again?**

Ashley: Never. They will keep growing differently.

Teacher: **Wow. This class has some high-level thinking going on. All in favor of what was just said, applaud loudly.**

There is loud applause.

Willy, you didn't applaud? Do you disagree?

Willy applauds with a smile.

So, what are you agreeing to?

Willy: I have no idea.

Teacher: I figured, which is okay. Let's back up. These three functions all have the same output when $x = 2$. The class is claiming they will have different outputs as x increases.

Willy applauds enthusiastically with an even bigger smile.

Okay, Willy now agrees with the rest of you, so let's see whether your collective thinking is accurate. What is 2 times 3, 2 to the third power, and 3^2? Each team, choose another representative to send to the clothesline for $x = 3$.

The representatives place the three new cards on the clotheslines for x = 3, *as shown in Figure 11.33.*

Figure 11.33 Placement for $x = 3$

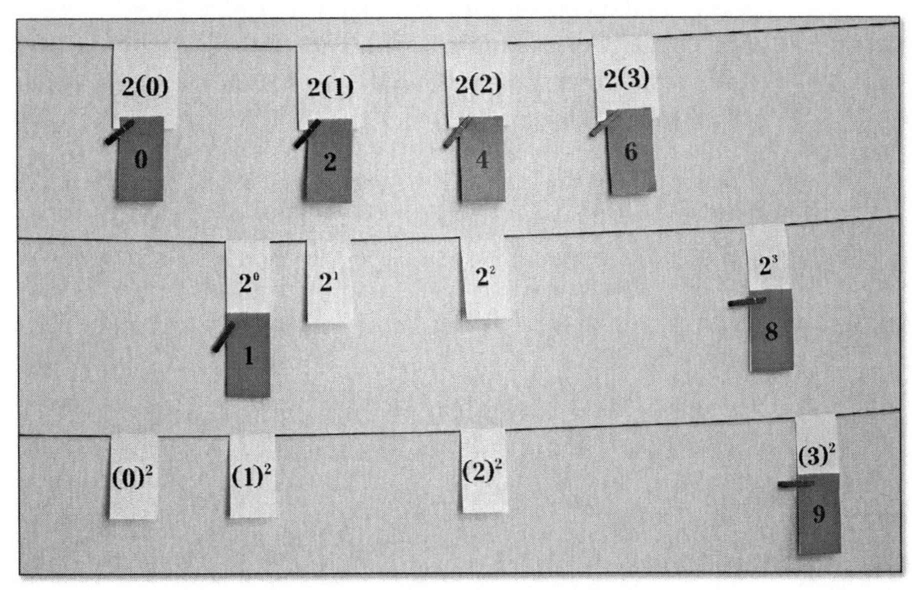

Your display appears to confirm your thinking, class. Give yourselves a big hand.

Students happily applaud themselves.

Now, I notice that all of our outputs are positive, so I am wondering whether we can ever get negative outputs from these functions. Discuss in your groups and prepare to vote. Are there any inputs that will yield a negative output for your functions? If it is possible for your team's function to generate negative outputs, please raise your hand.

Only the Linear Team raises their hands.

Quadratic Team, what makes you think that your function will not generate any negative values?

Quadratic Team: Because, when you square a negative, you still get a positive.

Teacher: Exponential Team, why do you claim that your function cannot generate a negative value?

Exponential Team: We argued about this a lot.

Teacher: That seems to be the theme of the day for the Exponential Team.

Exponential Team: No kidding. We were debating whether a negative exponent gives you a negative answer, but then we remembered that negative exponents give you fractions.

Teacher: Interesting. Let's see whether you are all correct and whether the linear function is the only one that will yield negative outputs. This time, I am giving two inputs to each team, $x = -1$ and $x = -2$. Each team, send two people to the clotheslines while the rest of the team gives feedback.

Students place all six cards on the clotheslines for the negative exponents (see Figure 11.34).

Figure 11.34 Placement for Negative Exponents

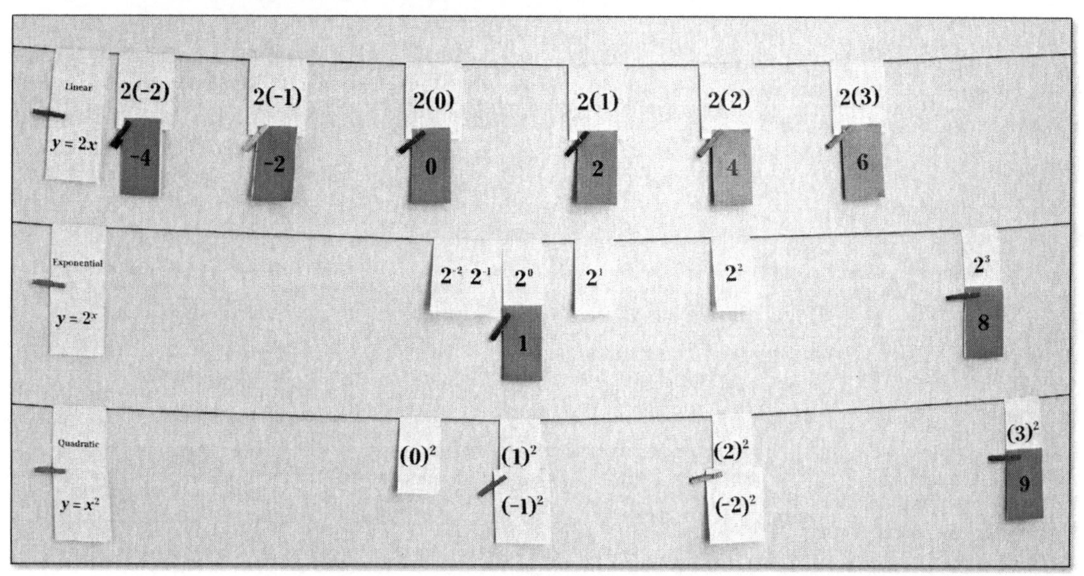

It appears that your thinking was correct, class. The linear function continues to decrease by 2 units into the negative range. What I find interesting is that while the exponential and quadratic functions "stay positive," they do so in different ways. The exponential keeps getting closer to 0 without reaching it, while the quadratic hits 0 and kind of rebounds in the other direction. That's what the graphs do in the coordinate plane.

Teacher: **Let's look back at our graph from the warm-up. See how the linear function has negative values for y as x continues to negative infinity? And, see how the exponential function keeps getting closer to the x-axis which represents a y-value of 0, but it never touches it? And see how the quadratic function decreases as we approach from the right, until $x = 0$? Then, the parabola increases again. In fact, the parabola is symmetrical, and we can even see that on the number line.**

Great work, crew. Let's write that on our activity sheet. Place the linear function on #1, the exponential on #2, and the quadratic on #3. Be sure to label each one with the name and equation like we did here, and to the right draw a quick sketch of the coordinate graph. Also, include what you learned from today's discussion.

Why These Numbers?

The functions of $y = 2x$, $y = 2^x$, and $y = x^2$ are the trifecta in Algebra 1. By the end of a first-year algebra course, students should be able to discuss the difference between these equations and their corresponding graphs. By placing the outputs on a number line, students can see why one graph is straight, another curves ever upward, and the third forms a parabola.

The Key Questions

- What do the equations and the coordinate graphs of these three types of functions look like?
- Why did your team have so much discussion?
- Will these functions continue to generate the same outputs?
- Which functions will not generate negative outputs?
- Will the exponential and quadratic functions continue to grow in the same manner as the linear function?
- What is 2 times _______, 2 to the _______ power, and ________ squared?
- Do the values on the number line justify your conjecture?

Analysis

I chose to unveil the functions simultaneously, one value at a time. I can't say that displaying one function at a time in its entirety wouldn't be better, but I wanted some predictions when these inputs are the same. Will all the other inputs be the same? This strategy worked nicely here, especially for the Exponential Team. Exponential functions and the association with limits seem to be the most challenging of the functions for students at this level.

There was very little procedural error in this lesson, because this came at the end of the year and after there was ample time to correct those issues. Apparently, some misconceptions came up in the team discussions, which I expected, but they were worked out among peers.

Having larger team discourse instead of smaller group talk was a risk. It is easier for students to stay silent and less engaged with the larger groups, but the intensity of students shouting to people at the clothesline kept everyone focused on what was going on. Engagement was high; however, individual sharing was diminished, which is why I implement this technique on only a few occasions. I really wanted to highlight the differences in the type of functions, so splitting the class into three teams created a focus. It was definitely a worthwhile trade-off.

Making the connections of the behavior on the number line to the shape of the graph on the coordinate plane was critical, so I made sure to formally summarize and insist the class record it in their notes.

The cards for this lesson are available in the Digital Resources (lesson36.pdf).

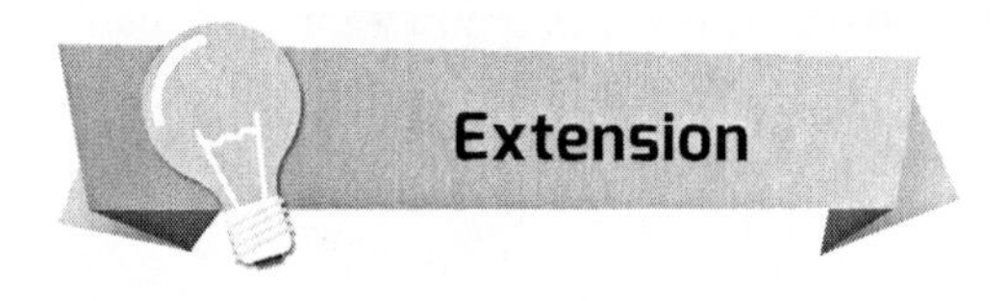

Extension

Instead of placing 2(1), 2^1, and $(1)^2$ on the number line, use function notation, placing L(1), E(1), and Q(1).

Lesson 37: Logarithms

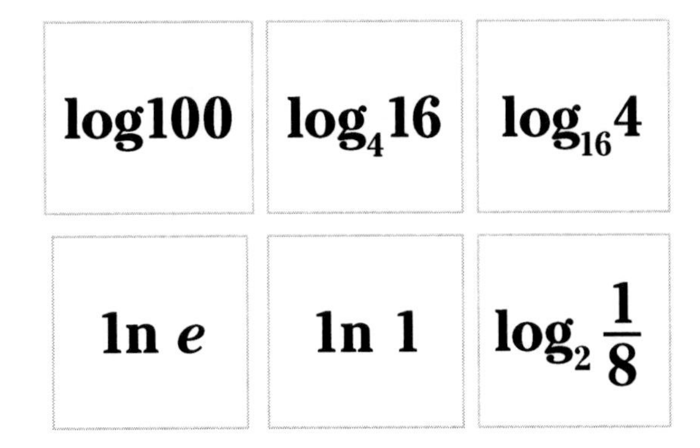

Objective: Accurately evaluate simple logs of various bases.

Teacher: **Good afternoon, class. Yesterday, we learned logarithms are the inverse functions of exponential functions. We also learned to evaluate logs, which we do by remembering that a log is an…**

Class: Exponent!

Teacher: **Good. So, for example, one of the logs to evaluate as a warm-up today is $\log_4 16$, which means 4 to what exponent is equal to 16? Today, you are going to the clothesline individually. Tilly, you are first since you haven't had a turn recently. Please choose one of these six values, and place it on the number line.**

Tilly picks $\log_4 16$ and pins it to 4 (see Figure 11.35).

Figure 11.35 Tilly's Placement

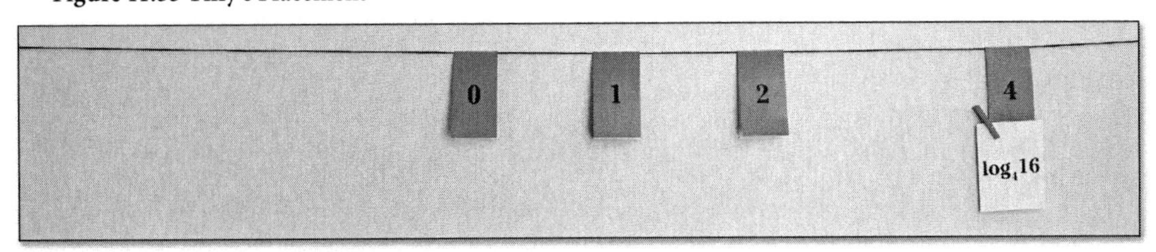

Tilly, please explain your reasoning.

Tilly: I chose 4, because 4 times 4 is 16.

Teacher: **Thank you, Tilly. You get to choose the next student to go.**

Tilly chooses Holly.

Holly, please either change the value already placed on the number line or choose a new one from the five remaining.

Holly chooses to change the value on the number line (see Figure 11.36).

Figure 11.36 Holly's Placement

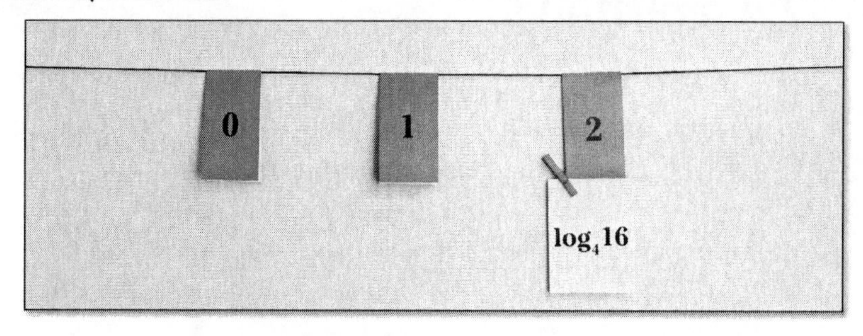

Teacher: Why did you change that one?

Holly: Because the log is an exponent, so you have to figure out 4 to what power is 16, not 4 times what is 16.

Teacher: And you chose 2?

Holly: Yes, because 4^2 is 16.

Tilly, I'm going to call you back. You may either change this back if you disagree with Holly, or you may keep her work and choose another card.

Tilly leaves Holly's work and adds a new card to the clothesline (see Figure 11.37).

Figure 11.37 Tilly's Second Placement

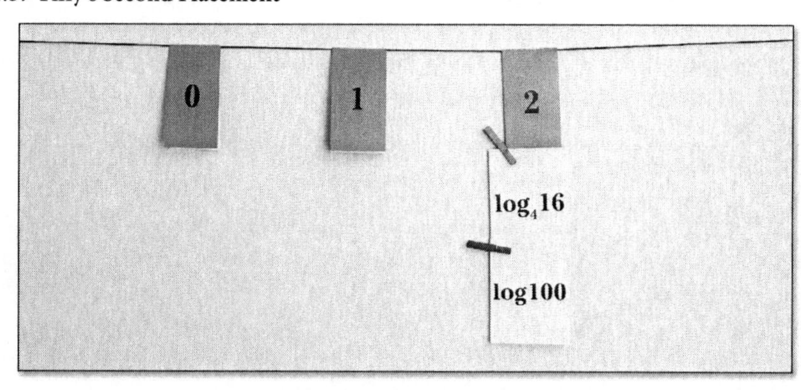

So, you agreed with Holly and you think log100 is also 2?

Tilly: Yes, Holly is right, and so 10 squared is 100.

Teacher: Ten? Where do you see 10?

Tilly: When the base of the log isn't showing it means base-10.

Teacher: Interesting, thank you. Will you choose our next person?

Tilly selects Jethro.

Jethro, you may change any of these cards or choose a new card from the four remaining.

Jethro: I don't know any of the others. All the easy cards are taken.

Teacher: That's okay. Pick one that looks the easiest, and I will walk you through it.

Jethro: Log of 16 and 4.

Teacher: We say, "log base-16 of 4." You say it.

Jethro: Log base-16 of 4.

Teacher: Good. So, 16 to what power equals 4?

Jethro: But 16 is bigger than 4, so how can we raise it to a power and make it smaller?

Teacher: Can we take a root?

Jethro: Oh yeah, square root. But I forget how to write that as an exponent.

Teacher: Give it a shot. What exponent means the same thing as taking a square root?

Jethro places his card on the -2 benchmark card, as shown in Figure 11.38.

Figure 11.38 Jethro's Placement

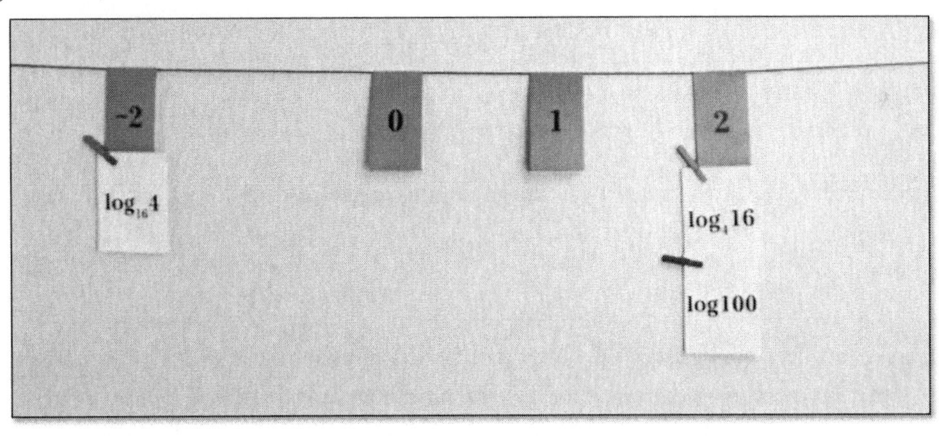

Are you saying 16 to the negative 2 power is 4?

Jethro: Yes.

Teacher: Thank you. Kelli, you seem excited to share. Come on up.

Kelli moves Jethro's card and places it between 0 and 1 (see Figure 11.39).

Figure 11.39 Kelli's Placement

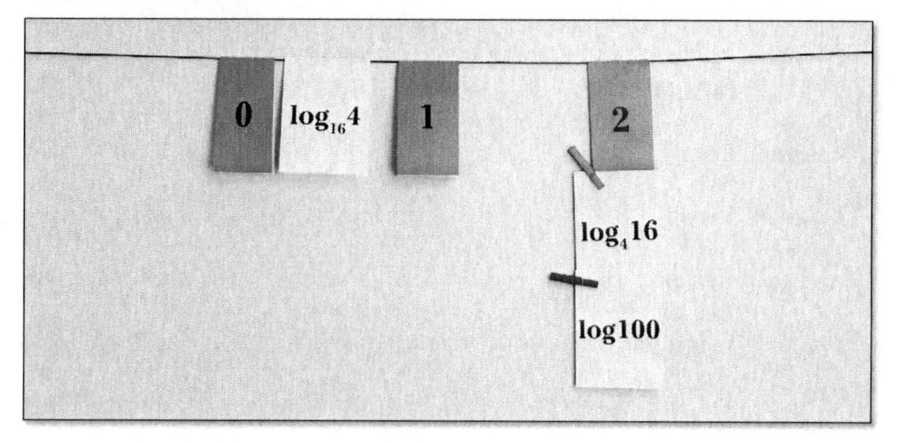

Teacher: **Now, that is interesting. Kelli, will you explain to Jethro why you changed his answer. Jethro, listen carefully because you will have a chance to defend your answer if you want. Kelli, go ahead.**

Kelli: Fractions as exponents are the same as roots.

Teacher: **We call those rational exponents.**

Kelli: One-half is a rational exponent, which means we take the square root. The square root of 16 is 4.

Teacher: **Jethro, Kelli is claiming the log base-16 of 4 equals $\frac{1}{2}$. If you disagree, change it back. If you agree, choose another one.**

Jethro: Why do I have to go again?

Teacher: **Because you are learning a lot from this and so is everyone else. Don't panic, just think.**

Jethro chooses a new card to place on the number line (see Figure 11.40).

Figure 11.40 Jethro's Second Placement

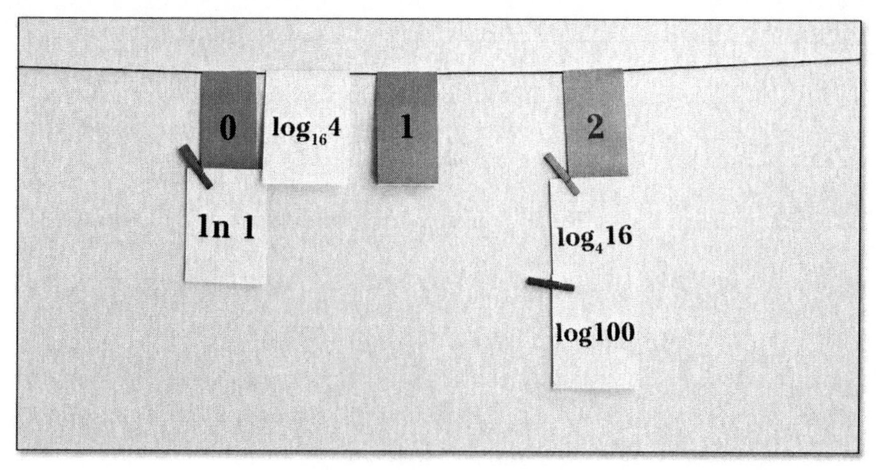

Teacher: For someone who was so reluctant, you didn't even hesitate on that.

Jethro: I remembered that anything to the 0 power is 1, and before we said that if we don't see a base it is 10. Ten to the 0 power is 1.

Teacher: You have the right thinking on the exponent, but not on the base. When Tilly said a base that is not shown means base-10, she meant for logs, but ln means natural logarithm, which is log base-e. Instead of writing log, we write ln instead. So, do you want to explain your answer again?

Jethro: Umm...*e* to the 0 power is 1?

Teacher: Are you sure?

Jethro: No, but I'm going with it.

Teacher: Then, choose our next person.

Jethro chooses Nate.

Nate.

Nate adds a card to the clothesline, as shown in Figure 11.41.

Figure 11.41 Nate's Placement

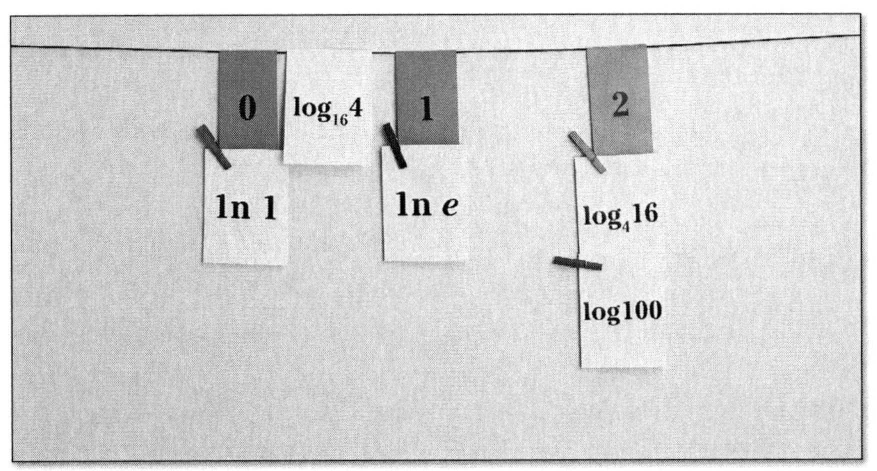

Why 1?

Nate: Because you just said that ln is base e, I just ask myself e to what power gives me e. That would be 1.

Teacher: Okay, thank you. Who is next?

Kylie raises her hand to volunteer.

Kylie, last one.

Kylie places the last card on the number line (see Figure 11.42).

Figure 11.42 Kylie's Placement

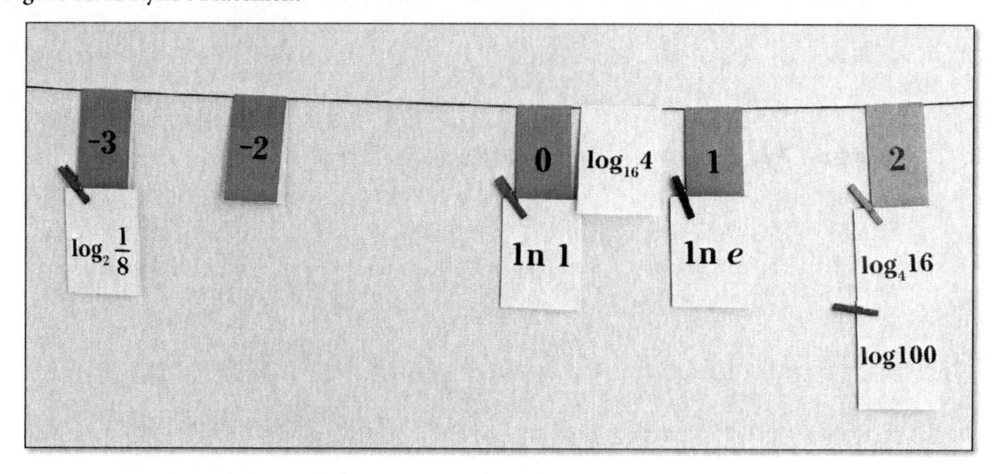

Teacher: Wow, we finally have a negative number. Why?

Kylie: Because when you have a negative exponent, you put the number underneath the fraction.

Teacher: In the denominator?

Kylie: Yes. Since 2 to the third power is 8, then 2^{-3} is 1 over 8.

Teacher: Let's read this last one. Repeat after me, everyone: Log base-2 of $\frac{1}{8}$ is -3...

Class: Log base-2 of $\frac{1}{8}$ is -3!

Teacher: Very good work. Let's record all six on one number line with a note about logarithms being exponents.

123 Why These Numbers?

The logarithmic expressions were chosen to show a variety of exponents (i.e., rational, negative, zero) while keeping the values simple to mentally compute. The card set also shows a variety of bases including base-10 and base-*e*.

The Key Questions

- Where do you put base-10 in the expression?
- Why did we get a negative number when taking the log of a fraction?
- A base of ________ raised to what power is equal to ________?
- What exponent means the same thing as taking a square root?
- What is the base of a natural log?
- Why did you change that response?

Analysis

The first student almost always chooses the simplest expression, $\log_4 16$, which lays the groundwork for the rest of the activity; namely, the logarithm yields an exponent. The error here of choosing a value of 4 because 4 times 4 equals 16 only further solidified that concept with the class. Finding the exponent of the base became the theme throughout the lesson.

Having students recognize a log with no base showing as log10 was my second biggest priority of the day. I did not anticipate students would extend that idea to the ln notation in the wrong fashion. This is an example of why explaining the reasoning is so important. The student who explained ln1 had the right answer but for the wrong reason. The notation of natural log, ln, has a base of e, not 10. This is an important distinction the class discussion brought forth. What I saw in the last two examples was more of an issue of rules of exponents than understanding logarithms. However, that is a common trend in math classes—students don't flunk current content, they flunk prior content. One of the greatest strengths of the clothesline is addressing those holes while simultaneously learning the current topic.

The cards for this lesson are available in the Digital Resources (lesson37.pdf).

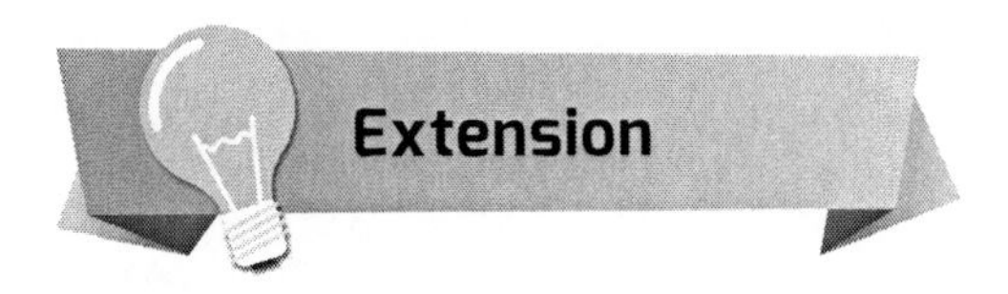

Extension

Place logarithmic expressions that have more complex answers requiring the change of case rule, such as $\log_8 16 = \frac{\log_2 16}{\log_2 8} = \frac{4}{3}$.

Lesson 38: Radians and Degrees

0° 30° 45° 60° 90° 120° 135° 150° 180°

210° 225° 240° 270° 300° 315° 330° 360° 0

Degrees Radians π 0π 2π $\frac{\pi}{2}$ $\frac{\pi}{3}$ $\frac{\pi}{4}$ $\frac{\pi}{6}$

$\frac{3\pi}{2}$ $\frac{2\pi}{3}$ $\frac{4\pi}{3}$ $\frac{5\pi}{3}$ $\frac{3\pi}{4}$ $\frac{5\pi}{4}$ $\frac{7\pi}{4}$ $\frac{5\pi}{6}$ $\frac{7\pi}{6}$ $\frac{11\pi}{6}$

Objective: Accurately convert degree measures of angles to radians.

Teacher: Good day, class. Please warm-up by calculating the circumference of a circle with a radius of 1 unit, and write it on your individual lapboards.

Nearly all students accurately compute 2π as the circumference.

Thank you. This circle is known as the unit circle because its radius is 1 unit. We also know there are 360° in a complete circle, so we can correlate 360° to 2π radians and 0° to 0 radians. Let me show you what this looks like on our double number line (see Figure 11.43).

Figure 11.43 Double Number Line—Degrees and Radians

Teacher: Today, we are going to match each major degree benchmark to its corresponding radian measure. I have all the radian measures we need on this desk. I will place the degree measures on the top line while you will place the corresponding radian measures directly underneath on the bottom line. All you need to do is remember that 360° equals 2π radians. I'm going to start with the most basic value and cut the circle in half. I have placed 180° on the top line. Juan, please point to the group that will place the corresponding radian measure.

Juan points to Lana's group.

Lana, your group is to place the radian equivalent of 180° on the second line. My hint to you is: 180 is half of 360, so what is half of 2π?

Lana places her card directly below 180°, as shown in Figure 11.44.

Figure 11.44 Lana's Placement

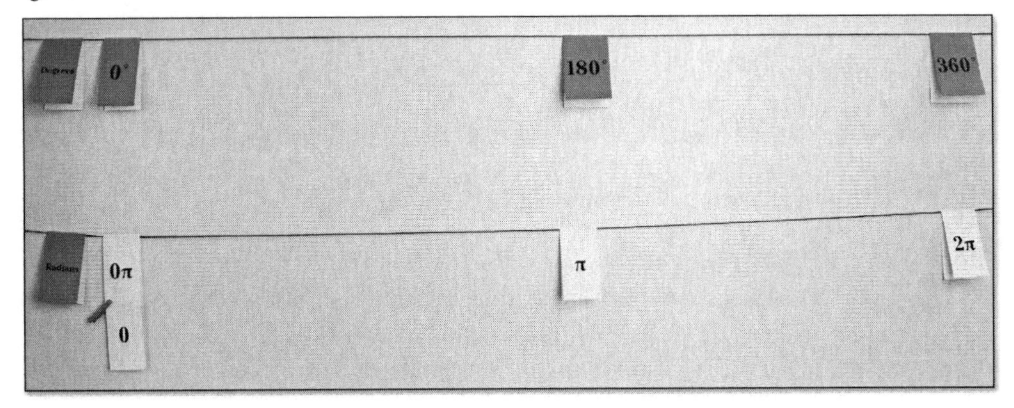

All in favor, clap twice.

All students clap twice.

Now, let's increase the challenge. I am going to place the other multiples of 90. Lana, which group is next?

Lana chooses Frankie's group.

Frankie, your group needs to head to the clothesline. Please find the radian measure of 90° and 270°.

Frankie's group places the radian measure for 90° and 270° (see Figure 11.45).

Figure 11.45 Frankie's Placement

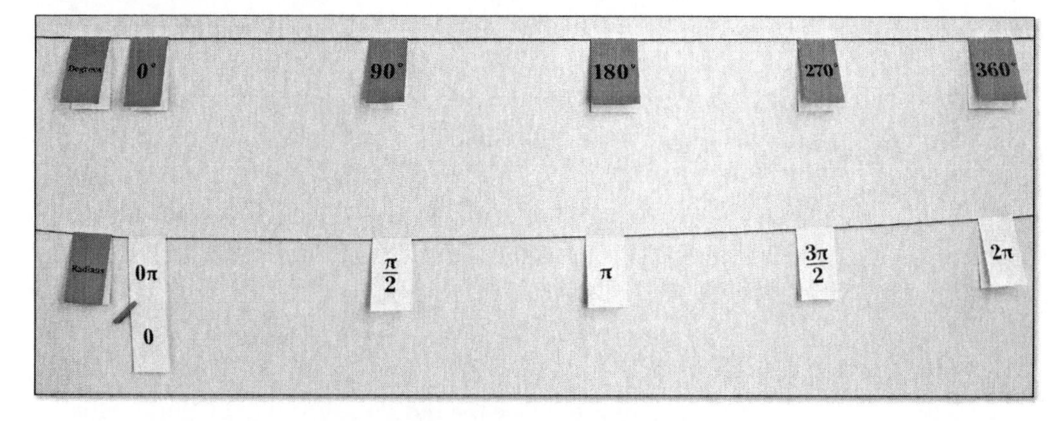

Teacher: Frankie, will your group please explain why you chose those values?

Frankie: Since 90 is half of 180, we chose half of π. Then, we had to find halfway between π and 2π, which is $1\frac{1}{2}\pi$, so we picked $\frac{3}{2}\pi$.

Teacher: So, let me see whether I understand you. We have $\frac{1}{2}\pi$, $\frac{2}{2}\pi$, $\frac{3}{2}\pi$, and $\frac{4}{2}\pi$.

Frankie: Yes.

Teacher: Matt, how is this $\frac{4}{2}\pi$ when it says right here that it is 2π?

Matt: Because $\frac{4}{2}$ is the same as 2.

Teacher: Well then, we should be able to apply that very intelligent reasoning to multiples of degrees that are not already on the line. Frankie, which group is next?

Frankie selects Suki's group.

Suki, have your group show us the radian equivalents of 45°, 135°, 225°, and 315°.

Suki and her group place the cards, as shown in Figure 11.46.

Figure 11.46 Suki's Placement

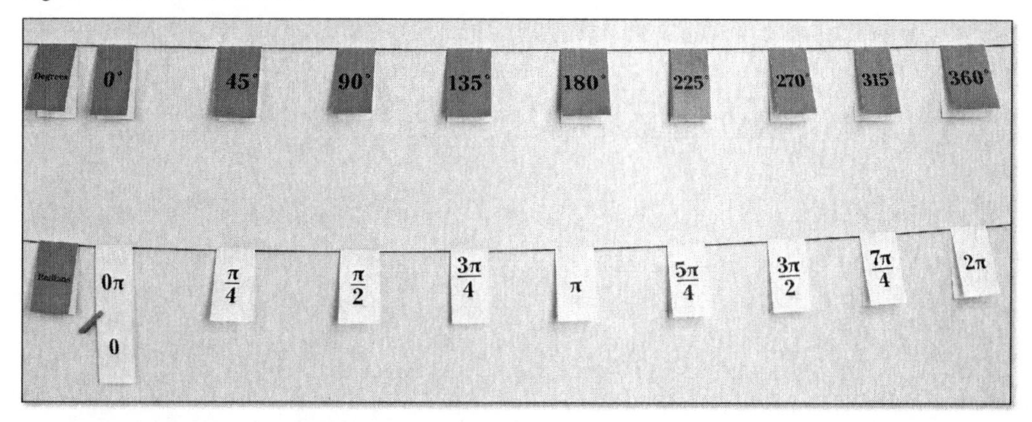

Teacher: Otto, what do you notice about all the values Suki's group chose?

Otto: They all have π?

Teacher: True. Is there something else?

Otto: I'm not sure.

Teacher: Take a look at the denominators.

Otto: They all have a 4.

Teacher: So Suki, Otto is claiming your group chose all fourths. Is that true?

Suki: Yes.

Teacher: Why?

Suki: We just kept cutting it in half. Half of 360°, then half of 180°, so half of 90°. And, 90° is $\frac{1}{2}$ of π, so we figured 45° would be $\frac{1}{4}$ of π.

Teacher: From there, how did you determine the other radian measures?

Suki: We just counted up $\frac{1}{4}\pi$ each time.

Teacher: Okay, let's do that. Everyone count, as I point. $\frac{1}{4}\pi$…

Class: $\frac{2}{4}\pi$…

Teacher: So, $\frac{2}{4}\pi$ is the same as $\frac{1}{2}\pi$. Keep counting!

Class: $\frac{3}{4}\pi$, $\frac{4}{4}\pi$…

Teacher: So, $\frac{4}{4}\pi$ is equivalent to 1π or simply π. Keep counting!

Class: $\frac{5}{4}\pi$, $\frac{6}{4}\pi$, $\frac{7}{4}\pi$, $\frac{8}{4}\pi$.

Teacher: Which is?

Class: Two π!

Teacher: Well done.

Randy: I get it!

Teacher: Great, Randy. Would you like to try the multiples of 60?

Randy: Sure.

Teacher: I will place them while you choose the radian values from the table.

Randy chooses cards from the table and places them on the clothesline (see Figure 11.47).

PART III – Chapter 11

Figure 11.47 Randy's Placement

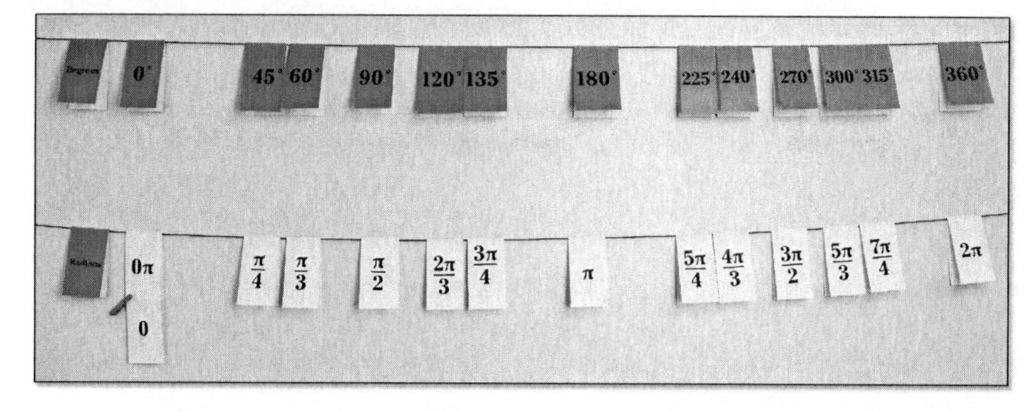

Teacher: I noticed you changed your mind while you were up there. What were you thinking?

Randy: I first divided 360° by 60 and got 6, so I chose fractions that have a 6 in them.

Teacher: Six in them?

Randy: Six for a denominator. But I realized that I did something wrong when I started counting up and had only $\frac{3}{4}\pi$. That's when I thought I should divide 180° by 60 since that is just π and 360° is actually 2π. So, I switched to thirds.

Teacher: Thank you, Randy. That was some tremendous thinking that we all learned from. I applaud your courage for walking us through that. Now, let's have the class count by $\frac{1}{3}\pi$ as you suggested. Everyone count, as I point. $\frac{1}{3}\pi$…

Class: $\frac{2}{3}\pi$, $\frac{3}{3}\pi$, $\frac{4}{3}\pi$, $\frac{5}{3}\pi$, $\frac{6}{3}\pi$.

Teacher: Well done! Now, we have one last set of values: multiples of 30. I am placing 30°, 150°, 210°, and 300° on the upper number line. All the other multiples of 30 have already been placed. Randy, you are choosing our next group.

Randy selects Justine's group.

Justine, your group has it easiest because there is only one set of cards left to choose from. Your only task, then, is to place them in the correct order. Come on up.

Justine's group places the final cards on the number line (see Figure 11.48).

Figure 11.48 Justine's Placement

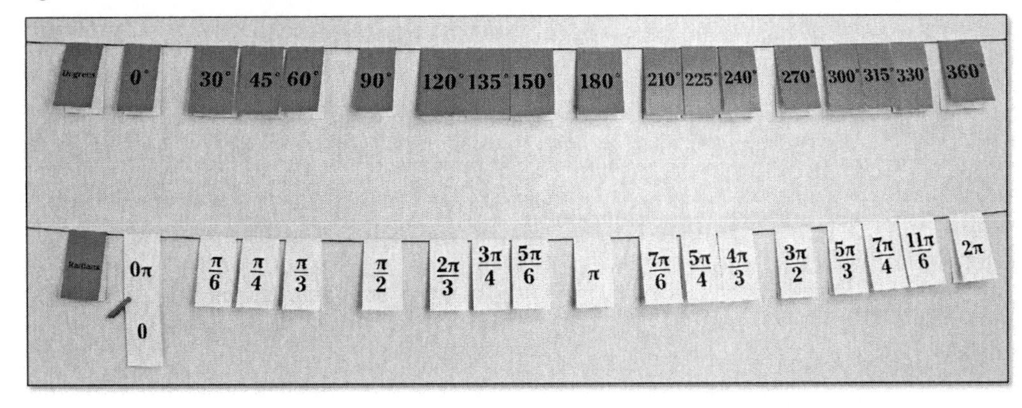

Teacher: Justine, can you explain why it makes sense that 30° is equivalent to $\frac{\pi}{6}$?

Justine: We did as Randy said and divided 180° by 30 to get 6. Sixths were the only cards left, so we knew that was right.

The class laughs.

Teacher: Very observant. Shall we count up to finish this? Everybody count, with me. $\frac{1}{6}\pi$...

Class: $\frac{2}{6}\pi$, $\frac{3}{6}\pi$, $\frac{4}{6}\pi$, $\frac{5}{6}\pi$, $\frac{6}{6}\pi$.

Teacher: Terrific, keep counting.

Class: $\frac{7}{6}\pi$, $\frac{8}{6}\pi$, $\frac{9}{6}\pi$, $\frac{10}{6}\pi$, $\frac{11}{6}\pi$, $\frac{12}{6}\pi$.

Teacher: Take a look at these values, and notice how the structure stems from the ratio of 360° to 2π. Record this carefully on your activity sheet, and make note of the ratio.

Why These Numbers?

The key to converting degrees to radians is the understanding of the ratio of 360° to 2π on the unit circle. From that starting point, the choice of the values to convert was all based on the multiples of the most common degree measures: 30°, 45°, 60°, and 90°.

The Key Questions

- How is this $\frac{4}{2}\pi$, when it says right here that it is 2π?
- Why did your group choose denominators of 4?
- Can you explain why it makes sense that 30° is equivalent to $\frac{\pi}{6}$?
- I noticed you changed your mind while you were at the clothesline. What were you thinking about?
- How did you determine the other radian measures?

Analysis

Surprisingly, students had few issues with generating the equivalent radian measures for the given degree measures. The warm-up of calculating the circumference, then making a strong connection between 360° and 2π radians, appeared to support students a great deal. Since there weren't many corrections to be made, I chose to emphasize the structure behind the progression of the fractions (e.g., $\frac{5}{4}\pi$, $\frac{6}{4}\pi$, $\frac{7}{4}\pi$, $\frac{8}{4}\pi$).

After this lesson, I left the completed double number line in front of the class for several days as we continued with other problems involving radians.

The cards for this lesson are available in the Digital Resources (lesson38.pdf).

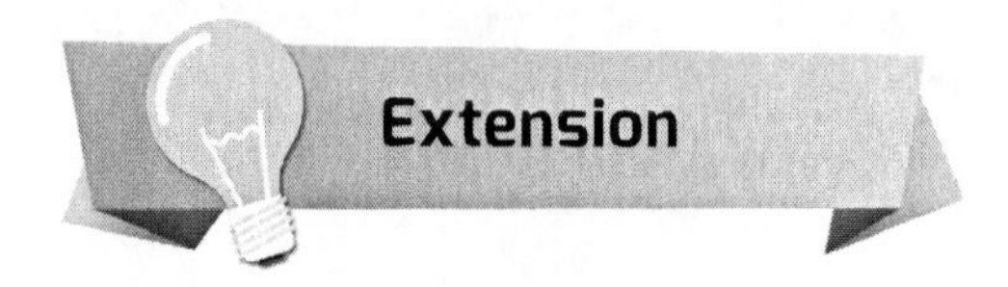

Extension

For a warm-up on a future day, start with radians, and have students choose corresponding degree measures.

Part IV

Appendixes

References Cited

Duncan, Greg. 2007. "School Readiness and Later Achievement." *Developmental Psychology* 43(6), 1428–1446.

Gersten, Russell and David J. Chard. 1999. "Number Sense: Rethinking Arithmetic Instruction for Students with Mathematical Disabilities." *The Journal of Special Education* 33.1: 18–28.

Jordan, Nancy. 2010. "Early Predictors of Mathematics Achievement and Mathematics Learning Difficulties." *Encyclopedia of Early Childhood Development.*

National Council of Teachers of Mathematics (NCTM). 2014. *Principles to Actions: Ensuring Mathematical Success for All.* Reston: National Council of Teachers of Mathematics.

National Governors Association (NGA) Center for Best Practices and Council of Chief State School Officers (CCSSO). 2010. *Common Core State Standards: Mathematics.* Washington, DC: National Governors Association Center for Best Practices, Council of Chief State School Officers. www.corestandards.org.

Smith, Margaret S., and Mary K. Stein. 2011. *5 Practices for Orchestrating Productive Mathematics Discussions.* Reston, VA: NCTM/Thousand Oaks, CA: Corwin Press.

Additional Resources

Suggested Reading

Doll, William E. 1993. *A Postmodern Perspective on Curriculum.* New York: Teacher's College.

Horn, Ilana. 2012. *Strength in Numbers: Collaborative Learning in Secondary Mathematics.* Reston: National Council of Teachers of Mathematics.

Kanold, Timothy. 2017. *Heart!: Fully Forming Your Professional Life as a Teacher and Leader.* Bloomington: Solution Tree Press.

Ma, Liping. 1999. *Knowing and Teaching Elementary Mathematics: Teachers' Understanding of Fundamental Mathematics in China and the United States (Studies in Mathematical Thinking and Learning Series),* 2nd ed. New York: Routledge.

Payne, Ruby. 2013. *Culture of Poverty: A Framework for Understanding Poverty, A Cognitive Approach,* 5th ed. Highlands: aha! Process, Inc.

Schmidt, William, Curtis McKnight, and Senta Raizen. 2002. *A Splintered Vision: An Investigation of U.S. Science and Mathematics Education.* Germantown: Wisconsin Teacher of Mathematics.

Senge, Peter. 1990. *The Fifth Discipline: The Art and Practice of the Learning Organization.* New York: Doubleday.

Shore, Chris. 2003. *MPJ's Ultimate Math Lessons.* Murrieta: The Math Projects Journal.

Stigler, James, and James Hiebert. 1999. *The Teaching Gap: Best Ideas From the World's Teachers for Improving Education in the Classroom.* New York: Free Press.

Additional Resources *(cont.)*

Clothesline Resources

Clothesline Math

www.clotheslinemath.com

Desmos

Desmos.com

Estimation 180

http://www.estimation180.com

Math Rock Stars

https://mathrockstarsblog.wordpress.com

The Mind of an April Fool

https://themindofanaprilfool.files.wordpress.com

Name: ______________________________ Date: ________________

Clothesline Math

Directions: For each set, record the given values, expressions, or drawings. After the discussion of their placement on the clothesline, record them on the number line.

1. __________, __________, __________

2. __________, __________, __________

3. __________, __________, __________

Name: ______________________________ Date: ________________

Clothesline Math *(cont.)*

4. __________, __________, __________

5. __________, __________, __________

6. __________, __________, __________

fold here

fold here

Clothesline Math
The Master Number Sense Maker
Shell Education

Clothesline Math
The Master Number Sense Maker
Shell Education

Clothesline Math
The Master Number Sense Maker
Shell Education

fold here

fold here

−6

−5

−4

Clothesline Math
The Master Number Sense Maker
Shell Education

Clothesline Math
The Master Number Sense Maker
Shell Education

d here

fold here

−3

−2

−1

Clothesline Math
The Master Number Sense Maker
Shell Education

Clothesline Math
The Master Number Sense Maker
Shell Education

Clothesline Math
The Master Number Sense Maker
Shell Education

fold here → ← fold here

0 1 2

Clothesline Math
The Master Number Sense Maker
Shell Education

Clothesline Math
The Master Number Sense Maker
Shell Education

Clothesline Math
The Master Number Sense Maker
Shell Education

fold here

fold here

3

4

5

fold here

fold here

0°

30°

45°

Clothesline Math
The Master Number Sense Maker
Shell Education

Clothesline Math
The Master Number Sense Maker
Shell Education

Clothesline Math
The Master Number Sense Maker
Shell Education

d here

fold here

60°

90°

180°

Clothesline Math
The Master Number Sense Maker
Shell Education

Clothesline Math
The Master Number Sense Maker
Shell Education

Clothesline Math
The Master Number Sense Maker
Shell Education

fold here

fold here

x

y

x°

Clothesline Math
The Master Number Sense Maker
Shell Education

Clothesline Math
The Master Number Sense Maker
Shell Education

Clothesline Math
The Master Number Sense Maker
Shell Education

d here

fold here

a

b

c

Clothesline Math
The Master Number Sense Maker
Shell Education

Clothesline Math
The Master Number Sense Maker
Shell Education

Clothesline Math
The Master Number Sense Maker
Shell Education

fold here

fold here

d

μ

$\bar{x}$

Clothesline Math
The Master Number Sense Maker
Shell Education

Clothesline Math
The Master Number Sense Maker
Shell Education

Clothesline Math
The Master Number Sense Maker
Shell Education

d here

fold here

Q_1 Q_2 Q_3

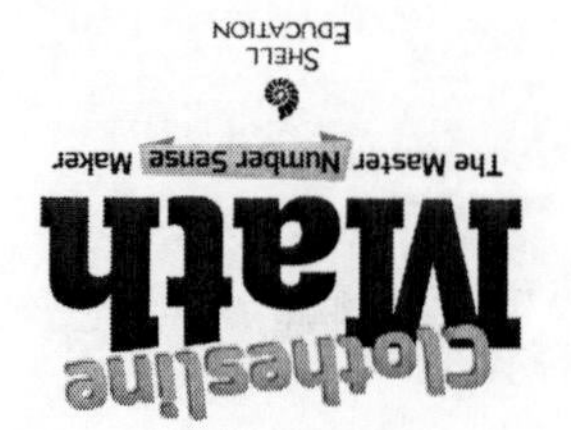

fold here

fold here

Median

Mean

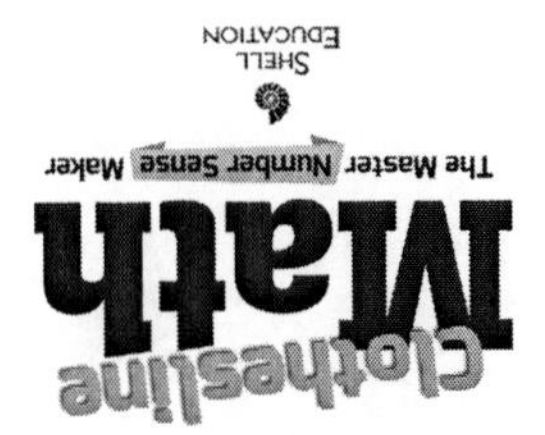

d here

Max

fold here

Min

fold here

fold here

Mode

Range

Clothesline Math
The Master Number Sense Maker
Shell Education

fold here

fold here

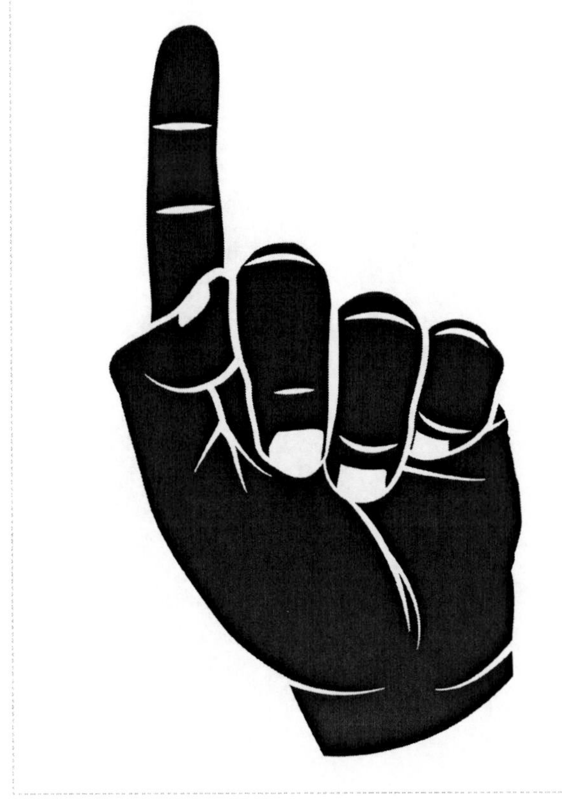

Clothesline Math
The Master Number Sense Maker
Shell Education

Clothesline Math
The Master Number Sense Maker
Shell Education

fold here

fold here

Clothesline Math
The Master Number Sense Maker
Shell Education

fold here

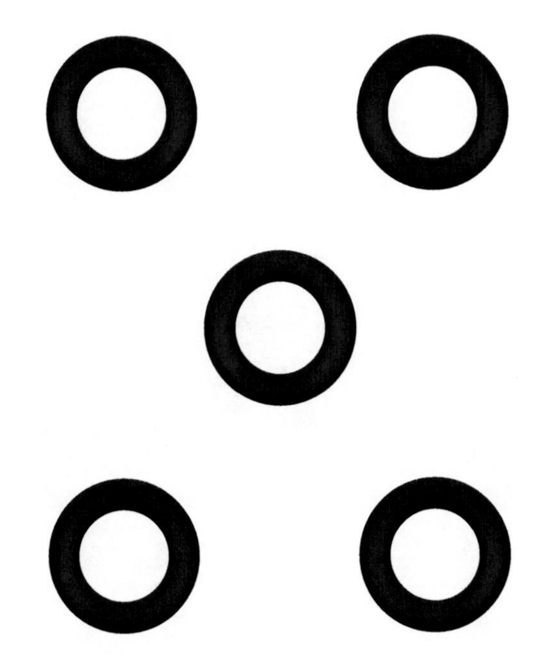

fold here

Clothesline Math
The Master Number Sense Maker
Shell Education

Clothesline Math
The Master Number Sense Maker
Shell Education

Clothesline Math
The Master Number Sense Maker
Shell Education

fold here

fold here

m_1

m_2

m_3

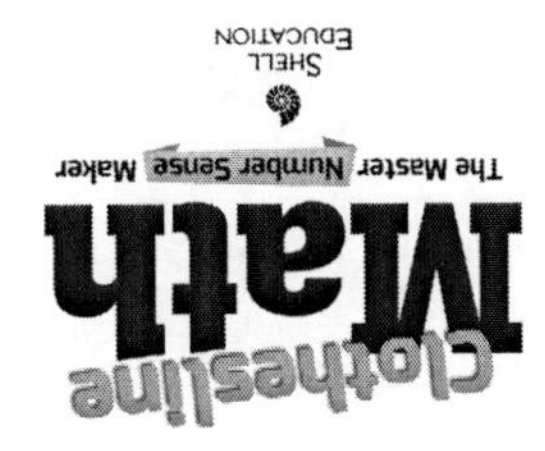

d here

fold here

b_1

b_2

b_3

Digital Resources

Lesson	Filename
Blank Cards	blankcards.pdf
Lesson 1 Card Set	lesson1.pdf
Lesson 2 Card Set	lesson2.pdf
Lesson 3 Card Set	lesson3.pdf
Lesson 4 Card Set	lesson4.pdf
Lesson 5 Card Set	lesson5.pdf
Lesson 6 Card Set	lesson6.pdf
Lesson 7 Card Set	lesson7.pdf
Lesson 8 Card Set	lesson8.pdf
Lesson 9 Card Set	lesson9.pdf
Lesson 10 Card Set	lesson10.pdf
Lesson 11 Card Set	lesson11.pdf
Lesson 12 Card Set	lesson12.pdf
Lesson 13 Card Set	lesson13.pdf
Lesson 14 Card Set	lesson14.pdf
Lesson 15 Card Set	lesson15.pdf
Lesson 16 Card Set	lesson16.pdf
Lesson 17 Card Set	lesson17.pdf
Lesson 18 Card Set	lesson18.pdf
Lesson 19 Card Set	lesson19.pdf
Lesson 20 Card Set	lesson20.pdf
Lesson 21 Card Set	lesson21.pdf
Lesson 22 Card Set	lesson22.pdf
Lesson 23 Card Set	lesson23.pdf
Lesson 24 Card Set	lesson24.pdf
Lesson 25 Card Set	lesson25.pdf
Lesson 26 Card Set	lesson26.pdf
Lesson 27 Card Set	lesson27.pdf
Lesson 28 Card Set	lesson28.pdf
Lesson 29 Card Set	lesson29.pdf
Lesson 30 Card Set	lesson30.pdf
Lesson 31 Card Set	lesson31.pdf
Lesson 32 Card Set	lesson32.pdf
Lesson 33 Card Set	lesson33.pdf
Lesson 34 Card Set	lesson34.pdf
Lesson 35 Card Set	lesson35.pdf
Lesson 36 Card Set	lesson36.pdf
Lesson 37 Card Set	lesson37.pdf
Lesson 38 Card Set	lesson38.pdf
Clothesline Math activity sheet	clotheslineactivity.pdf